W9-ABH-163

Linear Differential Transformations of the Second Order

Linear Differential Transformations of the Second Order

by

Otakar Borůvka
Mathematics Institute
Czechoslovak Academy of Sciences, Brno

translated by

F. M. Arscott
Professor of Mathematics
University of Surrey

The English Universities Press Ltd

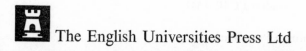

Dedicated in gratitude to the memory of my parents,
Emilie and Jan Borůvka.

ISBN 0 340 12408 3

First published in German under the title Lineare Differentialtransformationen 2. Ordnung by VEB Deutscher Verlag der Wissenschaften, Berlin, in 1967.

First published in English in 1971

Copyright of translation © 1971 F. M. Arscott
All rights reserved. No part of this publication may be reproduced, or transmitted in any form or by any means, electronic or mechanical, including photocopy, recording, or any information storage and retrieval system, without permission in writing from the publisher.

The English Universities Press Limited
St Paul's House, Warwick Lane, London EC4P 4AH

Made in Great Britain at the Pitman Press Bath

MATH.-SCI.
QA
372
.B6813
1971
1289277

Preface to the original edition

This book comprises a theory of transformations of ordinary linear homogeneous differential equations of the second order; the concern of this theory is to apply results relating to transformation of variables to the solutions of such differential equations. It is a qualitative theory, in the real domain, and is of global character.

The theory of transformations of such linear second-order equations was founded by E. E. Kummer, whose work in this field is distinguished particularly by his discovery of that non-linear third-order differential equation which is fundamental to the theory. Further discoveries in this direction led to wide-ranging research on transformations of linear differential equations of the n-th order, in connection with the equivalence problem; notable studies in this field are those of E. Laguerre, F. Brioschi, G. H. Halphen, A. R. Forsyth, S. Lie and P. Appell. Within the framework of these studies, there occurs from time to time work on the transformation of linear differential equations of the second order in the complex domain. Linear differential equations of the second order have an exceptional position among those of order n ($n \geqslant 2$) since only in the case $n = 2$ are two differential equations always equivalent.

The transformation theory expounded in this book is a far-reaching development of certain new fundamental ideas, and consists essentially of two parts. The first part comprises the "theory of dispersions," named after its basic idea; it relates to oscillatory differential equations and proceeds from the concept of "central dispersion" to a constructive theory for integration of the Kummer differential equation. The other part consists of "general transformation theory", in which we study properties of the solutions of the Kummer differential equation under general conditions, in connection with the process of transforming linear differential equations of the second order. One section of this theory is devoted to questions regarding complete solutions of the Kummer equation; such complete solutions are distinguished by the fact that they provide functions which transform integrals of linear differential equations of the second order into each other over their entire domain.

The methods used in building up this transformation theory made it clear that much can be gained by a deepening and widening of certain concepts in the classical theory of linear second-order differential equations. This is particularly true of the concept of "phase" which one can recognize, in retrospect, as one of the most important ideas used in transformation theory. The significance of this concept is brought out in the chapter on "phase theory" which develops a body of results related to this notion in preparation for later use. Other theories, also of a preparatory character, are built up as well; these relate to the subjects of "conjugate numbers" and of centro-affine properties of plane curves.

My concern has been to give the book a form which shall be fresh and stimulating, and at the same time be a complete and unified whole. The material on which I have worked has not only called for the methods of classical analysis but has also given scope, in many directions, for use of the apparatus of modern algebra, particularly group theory; in this way has come the discovery of some deep results. The reader must judge how far I have succeeded in achieving the objectives I set myself; how far, indeed, potentialities have been turned into fact.

I take this opportunity of expressing warmest thanks to my colleagues, Dozents E. Barvínek and Fr. Neuman, for their careful reading of the manuscript and for their kindly advice. Dr. Neuman has, moreover, compiled the bibliography on work carried out by participants in my seminar and arising from the subject-matter of this book. I am grateful also to Mrs. H. Fendrychová for her careful preparation of the diagrams; finally, I am greatly indebted to the VEB Deutscher Verlag der Wissenschaften in Berlin for their generous and accurate collaboration.

Brno *O. Borůvka*

Translator's note

The greater part of this book, sections 1 to 27, was published in German in 1967 by VEB Deutscher Verlag Der Wissenschaften in Berlin, under the title *Lineare Differentialtransformationen 2. Ordnung*; for this edition, Professor Borůvka has contributed two further sections, 28 and 29, giving recent results in the field and bringing the work right up to date. The material in these two sections is published here for the first time.

In making the translation, a few minor changes of format have been effected. A decimal system of section and paragraph numbering has been adopted, so that sections run from 1 to 29 and paragraphs from 1.1 to 29.4. Formulae, however, are numbered consecutively within each section, paragraphs being ignored; thus the formulae in section 1 are numbered (1.1) to (1.23). Where reference is made to a formula in the same section, only the decimal part of the reference number is used.

I have followed the German text closely, and only in a few places, where the interwoven structure of the original did not go easily into English, have I rewritten any substantial portion. The only notational change has been the use of θ in place of ϑ for the polar function.

Many of the ideas and concepts in this book are sufficiently novel for the German technical terms not yet to have English equivalents which can be regarded as standard. I have therefore set out below a short glossary of such terms, with the translation here employed. On this, and on many other technical problems, I have had the inestimable benefit of the author's advice, help and encouragement. This translation is my tribute to the fine work both of Academician Otakar Borůvka himself and also of his coworkers in Czechoslovakia; it is made in the hope that thereby the work will be accessible to that wider circle of mathematical readers which its beauty and originality so richly deserve.

Finally, my warmest thanks are due to the three secretaries, Mrs. P. J. Goodwin and Mrs. J. Hoogendyck of the University of Surrey and Miss Marian Martin of the University of Calgary for their invaluable patience and skill in producing the typescript, and to my daughter, Miss Christine Arscott, for her assistance in transcribing formulae from the original text to this.

University of Surrey
Guildford, Surrey

F. M. Arscott

Glossary

Art	kind	Gattung	category
ausgezeichnet	singular	grund-	fundamental
begleitend	associated	Knotenpunkt	node
Büschel	bunch	Repräsentant	representative
Darstellung	representation	Träger	carrier
eigentlich	proper	Transformierende	transforming function
Erzeugende	generator	Typus	type

Table of contents

Survey

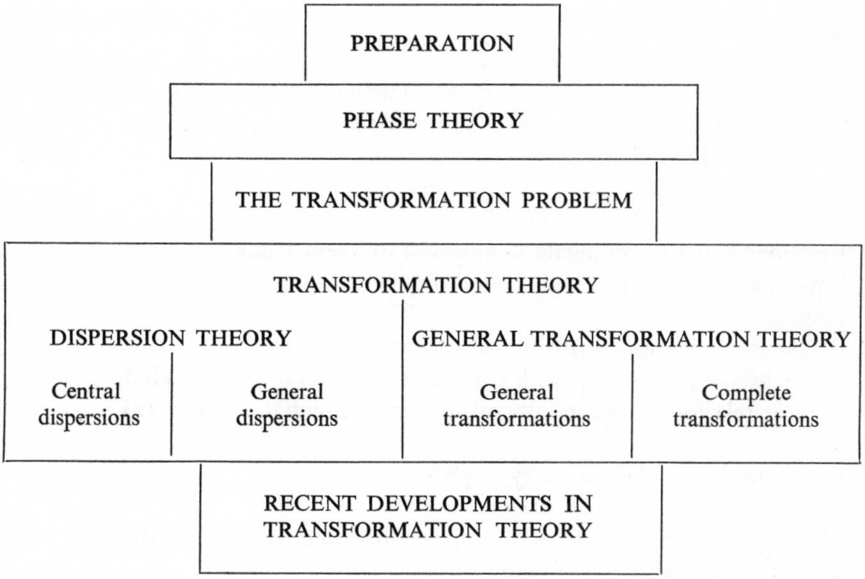

I Foundations of the theory

In this first chapter our aim is to assemble certain information needed in our subsequent construction of the theory. Well known facts will be set out in an abbreviated form only and, in the case of theorems, without proof. We can assure the reader that the following account provides a careful preparation for the problems which are going to be discussed, and without it he might well find the rest of the book wearisome to read.

General properties of ordinary linear homogeneous differential equations of the second order

1 Introduction

1.1 Preliminaries

In this book we shall be concerned with ordinary linear differential equations of the second order of the form

$$y'' = q(t)y \tag{q}$$

We shall suppose that the function q, which for brevity we call the *carrier* of the differential equation (q), is defined in an open bounded or unbounded interval $j = (a, b)$ and belongs to the class C_0. When necessary, the function q will naturally be required to have further properties. The symbol C_0 means as usual the class of functions which are continuous in the interval considered while C_k denotes the class of functions with continuous derivatives up to and including the k-th order (k=1, 2,. . .).

Differential equations of the form (q) are called *Sturm-Liouville* or *Jacobian* differential equations. We shall use the latter term.

A linear differential equation of the second order of the form

$$Y'' + a(x)Y' + b(x)y = 0, \tag{a}$$

whose coefficients a, b are defined in an interval J and belong to the class C_0, can be brought into the Jacobian form (q) by means of a transformation of the independent variable of the form

$$t = t_0 + t_0' \int_{x_0}^{x} \exp\left(-\int_{x_0}^{\sigma} a(\tau)\,\mathrm{d}\tau\right) \mathrm{d}\sigma \tag{1.1}$$

with arbitrary numbers $x_0 \in J$, t_0, $t_0' \neq 0$. The coefficient q is defined and continuous in the interval j, given by the range of the function $t(x)$, $x \in J$ and we have

$$q(t) = -\frac{1}{t_0'^2} b(x) \cdot \exp 2\int_{x_0}^{x} a(\tau)\,\mathrm{d}\tau. \tag{1.2}$$

The connection between solutions of the differential equations (a), (q) is given by the formula

$$Y(x) = y(t) \tag{1.3}$$

in which $x \in J$ and $t \in j$ are homologous values, that is to say connected by the relation (1.1). Hence, if $Y(x)$ is a solution of the differential equation (a) then the function $y(t)$ defined by (3) represents a solution of the differential equation (q), and conversely.

There is another possible way of putting the differential equation (a) into Jacobian form, when the coefficient $a \in C_1$. In this case the transformation

$$Y = \exp\left(-\frac{1}{2}\int_{x_0}^{x} a(\tau)\,d\tau\right) \cdot y$$

of the dependent variable leads to the Jacobian differential equation (q) with the carrier

$$q(x) = \frac{1}{4} a^2(x) + \frac{1}{2} a'(x) - b(x)$$

in the interval J. By a solution of the differential equation (q) we mean a function $y \in C_2$, defined in an interval $i \subset j$ and satisfying (q). In the case when $i = j$ we shall generally use the term *integral* instead of the term *solution.*

It is known that there is precisely one integral y of the differential equation (q) passing through an arbitrary point (t_0, y_0), $t_0 \in j$, with arbitrary gradient y'_0; that is to say, such that $y(t_0) = y_0$, $y'(t_0) = y'_0$. The differential equation (q) is always satisfied by the identically zero function, but usually we shall leave this solution out of our consideration. Sometimes, for convenience, we associate ideas relating to the differential equation (q) with the carrier q itself, for instance we shall speak of solutions or integrals of the carrier q.

1.2 The Wronskian determinant

Given an ordered pair of solutions u, v of the differential equation (q), with the same interval of definition $i \subset j$, there is associated with them the Wronskian determinant or (simply) the Wronskian $w = uv' - u'v$ whose value is constant. The solutions u, v are linearly dependent on or independent of each other according as the Wronskian of the ordered pair u, v or of the pair v, u is equal to zero or different from zero. If the solutions u, v or their first derivatives possess zeros, they are dependent if and only if they or their first derivatives have a common zero. In this case the solutions u, v have all their zeros in common and the same is true for their derivatives u', v'.

Two functions $u, v \in C_2$ and defined in the interval j are independent integrals of a differential equation (q) if and only if the Wronskian $w = uv' - u'v$ is constant and not zero. The function q is then given by

$$q = \frac{1}{w}(u'v'' - u''v'). \tag{1.4}$$

If a solution u of the differential equation (q) is everywhere non-zero then the function $v(t)$ given by

$$v(t) = u(t) \cdot \int_{t_0}^{t} \frac{d\sigma}{u^2(\sigma)} \quad (t_0 \in j) \tag{1.5}$$

represents a further solution of (q). The solutions u, v are linearly independent: $w = 1$.

1.3 Bases

The set of all integrals of the differential equation q forms a two-dimensional linear space, the so-called *integral space* r of the differential equation (q). Every ordered pair of linearly independent integrals u, v of the differential equation (q) forms a *basis* (u, v) of r; an arbitrary integral $y \in r$ has uniquely determined constant coordinates c_1, c_2 with respect to the basis (u, v); that is to say $y = c_1 u + c_2 v$. Conversely, given any ordered pair of constants c_1, c_2 there is precisely one corresponding integral y of the differential equation (q) with the coordinates c_1, c_2. Any basis of r will also be called a basis of the differential equation (q). The bases (u, v), (v, u) will be called *inverse*, and two bases (u, v), (ku, kv), in which k $(\neq 0)$ is an arbitrary constant, will be called *proportional*.

1.4 Integral curves

A basis (u, v) of the differential equation (q) determines a plane curve with parametric coordinates $u(t)$, $v(t)$. The curve can therefore be defined by means of the parametric representation $x_1 = u(t)$, $x_2 = v(t)$, with respect to a fixed coordinate system with origin O. We shall call such a curve an *integral curve* of the differential equation (q). If the variable t represents time then the integral curve can be considered as the trajectory of the point $P(t) = P[u(t), v(t)]$. The oriented area traced out by the radius vector $\overrightarrow{OP}$ in the time interval $t_1 < t < t_2$ has the value $-\frac{1}{2}w(t_2 - t_1)$, where w represents of course the Wronskian of u, v. An element of the two-dimensional linear homogeneous transformation group, that is to say a centroaffine plane transformation, transforms any integral curve of the differential equation (q) into another such integral curve. Those properties of an integral curve of the differential equation (q) which are invariant under such transformations, hold for all integral curves of (q) and are determined by appropriate properties of the carrier q of the differential equation (q). Conversely, those properties of the integral curves of the differential equation (q) which arise from special properties of the carrier q, have an invariant character with respect to centroaffine plane transformations.

1.5 Kinematic interpretation of integrals

Occasionally it is useful to regard the value $u(t)$ of an integral u of the differential equation (q) as the directed distance of a point P, moving on an oriented line G, from a fixed point or origin O on the line G. At any instant t $(\in j)$ the point P is at a distance $u(t)$ units from the origin O, and lies in the positive or negative direction on the line G according as $u(t) > 0$ or < 0. The instants when P passes through the origin O are given precisely by the zeros of u. We say that the motion of the point P *follows* the integral u of the differential equation (q).

1.6 Types of differential equations (q)

All integrals of the differential equation (q) have the same oscillatory character, that is to say they all have either a finite or an infinite number of zeros in the interval j.

In the first case the differential equation (q) is said to be *of finite type* or *non-oscillatory*. More precisely, it is said to be *of type* (m), m integral, $m \geqslant 1$, if it possesses integrals with m zeros in the interval j but none with $m + 1$ zeros. In the second case the differential equation (q) is said to be *of infinite type*; specifically, it is described as *left* or *right oscillatory* according as the zeros of its integral cluster towards the left-hand end point or the right-hand end point of the interval j, and *oscillatory* if the zeros cluster towards both end points. Alternatively, we describe a differential equation (q) of infinite type as being of the *first*, *second* or *third category*.

Later (§§ 3.6, 3.10, 7.2) we shall have occasion to separate differential equations (q) of finite type (m), $m \geqslant 1$, into general and special differential equations. The term *kind* of a differential equation (q) will have two meanings; if (q) is of finite type, then its "kind" denotes whether it is general or special; if (q) is of infinite type than its "kind" is its category.

All zeros of integrals of the differential equation (q) are isolated.

1.7 The Schwarzian derivative

We shall now consider a bi-rational transformation T defined in a three-dimensional real coordinate space S_3, specified by the formulae*

$$\left.\begin{aligned} X' &= \frac{1}{\dot{x}}, & \dot{x} &= \frac{1}{X'}, \\ X'' &= -\frac{\ddot{x}}{\dot{x}^3}, & \ddot{x} &= -\frac{X''}{X'^3}, \\ X''' &= 3\frac{\ddot{x}^2}{\dot{x}^5} - \frac{\dddot{x}}{\dot{x}^4}, & \dddot{x} &= 3\frac{X''^2}{X'^5} - \frac{X'''}{X'^4}. \end{aligned}\right\} \tag{1.6}$$

It thus associates with each other two points X' ($\neq 0$), X'', X'''; and $\dot{x}$ ($\neq 0$), $\ddot{x}$, $\dddot{x}$. This transformation T leaves the function

$$K(X) = \frac{X''^2}{X'^3} \tag{1.7}$$

invariant, that is to say

$$K(X) = K(x). \tag{1.8}$$

The so-called Schwarzian function

$$S(X) = \frac{1}{2}\frac{X'''}{X'} - \frac{3}{4}\frac{X''^2}{X'^2} \tag{1.9}$$

is transformed as follows:

$$\frac{S(X)}{X'} + \frac{S(x)}{\dot{x}} = 0. \tag{1.10}$$

* It is useful to note the convention, adopted throughout, that differentiation with respect to t is indicated by a prime, and with respect to T by a dot. In (1.6), however, primes and dots serve only to label the coordinates X', $\dot{x}$, etc. (Trans.)

The transformation T is of particular importance in the study of relationships between the values of two (mutually) inverse functions of one variable and of their derivatives.

Let $X(t)$, $x(T)$ be two inverse functions whose intervals of definition we shall denote by i, I respectively. We naturally suppose that the functions X, x are monotonic in the intervals $i = x(I)$, $I = X(i)$. For convenience of terminology we call the numbers t, T *homologous* if they are related by the formulae $T = X(t)$, $t = x(T)$, sometimes the number $t(T)$ will be described as the number homologous to $T(t)$.

We assume that the functions X, x are three times differentiable in the intervals i, I, and $X' \neq 0$, $\dot{x} \neq 0$. Then the rules of differentiation give the following formulae, holding at two homologous numbers $t \in i$, $T \in I$:

$$\left.\begin{aligned} X'\dot{x} &= 1, \\ X''\dot{x} + X'^2\ddot{x} = 0; \qquad \ddot{x}X' + \dot{x}^2X'' &= 0, \\ X'''\dot{x}^2 + 3X''\ddot{x} + \dddot{x}X'^2 &= 0. \end{aligned}\right\} \tag{1.11}$$

Thus the bi-rational transformation T has the property that the values of the derivatives of two inverse functions X, x go over into each other at homologous points.

1.8 Projective property of the Schwarzian derivative

Let $X(t)$ be a three times differentiable function in the interval j, whose derivative X' is everywhere non-zero, i.e. $X'(t) \neq 0$ for $t \in j$. By the *Schwarzian derivative* of the function X we mean the Schwarzian function $S(X)$ formed with the derivatives X', X'', X'''. The value of the Schwarzian derivative of X at the point $t \in j$ will be denoted by $\{X, t\}$, that is to say

$$\{X, t\} = \frac{1}{2}\frac{X'''(t)}{X'(t)} - \frac{3}{4}\frac{X''^2(t)}{X'^2(t)}. \tag{1.12}$$

A simple calculation yields the relationship

$$\{X, t\} = -\sqrt{|X'|}\left(\frac{1}{\sqrt{|X'|}}\right)'', \tag{1.13}$$

in which on the right-hand side we take the value of the function at the point t. Schwarzian derivatives are of particular value in the linear transformation of functions. A fundamental theorem is the following:

Theorem. The Schwarzian derivatives $\{X, t\}$, $\{Y, t\}$ of two functions X, $Y \in C_3$ in an interval j are identical if and only if X and Y are related projectively, that is

$$Y(t) = \frac{c_{11}X(t) + c_{10}}{c_{21}X(t) + c_{20}}, \tag{1.14}$$

$t \in j$, c_{10}, c_{11}, c_{20}, $c_{21} = \text{const.}$

Proof. Simple calculation shows that the condition for the identity of the two Schwarzian derivatives is sufficient; we have therefore only to prove its necessity.

Let $\{X, t\} = \{Y, t\}$ in the interval j; for brevity, set $\{X, t\} = \{Y, t\} = q(t)$, then $q(t)$ is a continuous function. The formulae (13) shows that the functions $1/\sqrt{|X'|}$, $1/\sqrt{|Y'|}$ are integrals of the differential equation (−q). Since these functions are everywhere non-zero then (from (5)), $X/\sqrt{|X'|}$, $Y/\sqrt{|Y'|}$ are also integrals of the same differential equation (−q), the integrals $1/\sqrt{|X'|}$, $X/\sqrt{|X'|}$ and also the integrals $1/\sqrt{|Y'|}$, $Y/\sqrt{|Y'|}$ being linearly independent. Consequently there exist constants $c_{10}, c_{11}, c_{20}, c_{21}$ such that

$$\frac{Y}{\sqrt{|Y'|}} = c_{11}\frac{X}{\sqrt{|X'|}} + c_{10}\frac{1}{\sqrt{|X'|}},$$
$$\frac{1}{\sqrt{|Y'|}} = c_{21}\frac{X}{\sqrt{|X'|}} + c_{20}\frac{1}{\sqrt{|X'|}},$$

and from this the relation (14) follows. This completes the proof.

We must now mention two further properties of the Schwarzian derivative which are important for our further studies. In this, $X(t)$, $x(T)$ are three times differentiable functions, inverse to each other in the intervals i, I.

1. By easy calculation we find that the following relationship holds at arbitrary points $t \in i$, $T \in I$;

$$\left.\begin{aligned} \frac{\{X, t\}}{X'} &= \frac{1}{4}\frac{X''^2}{X'^3} - \frac{1}{2}\left(\frac{1}{X'}\right)'', \\ \frac{\{x, T\}}{\dot{x}} &= \frac{1}{4}\frac{\ddot{x}^2}{\dot{x}^3} - \frac{1}{2}\left(\frac{1}{\dot{x}}\right)^{\cdot\cdot}. \end{aligned}\right\} \tag{1.15}$$

In particular, if the numbers t, T are homologous, then formulae such as (6), (8), are valid and we obtain the following result:

At two homologous points $t \in i$, $T \in I$ there holds the symmetric relationship

$$\frac{\{X, t\}}{X'} + \frac{1}{2}\left(\frac{1}{X'}\right)'' = \frac{\{x, T\}}{\dot{x}} + \frac{1}{2}\left(\frac{1}{\dot{x}}\right)^{\cdot\cdot}. \tag{1.16}$$

2. Let Z be a three times differentiable function in an interval h, such that $Z(t) \subset i$ and $Z'(t) \neq 0$ for $t \in h$.

Then the composite function XZ exists in the interval h; its Schwarzian derivative also exists there and we have the relationship

$$\{XZ, t\} = \{X, Z(t)\}\, Z'^2(t) + \{Z, t\}. \tag{1.17}$$

1.9 Associated differential equations

In this section we assume that the carrier q of the differential equation (q) in the interval j is everywhere non-zero and $\in C_2$. We then define in the interval j the following differential equation, which we call the *first associated differential equation* ($\hat{q}_1$) of (q):

$$y'' = \hat{q}_1(t) y_1, \tag{q$_1$}$$

where

$$\hat{q}_1(t) = q(t) + \sqrt{|q(t)|}\left(\frac{1}{\sqrt{|q(t)|}}\right)''. \tag{1.18}$$

The function $\hat{q}_1$, the so-called *first associated carrier of* q, can obviously be put into the form

$$\hat{q}_1(t) = q(t) - \frac{1}{2}\frac{q''(t)}{q(t)} + \frac{3}{4}\frac{q'^2(t)}{q^2(t)} \tag{1.19}$$

or alternatively

$$\hat{q}_1(t) = q(t) - \left\{\int_{t_0}^{t} q(\sigma)\, d\sigma, t\right\} \quad (t_0 \in j). \tag{1.20}$$

The significant connection between the differential equations (q), ($\hat{\mathrm{q}}_1$) lies in the fact that given any integral y of the differential equation (q) the function

$$y_1(t) = \frac{y'(t)}{\sqrt{|q(t)|}} \tag{1.21}$$

is an integral of the differential equation ($\hat{\mathrm{q}}_1$). We have also the converse relationship:

For every integral y_1 of the differential equation ($\hat{\mathrm{q}}_1$) *the function $y_1\sqrt{|q(t)|}$ represents the derivative y' of precisely one integral y of* (q).

Proof. Let y_1 be an integral of ($\hat{\mathrm{q}}_1$). We choose an arbitrary number $t_0 \in j$.

(a) We suppose that there is an integral y of the differential equation (q) such that in the interval j

$$y_1\sqrt{|q|} = y'. \tag{1.22}$$

At the point t_0 the integral y and its derivative y' obviously take the values

$$y_0 = \frac{1}{q(t_0)}[y_1\sqrt{|q|}]'_{t=t_0}; \qquad y'_0 = y_1(t_0)\sqrt{|q(t_0)|}. \tag{1.23}$$

We see that there is at most one integral y of (q) satisfying the relation (22), namely that integral of (q) determined by the initial values (23).

(b) We now define, in the interval j, the function y as follows

$$y(t) = \frac{1}{q(t_0)}[y_1\sqrt{|q|}]'_{t=t_0} + \int_{t_0}^{t} y_1(\sigma)\sqrt{|q(\sigma)|}\, d\sigma.$$

The function y and its derivative obviously take the values (23) at the point t_0. Moreover, the condition (22) clearly holds in the interval j. Then it follows easily that y_1 satisfies the differential equation ($\hat{\mathrm{q}}_1$), and by application of the formula (19) we see that it also satisfies the equation

$$y''' - y''\frac{q'}{q} - y'q = 0$$

which may be written

$$\left[\frac{y'' - qy}{q}\right]' = 0.$$

Consequently we have

$$y'' - qy = kq \qquad (k = \text{const}),$$

and then equation (22) and the initial values given in (23) show that $k = 0$. The function y is consequently an integral of the differential equation (q), and the proof is complete.

The mapping P of the integral space r of (q) on the integral space r_1 of ($\hat{q}_1$), by which each integral $y \in r$ is mapped into the integral $y_1 = y'/\sqrt{|q|} \in r_1$, is called the *projection* of the integral space r onto the integral space r_1. We also say that y_1 ($= Py$) is the projection of y, and call the integrals y, y_1 *associated*. The reader may easily verify that the Wronskians of two bases (u, v), (Pu, Pv), of r and r_1 respectively, have the same value.

The differential equation ($\hat{q}_1$) represents the first associated differential equation of (q). The n-th *associated differential equation* ($\hat{q}_n$) of (q) is defined as the first associated differential equation of ($\hat{q}_{n-1}$). For example, if we take the Bessel differential equation

$$y'' = -\left(1 + \frac{1 - 4\nu^2}{4t^2}\right) y \quad (j = (0, \infty),\ \nu = \text{const}) \tag{1.24}$$

then the first associated differential equation belonging to this is

$$y'' = -\left(1 + \frac{1 - 4\nu^2}{4t^2} + \frac{12(1 - 4\nu^2)}{(4t^2 + 1 - 4\nu^2)^2}\right) y. \tag{1.25}$$

2 Elementary properties of integrals of the differential equation (q)

2.1 Relative positions of zeros of an integral and its derivative

Between two zeros of an integral y of the differential equation (q) there always lies at least one zero of its derivative y'. Between two zeros of the derivative y' there always lies at least one zero of y or one zero of q. It follows that:

Between two neighbouring zeros of an integral y of the differential equation (q) *lies precisely one zero of y', if q does not vanish in this interval. Between two neighbouring zeros of the derivative y' lies precisely one zero of y if $q \neq 0$ in this interval.*

In this statement, the inequality $q \neq 0$ can without loss of generality be replaced by $q < 0$, in consequence of the following theorem.

Theorem. If, between two neighbouring zeros of an integral y of the differential equation (q), *or between two zeros of its derivative y', or between a zero of y and a zero of y', the function q does not vanish then it must be negative, i.e. $q < 0$.*

Proof. Obviously, it is sufficient only to consider the third case. Let $t_1, x_1 \in j$, with $t_1 < x_1$, and assume, for example, that $y(t_1) = y'(x_1) = 0$, while $y(t) > 0$, $y'(t) > 0$ for $t \in (t_1, x_1)$. If possible, let $q > 0$ in the interval (t_1, x_1). Then in this interval $y'' > 0$, the function y' is increasing and since $y'(x_1) = 0$, y' is negative, which contradicts our hypothesis and so proves the theorem.

2.2 Ratios of integrals and their derivatives

For two integrals u, v of the differential equation (q) the following formulae hold in the interval j, with the exception, naturally, of points where the denominators vanish:

$$\left(\frac{u}{v}\right)' = -\frac{w}{v^2}, \quad \left(\frac{u'}{v'}\right)' = \frac{wq}{v'^2}, \quad \left(\frac{uu'}{vv'}\right)' = w\,\frac{quv - u'v'}{v^2v'^2} \tag{2.1}$$

$$(w = uv' - u'v).$$

We obtain from these the following results:

Let the integrals u, v be linearly independent; then in every interval $i \subset j$ containing no zeros of v, the ratio u/v is either an increasing or a decreasing function, according as $w < 0$ or $w > 0$. On the same assumption, in every interval $i \subset j$ which contains no zeros of v', the ratio u'/v' is an increasing or decreasing function according as $wq > 0$ or $wq < 0$. A similar statement also holds for the function uu'/vv'.

By integration of the above formulae over an interval $(t, x) \subset j$, in which the denominators involved are not zero, we obtain

$$\left.\begin{aligned} \frac{u(x)}{v(x)} - \frac{u(t)}{v(t)} = -w\int_t^x \frac{d\sigma}{v^2}, \qquad \frac{u'(x)}{v'(x)} - \frac{u'(t)}{v'(t)} = w\int_t^x \frac{q\,d\sigma}{v'^2}, \\ \frac{u(x)u'(x)}{v(x)v'(x)} - \frac{u(t)u'(t)}{v(t)v'(t)} = w\int_t^x \frac{quv - u'v'}{v^2v'^2}\,d\sigma. \end{aligned}\right\} \tag{2.2}$$

If the numbers t, x are zeros of the function u or of the function u', or if one of them is a zero of u and the other a zero of u', then the integral on the right hand side of the corresponding formula is zero.

2.3 The ordering theorems

There are several important laws governing the location of zeros of two independent integrals of the differential equation (q) and of their derivatives. These are described in the following four theorems, the so-called *ordering theorems*. Proofs follow from the formulae (2) above.

Let u, v be independent integrals of the differential equation (q) *and t_1, x_1 be numbers in the interval j with $t_1 < x_1$.*

(1) *Let $u(t_1) = u(x_1) = 0$, $u(t) \neq 0$ for $t \in (t_1, x_1)$, then the integral v has precisely one zero in the interval (t_1, x_1).*

We now make the additional assumption that $q \neq 0$ for $t \in j$.

(2) *Let $u'(t_1) = u'(x_1) = 0$, $u'(t) \neq 0$ for $t \in (t_1, x_1)$, then the function v' has precisely one zero in the interval (t_1, x_1).*

(3) *Let $u'(t_1) = u(x_1) = 0$, $u(t) \neq 0$ for $t \in (t_1, x_1)$. If $t_2 < t_1$ and $v'(t_2) = 0$, then the integral v has a zero $x_2 \in (t_2, x_1)$. If $x_2 > x_1$ and $v(x_2) = 0$, then the function v' has a zero $t_2 \in (t_1, x_2)$.*

(4) *Let $u(t_1) = u'(x_1) = 0$, $u(t) \neq 0$ for $t \in (t_1, x_1)$. If $t_2 < t_1$ and $v(t_2) = 0$, then the function v' has a zero $x_2 \in (t_2, x_1)$. If $x_2 > x_1$ and $v'(x_2) = 0$, then the function v has a zero $t_2 \in (t_1, x_2)$.*

Proof. We shall give only the proof of the first part of (3). Assuming the contrary, we suppose that $v(t) \neq 0$ for $t \in (t_2, x_1)$. Then $v(t)v'(t) \neq 0$ for $t \in (t_1, x_1)$ and, indeed, even for $t \in [t_1, x_1]$. We can thus apply the last formula (2) to the integrals u, v in the interval $[t_1, x_1]$ from which it follows that

$$\int_{t_1}^{x_1} \frac{quv - u'v'}{v^2v'^2}\,d\sigma = 0.$$

Obviously we can assume that $v'(t_1) < 0$, $u'(x_1) < 0$. Then in the interval (t_1, x_1) we have $u > 0$, $u' < 0$, $v > 0$, $v^1 < 0$. This, however, is inconsistent with the above integral relationship, so our assumption is false and the proof is completed.

2.4 The (Riemann) integrals $\int_{x_0}^{x_1} \frac{d\sigma}{y^2(\sigma)}, \int_{x_0}^{x_1} \frac{q(\sigma)}{y'^2(\sigma)}\, d\sigma$ in the neighbourhood of a singular point

Let y be an integral of the differential equation (q) with a zero at the point c. We consider a left or right neighbourhood j_{-1} or j_0 of c, in which the integral y does not vanish, and choose first a number $x_0 \in j_{-1}$. We wish to study the behaviour of the integral $\int_{x_0}^{t} d\sigma/y^2(\sigma)$, $t \in j_{-1}$, in the neighbourhood of the singular point c.

Obviously, for $\sigma \in j_{-1}$ we have the formula

$$y(\sigma) = y'(c)\,(\sigma - c) + \frac{(\sigma - c)^2}{2}\, y''(\tau),$$

where $\sigma < \tau < c$. From this it follows that

$$y^2(\sigma) = y'^2(c)\,(\sigma - c)^2 \left[1 + \frac{\sigma - c}{2} \cdot \frac{y''(\tau)}{y'(c)}\right]^2,$$

hence

$$\frac{1}{y^2(\sigma)} = \frac{1}{y'^2(c)\,(\sigma - c)^2} \cdot \frac{1}{\left[1 + \frac{\sigma - c}{2} \cdot \frac{y''(\tau)}{y'(c)}\right]^2}.$$

Now we apply the Taylor expansion formula to obtain

$$\frac{1}{\left[1 + \frac{\sigma - c}{2} \cdot \frac{y''(\tau)}{y'(c)}\right]^2} = 1 - (\sigma - c)\frac{y''(\tau)}{y'(c)} + \frac{(\sigma - c)^2}{4} \cdot \frac{y''^2(\tau)}{y'^2(c)} \cdot \frac{3}{\left[1 + \Theta \frac{\sigma - c}{2} \cdot \frac{y''(\tau)}{y'(c)}\right]^4}$$

with $0 < \Theta < 1$. Consequently

$$\frac{1}{y^2(\sigma)} = \frac{1}{y'^2(c)} \left[\frac{1}{(\sigma - c)^2} - \frac{q(\tau)}{y'(c)} \cdot \frac{\tau - c}{\sigma - c} \cdot \frac{y(\tau) - y(c)}{\tau - c}\right] + O(1),$$

in which the symbol O naturally relates to the left neighbourhood of c. For $\sigma \in j_{-1}$, let

$$g(\sigma) = \frac{1}{y^2(\sigma)} - \frac{1}{y'^2(c)} \cdot \frac{1}{(\sigma - c)^2}; \tag{2.3}$$

we then have

$$g(\sigma) = -\frac{q(\tau)}{y'^3(c)} \cdot \frac{\tau - c}{\sigma - c} \cdot \frac{y(\tau) - y(c)}{\tau - c} + O(1). \tag{2.4}$$

By the formula (3) the function g is continuous in the interval j_{-1}, while (4) shows that it is bounded there. From this follows the existence of the Riemann integral

$\int_{x_5}^{c} g(\sigma)\, d\sigma$. We now make use of the formula (3) to extend the definition of the function g over the interval j_0. An argument similar to that used above shows that for every $x_1 \in j_0$ the integral $\int_{c}^{x_1} g(\sigma)\, d\sigma$ exists.

For every two numbers $x_0 \in j_{-1}$, $x_1 \in j_0$, the integral

$$\int_{x_0}^{x_1} g(\sigma)\, d\sigma = \int_{x_0}^{x_1} \left[\frac{1}{y^2(\sigma)} - \frac{1}{y'^2(c)} \cdot \frac{1}{(\sigma - c)^2}\right] d\sigma$$

exists. Now let x_0, x_m ($x_0 < x_m$) be arbitrary numbers in the interval j, which are not zeros of y and between which lie precisely m ($\geqslant 1$) zeros $c_1, \ldots, c_m$ of y, ordered so that $x_0 < c_1 < \ldots < c_m < x_m$.

Now we define the function g_m as follows:

$$g_m(\sigma) = \frac{1}{y^2(\sigma)} - \sum_{\mu=1}^{m} \frac{1}{y'^2(c_\mu)} \cdot \frac{1}{(\sigma - c_\mu)^2},$$

this definition being valid in the interval $[x_0, x_m]$ with the exception of the points c_μ. We choose a number x_μ in each interval $(c_\mu, c_{\mu+1})$, $\mu = 1, \ldots, m - 1$. From the above result, the following integral exists for $\nu = 1, \ldots, m$:

$$\int_{x_{\nu-1}}^{x_\nu} g_m(\sigma)\, d\sigma = \int_{x_{\nu-1}}^{x_\nu} \left[\frac{1}{y^2(\sigma)} - \frac{1}{y'^2(c_\nu)} \cdot \frac{1}{(\sigma - c_\nu)^2}\right] d\sigma + \sum_{\nu \neq \mu = 1}^{m} \frac{1}{y'^2(c_\mu)} \left[\frac{1}{x_\nu - c_\mu} - \frac{1}{x_{\nu-1} - c_\mu}\right].$$

Then by summation we obtain the formula

$$\begin{aligned}\int_{x_0}^{x_m} g_m(\sigma)\, d\sigma = \sum_{\nu=1}^{m} \int_{x_{\nu-1}}^{x_\nu} &\left[\frac{1}{y^2(\sigma)} - \frac{1}{y'^2(c_\nu)} \cdot \frac{1}{(\sigma - c_\nu)^2}\right] d\sigma \\ &- \sum_{\mu=1}^{m} \frac{1}{y'^2(c_\mu)} \left[\frac{1}{c_\mu - x_{\mu-1}} + \frac{1}{x_\mu - c_\mu}\right] \\ &+ \sum_{\mu=1}^{m} \frac{1}{y'^2(c_\mu)} \left[\frac{1}{c_\mu - x_0} + \frac{1}{x_m - c_\mu}\right].\end{aligned}$$

2.5 Application to the associated equation

Now we assume that $q \in C_2$ and does not vanish in the interval j. Then we can apply the above results to the first associated differential equation ($\hat{q}_1$) of (q) (§ 1.9).

Let y be an integral of (q) and $e \in j$ a zero of its derivative y'. We define the function $h(\sigma)$ in a neighbourhood of e, $\sigma \neq e$, by

$$h(\sigma) = \frac{q(\sigma)}{y'^2(\sigma)} - \frac{1}{q(e) y^2(e)} \cdot \frac{1}{(\sigma - e)^2};$$

then for every two numbers x_0, $x_1 \in j$, which are not zeros of the derivative y' and between which lies precisely the one zero e of y', there exists the integral

$$\int_{x_0}^{x_1} h(\sigma)\, d\sigma = \int_{x_0}^{x_1} \left[\frac{q(\sigma)}{y'^2(\sigma)} - \frac{1}{q(e)y^2(e)} \cdot \frac{1}{(\sigma - e)^2}\right] d\sigma.$$

More generally; let x_0, x_m ($x_0 < x_m$) be arbitrary numbers in the interval j which are not zeros of the derivative y' and between which lie precisely m ($\geqslant 1$) zeros $e_1, \ldots, e_m$ of y' ordered such that $x_0 < e_1 < \ldots < e_m < x_m$. In the interval $[x_0, x_m]$ with the exception of the numbers e_μ we define the function h_m as:

$$h_m(\sigma) = \frac{q(\sigma)}{y'^2(\sigma)} - \sum_{\mu=1}^{m} \frac{1}{q(e_\mu)y^2(e_\mu)} \cdot \frac{1}{(\sigma - e_\mu)^2},$$

and in every interval $(e_\mu, e_{\mu+1})$ we choose a number x_μ, $\mu = 1, \ldots, m - 1$. Then the integral of the function h_m exists between the limits x_0, x_m, and we have the following formula:

$$\begin{aligned}\int_{x_0}^{x_m} h_m(\sigma)\, d\sigma = {} & \sum_{\nu=1}^{m} \int_{x_{\nu-1}}^{x_\nu} \left[\frac{q(\sigma)}{y'^2(\sigma)} - \frac{1}{q(e_\nu)y^2(e_\nu)} \cdot \frac{1}{(\sigma - e_\nu)^2}\right] d\sigma \\ & - \sum_{\mu=1}^{m} \frac{1}{q(e_\mu)y^2(e_\mu)} \left[\frac{1}{e_\mu - x_{\mu-1}} + \frac{1}{x_\mu - e_\mu}\right] \\ & + \sum_{\mu=1}^{m} \frac{1}{q(e_\mu)y^2(e_\mu)} \left[\frac{1}{e_\mu - x_0} + \frac{1}{x_m - e_\mu}\right].\end{aligned}$$

2.6 Basis functions

We now consider two differential equations

$$y'' = q(t)y, \qquad \text{(q)}$$

$$\ddot{Y} = Q(T)Y \qquad \text{(Q)}$$

on the intervals j, J, i.e. $t \in j$, $T \in J$. We do not exclude the possibility that these two differential equations coincide.

Let (u, v), (U, V) be an ordered pair of arbitrary bases for (q), (Q) respectively. By a *basis function* belonging to this ordered pair of bases we mean a function $F(t, T)$ defined on the region $j \times J$ by one of the following four formulae:

1. $u(t)V(T) - v(t)U(T)$,
2. $u'(t)\dot{V}(T) - v'(t)\dot{U}(T)$,
3. $u(t)\dot{V}(T) - v(t)\dot{U}(T)$,
4. $u'(t)V(T) - v'(t)U(T)$.

Thus there are four basis functions corresponding to the above basis pair for the differential equations (q), (Q) (and consequently to every such basis pair). If the differential equations (q). (Q) coincide, then we speak of *basis functions of the differential equation* (q).

We consider a basis function $F(t, T)$. Let $t_0 \in j$, $X_0 \in J$ be arbitrary numbers for which $F(t_0, X_0) = 0$ and in the cases 2 and 3 assume also that $Q(X_0) \neq 0$. We wish to show that *there is precisely one function $X(t)$ defined in a neighbourhood of t_0, which takes the value X_0 at the point t_0, is continuous in its interval of definition, and satisfies the equation $F[t, X(t)] = 0$. This function X has moreover, in its interval of definition, the continuous derivative*

$$X'(t) = -\frac{F'[t, X(t)]}{\dot{F}[t, X(t)]}.$$

In the individual cases the derivative $X'(t)$ is therefore given by the following expressions

$$1.\quad -\frac{u'(t)V[X(t)] - v'(t)U[X(t)]}{u(t)\dot{V}[X(t)] - v(t)\dot{U}[X(t)]},$$

$$2.\quad -\frac{q(t)}{Q[X(t)]}\cdot\frac{u(t)\dot{V}[X(t)] - v(t)\dot{U}[X(t)]}{u'(t)V[X(t)] - v'(t)U[X(t)]},$$

$$3.\quad -\frac{1}{Q[X(t)]}\cdot\frac{u'(t)\dot{V}[X(t)] - v'(t)\dot{U}[X(t)]}{u(t)V[X(t)] - v(t)U[X(t)]},$$

$$4.\quad -q(t)\frac{u(t)V[X(t)] - v(t)U[X(t)]}{u'(t)\dot{V}[X(t)] - v'(t)\dot{U}[X(t)]}.$$

To illustrate the method of proof, take the function

$$F(t, T) = u(t)V(T) - v(t)U(T).$$

According to our assumption we have $F(t_0, X_0) = 0$ and the function F obviously possesses continuous partial derivatives

$$F'(t, T) = u'(t)V(T) - v'(t)U(T),$$
$$\dot{F}(t, T) = u(t)\dot{V}(T) - v(t)\dot{U}(T).$$

at every point $(t, T) \in j \times J$.

Further, $\dot{F}(t_0, X_0) \neq 0$, for otherwise we would have

$$(F(t_0, X_0) =)\quad u(t_0)V(X_0) - v(t_0)U(X_0) = 0,$$
$$(\dot{F}(t_0, X_0) =)\quad u(t_0)\dot{V}(X_0) - v(t_0)\dot{U}(X_0) = 0,$$

and these two relations (when we recall that $u^2(t_0) + v^2(t_0) \neq 0$) contradict the linear independence of the integrals U, V of (Q). Now we only need to apply the classic implicit function theorem, and the proof is complete.

We observe that: if two functions $(z =)\ x$, X are continuous in an interval $i \subset j$, take the same value at a point of the interval i, and satisfy in this interval the equation $F(t, z) = 0$, then they coincide in the interval i. For, if this were not so, there would be numbers $t_1 < t_2$ in the interval i such that, for instance, $x(t_1) = X(t_1)$ and $x(t) \neq X(t)$ for $t_1 < t \leqslant t_2$. This, however, contradicts the above theorem.

3 Conjugate numbers

Conjugate numbers, the concept and properties of which will be described in this paragraph, play a very important role in the theory of linear differential equations of the second order and particularly in the transformation theory of such differential equations. We will consider a differential equation (q).

3.1 The concept of conjugate numbers

Let $t \in j$ be arbitrary, and let u, v be arbitrary integrals of the differential equation (q), such that $u(t) = 0$, $v'(t) = 0$.

We describe a number $x \in j$ $(x \neq t)$ as being *conjugate* with the number t (with respect to the differential equation (q)), and more precisely of the

1st kind, 2nd kind, 3rd kind, 4th kind,

according as

$$u(x) = 0, \quad v'(x) = 0, \quad u'(x) = 0, \quad v(x) = 0.$$

If the number x is conjugate with t of the κ-th kind, then we call it κ-conjugate with t; $\kappa = 1, 2, 3, 4$. A number x which is κ-conjugate with t is called *left* or *right conjugate* (of the κ-th kind), according as $x < t$ or $x > t$. We see that the number x is a conjugate number with t of the 1st, 2nd, 3rd or 4th kind, according as it is a zero of the function u, v', u', v respectively. Let the number x be the n-th zero $(n \geqslant 1)$ of this function lying on the left or right of t; then we call it the n-th *left* or *right conjugate number with* t, of the corresponding kind.

Let x be the n-th left or right conjugate number with t of the 1st or 2nd kind, then t is the n-th right or left conjugate number with x of the 1st or 2nd kind. Because of this symmetry we can refer simply to 1st or 2nd kind conjugate numbers, and particularly in the case $n = 1$ to *neighbouring* 1st or 2nd conjugate numbers t, x. On the assumption that between two neighbouring zeros of an integral of (q), or of its derivative, there lies precisely one zero of this derivative or of the integral respectively, we can make the further statement: Let x be the n-th left or right conjugate number with t of the 3rd or 4th kind; then t is respectively the n-th right or left conjugate number with x of the 4th or 3rd kind.

We know that two integrals of the differential equation (q) which have a common zero, or whose derivatives have a common zero, are merely constant multiples of each other and it follows that all their zeros and all the zeros of their derivatives coincide. Hence the concept of conjugate numbers is inherent in the differential equation (q) itself and does not depend on the particular choice of the integrals u, v used in their definition.

3.2 Classification of differential equations (q) with respect to conjugate numbers

By consideration of conjugate numbers of the κ-th kind ($\kappa = 1, 2, 3, 4$) we can separate differential equations of the type (q) into two classes according as there exist, in the interval j, conjugate numbers of the κ-th kind or not. For example, taking the differential equation (q) with the carrier $q = -1$, consider an open interval j of length $|j|$. In the case $0 < |j| \leqslant \frac{1}{2}\pi$ there are no conjugate numbers, in the case $\frac{1}{2}\pi < |j| \leqslant \pi$ there are conjugate numbers only of the 3rd and 4th kinds, while for $\pi < |j|$ there are conjugate numbers of all kinds.

When we have differential equations of the type (q) for which κ-conjugate numbers exist, we call them *differential equations with conjugate numbers (points) of the κ-th kind*, or *differential equations with κ-conjugate numbers (points)*. We also express this fact by saying that the differential equation (q) admits of or possesses conjugate numbers (points) of the κ-th kind, or κ-conjugate numbers.

Differential equations of the type (q) for which no κ-conjugate numbers exist we call *differential equations without conjugate numbers (points) of the κ-th kind* or *differential equations without κ-conjugate numbers (points)*. In this case we also say that the differential equation (q) admits of or possesses no conjugate numbers (points) of the κ-th kind or no κ-conjugate numbers (points).

3.3 Properties of differential equations (q) with κ-conjugate numbers

In this and in §§ 3.4 to 3.9 we shall concern ourselves with differential equations (q_κ) with conjugate numbers of the κ-th kind ($\kappa = 1, 2, 3, 4$). In the cases $\kappa = 2, 3, 4$ it is convenient to suppose that $q_\kappa(t) < 0$ for $t \in j$, in order to simplify our study; in particular, this assumption makes it possible to apply the ordering theorems (§ 2.3) in the interval j.

Let (q_κ) be a differential equation with conjugate numbers of the κ-th kind. First we have the following theorems:

Given any number $x_1 \in j$, which possesses left conjugate numbers of the κ-th kind, there exist smaller numbers which also possess left conjugate numbers of the κ-th kind. All numbers greater than x_1 lying in the interval j possess left conjugate numbers of the κ-th kind. Given any number $x_1 \in j$ which possesses right conjugate numbers of the κ-th kind, there exist greater numbers which also possess right conjugate numbers of the κ-th kind. All numbers smaller than x_1 lying in the interval j possess right conjugate numbers of the κ-th kind.

Proof. We shall only give the proof of the first theorem. Let t_1 be the first left conjugate number with x_1 of the κ-th kind. We choose an arbitrary number $t_2 < t_1$, $t_2 \in j$. Then by the ordering theorems there is a number x_2 which is right conjugate with t_2 of the kind κ' ($\kappa' = \kappa$ or $\kappa' = \kappa \pm 1$) and $x_2 < x_1$. This number x_2 then possess the left

κ-conjugate number t_2, which proves the first part of our result. The second part follows immediately from the ordering theorems.

3.4 Fundamental numbers

We denote by R_κ, S_κ respectively the sets of numbers lying in the interval j which possess left (or right) conjugate numbers of the κ-th kind. Since, on our hypothesis, κ-conjugate numbers exist, in the cases $\kappa = 1, 2$ both sets R_κ, S_κ are non-empty, while in the cases $\kappa = 3, 4$ at least one of these is non-empty.

Let $R_\kappa \neq \varnothing$. If the set R_κ is bounded below, then it possesses a greater lower bound r_κ, which either belongs to the interval j, that is $a < r_\kappa$, or coincides with its end point a: $a = r_\kappa$. If the set R_κ is unbounded below, which is obviously only possible in the case $a = -\infty$, then we define the number r_κ as: $r_\kappa = a = -\infty$. The number r_κ we call the *left fundamental number of the κ-th kind* or the *left κ-fundamental number of the κ-th kind* or *the left κ-fundamental number* of the differential equation (q_κ). If $a < r_\kappa$, we call r_κ *proper*, and if $a = r_\kappa$ we call r_κ *improper*.

From the theorems in Section 3 we conclude that if the left κ-fundamental number r_κ of the differential equation (q_κ) is proper, then it is the largest number $\in j$ for which there are no left κ-conjugate numbers. In this case the interval j separates into two sub-intervals $(a, r_\kappa]$, (r_κ, b), of which the first is composed of numbers which do not possess any left κ-conjugate numbers, while there are left κ-conjugate numbers for every member of the second sub-interval

Similarly, if $S_\kappa \neq \varnothing$ and S_κ is bounded above, it has a least upper bound s_κ, with $s_\kappa < b$ or $s_\kappa = b$. If S_κ is unbounded above, we define $s_\kappa = b = \infty$. This number s_κ is called the *right fundamental number of the κ-th kind* or the *right κ-fundamental number* of the differential equation (q_κ), and is *proper* if $s_\kappa < b$, *improper* if $s_\kappa = b$. If s_κ is proper the interval j is composed of the two sub-intervals (a, s_κ), $[s_\kappa, b)$; right κ-conjugate numbers are possessed by every number in (a, s_κ) and by none in $[s_\kappa, b)$.

3.5 Fundamental integrals and fundamental sequences

In this and the following paragraphs we are concerned principally with 1-conjugate and 2-conjugate numbers, and it is convenient to use λ in place of κ, allowing λ to take the values 1, 2 only. We consider a differential equation (q_λ) whose left λ-fundamental number r_λ is proper; $a < r_\lambda$.

Let u_λ be an integral of (q_λ) which vanishes or has its derivative vanishing at the point r_λ, according as $\lambda = 1$ or $\lambda = 2$, that is $u_1(r_1) = 0$, $u_2'(r_2) = 0$. Such an integral u_λ we shall call a *left fundamental integral* of the λ-th kind or a *left λ-fundamental integral*, of (q_λ). Obviously any integral of (q_λ) which is dependent on the integral u_λ is also a left λ-fundamental integral of (q_λ).

Now let $\nu = 1, 2, \ldots$. In the case $\lambda = 1$ let $r_{1,\nu}$ be the ν-th zero of the fundamental integral u_1 lying on the right of r_1. Moreover, if $q_1(t) < 0 \;\forall\; t \in j$, let $r_{4,\nu}$ be the ν-th zero of the derivative u_1' of u_1 on the right of r_4 (assuming, of course, that this zero exists.) The fundamental number r_4 definitely exists (§ 3.7). We set $r_1 = r_{1,0}$, $r_4 = r_{4,0}$.

In the case $\lambda = 2$ let $r_{2,\nu}$ be the ν-th zero of the derivative u_2' of u_2 on the right of r_2. Moreover, let $r_{3,\nu}$ be the ν-th zero of the fundamental integral u_2 on the right of r_3, (again assuming that these zeros exist.) The fundamental number r_3 does exist (§ 3.7). We set $r_2 = r_{2,0}$, $r_3 = r_{3,0}$.

The sequence (finite or infinite) of numbers $\in j$

$$r_{\kappa,0} < r_{\kappa,1} < r_{\kappa,2} < \cdots$$

is known as the *left fundamental sequence of the κ-th kind* or the *left κ-fundamental sequence* of the differential equation; $\kappa = 1, 2, 3, 4$. We denote this sequence by $\boldsymbol{R}_\kappa$. Obviously the fundamental sequence is an entity characteristic of the differential equation itself and does not depend on the choice of the particular fundamental integral used in its definition.

Similarly, let us consider a differential equation (q_λ) $(\lambda = 1, 2)$ whose right λ-fundamental number s_λ is proper; $s_\lambda < b$. We define a *right fundamental integral of the λ-th kind* or a *right fundamental λ-integral* (of q_λ) as an integral v_λ such that $v_1(s_1) = 0$ or $v_2'(s_2) = 0$, according as $\lambda = 1, 2$. When $\lambda = 1$ we define $s_{1,\nu}$ as the ν-th zero of v_1 to the left of s_1 and if $q_1(t) < 0$ in j we define $s_{4,\nu}$ as the ν-th zero of v_1' to the left of s_4, (provided it exists). Further, when $\lambda = 2$ we define $s_{2,\nu}$ and $s_{3,\nu}$ correspondingly as the ν-th zeros of v_2' and v_2 to the left of s_2 and s_3; assuming again that these numbers exist (s_4 and s_3 always do). We write $s_2 = s_{2,0}$, $s_3 = s_{3,0}$.

We then have the *right fundamental sequence of the κ-th kind*, or the *right κ-fundamental sequence*, $\boldsymbol{S}_\kappa$

$$s_{\kappa,0} > s_{\kappa,1} > s_{\kappa,2} > \cdots,$$

The elements of $\boldsymbol{R}_\kappa$, $\boldsymbol{S}_\kappa$ in the interval j are called *singular numbers of the κ-th kind* of the differential equation.

Every two distinct terms ($\neq r_3$ or s_3) of the fundamental sequences $\boldsymbol{R}_1$, $\boldsymbol{R}_3$; $\boldsymbol{S}_1$, $\boldsymbol{S}_3$ are 1-conjugate with each other; every two distinct terms, ($\neq r_4$ or s_4) of each of the fundamental sequences $\boldsymbol{R}_2$, $\boldsymbol{R}_4$; $\boldsymbol{S}_2$, $\boldsymbol{S}_4$ are 2-conjugate with each other. The subintervals of j

$$i_{\kappa,\nu} = (r_{\kappa,\nu-1}, b) \qquad (j_{\kappa,\nu} = (a, s_{\kappa,\nu-1}))$$
$$(\kappa = 1, 2, 3, 4; \qquad \nu = 1, 2, \ldots),$$

consist precisely of those numbers $t \in j$ which possess the ν-th left or right κ-conjugate number respectively.

3.6 General and special equations of finite type

Following on from the above, let (q_λ) be a differential equation, for which both fundamental number r_λ and s_λ are proper $(\lambda = 1, 2)$. Then the differential equation (q_λ) is of finite type (m) (naturally with $m \geqslant 2$). For, if the fundamental numbers r_λ, s_λ are proper, then the left λ-fundamental integral u_λ has on the left of r_λ either no zero (if $\lambda = 1$) or at most one zero (if $\lambda = 2$), and v_λ has similar properties to the right of s_λ.

Generally, the fundamental numbers r_λ, s_λ are not λ-conjugate. In this case the differential equation (q_λ) is called *general*, or *non-special, of the λ-th kind*, more briefly *λ-general* or *λ-non-special.* In the case of a λ-general differential equation (q_λ), the two λ-fundamental integrals u_λ and v_λ are independent, and the two κ-fundamental sequences $\boldsymbol{R}_\kappa$, $\boldsymbol{S}_\kappa$, $\kappa = 1, 2, 3, 4$, have no common terms in the interval j.

If the fundamental numbers r_λ, s_λ are λ-conjugate, then the differential equation (q_λ) is called *special of the λ-th kind* or briefly *λ-special.* In the case of a λ-special differential equation (q_λ) the two λ-fundamental integrals u_λ and v_λ are dependent, and the two fundamental sequences $\boldsymbol{R}_\lambda$, $\boldsymbol{S}_\lambda$ coincide, forming the so-called *fundamental sequence of the λ-th kind*, or briefly the *λ-fundamental sequence* of the differential equation (q_λ).

3.7 Relations between conjugate numbers of different kinds

We now wish to consider relationships between conjugate numbers of different kinds of the same differential equation (q).

We take $q(t) < 0$ for $t \in j$, and let the symbols R_κ, S_κ, $\kappa = 1, 2, 3, 4$ have the same meaning as in § 3.4. Let $t \in R_1$; then every integral u of (q) which vanishes at t has a first zero x lying to the left of t, and between t, x there is precisely one zero x' of u'. The number x' is obviously left 3-conjugate with t, and t is right 4-conjugate with x'. Hence

$$R_1 \subset R_3, \qquad S_1 \subset S_3.$$

Moreover, from $R_1 \neq \varnothing$ or $S_1 \neq \varnothing$ it follows that $R_\kappa \neq \varnothing$, $S_\kappa \neq \varnothing$ for $\kappa = 1, 3, 4$. Similarly we obtain

$$R_2 \subset R_4, \qquad S_2 \subset S_4$$

and from $R_2 \neq \varnothing$ or $S_2 \neq \varnothing$ the conclusion that $R_\kappa \neq \varnothing$, $S_\kappa \neq \varnothing$ for $\kappa = 2, 3, 4$.

If, therefore, the differential equation (q) possesses conjugate numbers of the first or second kind then the fundamental numbers r_κ, s_κ exist for $\kappa = 1, 3, 4$ or $\kappa = 2, 3, 4$, respectively. We now assume that the differential equation (q) has conjugate numbers of all four kinds.

First, the above relations between the sets R_κ, S_κ give the following inequalities for the fundamental numbers:

$$r_3 \leqslant r_1, \qquad r_4 \leqslant r_2; \qquad s_1 \leqslant s_3, \qquad s_2 \leqslant s_4. \tag{3.1}$$

Obviously, if r_1 or r_2 is improper, then r_3 or r_4 is improper also, and if s_1 or s_2 is improper, so is s_3 or s_4. We have, moreover, the following theorem:

Theorem. Let one of the two fundamental numbers r_λ, $\lambda = 1, 2$ be improper, then the left fundamental numbers of all four kinds are improper. In this case the left end point a of the interval j is a cluster point of zeros of the individual integrals of (q).

If one of the two fundamental numbers r_λ, $\lambda = 1, 2$ is proper, so is the other. In this case the left end point a of j is not a cluster point of zeros of the individual integrals of (q).

The same is true with r_λ, "left" and a replaced by s_λ, "right" and b.

Proof. (a) For definiteness, let r_1 be improper; consider an integral u of (q). Assume a is not a cluster point of zeros of the integrals of (q); u has therefore a least zero $x \in j$. Then no number $t \in (a, x)$ possesses a left 1-conjugate number. We have therefore $a < x \leqslant r_1$, contrary to our assumption. Consequently, in every right neighbourhood of a there lie infinitely many zeros of u and consequently (§ 2.1) also of u', hence $r_2 = a$.

(b) The first part of the second statement follows from (a). Moreover if a is a cluster point of zeros of the individual integrals of (q), then in every right neighbourhood of a there are numbers which possess left conjugate numbers of the λ-th kind; hence $r_\lambda = a$, $\lambda = 1, 2$. Other cases are proved similarly.

3.8 Equations with proper fundamental numbers

Consider now a differential equation (q) with proper λ-fundamental numbers r_λ, s_λ $\lambda = 1, 2$. First we observe that every integral of (q) has therefore at least one zero. We recall also our assumption that $q(t) < 0 \ \forall \, t \in j$.

We now state and prove a theorem relating to left fundamental numbers and integrals. An exactly similar theorem holds for right fundamental numbers and integrals, which is obtained merely by replacing "left" by "right", u by v, a by b, r by s and $<$ by $>$, while the proof is entirely analogous. We shall not therefore write down this theorem or its proof explicitly.

Let u_λ be a left λ-fundamental integral of (q). Our theorem is:

1. *Let the fundamental number r_4 be proper, so that $a < r_4$, then the derivative u_1' of u_1 has precisely one zero to the left of r_1 and this latter zero is r_4. If the fundamental number r_4 is improper so that $a = r_4$, then the derivative u_1' of u_1 has no zero to the left of r_1. In this case $r_3 = r_1$ and the first zero of u_1' to the right of r_1 is the fundamental number r_2. Hence $r_1 < r_2$ and the two fundamental integrals u_1, u_2 are dependent.*

2. *Let the fundamental number r_3 be proper, so that $a < r_3$; then the fundamental integral u_2 has precisely one zero to the left of r_2, and this is r_3. If the fundamental number r_3 is improper, so that $a = r_3$, then the fundamental integral u_2 has no zero to the left of r_2. In this case $r_4 = r_2$ and the first zero of u_2 lying to the right of r_2 is the fundamental number r_1. Thus $r_2 < r_1$ and the two fundamental integrals u_1, u_2 are dependent.*

Proof. We shall restrict ourselves to the proof of the first statement.
(a) Let the fundamental number r_4 be proper. We consider a number $t_1 \in (a, r_4)$ and an integral u of (q) whose derivative vanishes at t_1; $u'(t_1) = 0$. On account of the fact that $t_1 < r_4$ the integral u has no zeros to the left of t_1. It therefore possesses zeros to the right of t_1; we consider the least of these, namely x_1. We must have $x_1 \leqslant r_1$, because otherwise x_1 would possess a left 1-conjugate number and consequently there would be a zero of u to the left of t_1. If $x_1 = r_1$, then the two integrals u, u_1 are dependent, and consequently their derivatives u', u_1' have the same zeros. In this case it follows that u_1' has at least one zero t_1 to the left of r_1. If $x_1 < r_1$, then we have the following situation: $t_1 < x_1$, $u'(t_1) = u(x_1) = 0$, $u(t) \neq 0$ for $t \in (t_1, x_1)$; $r_1 > x_1$, $u_1(r_1) = 0$. Then the third ordering theorem § 2.3 (3) shows that the function u_1 has a zero to the left of r_1.

The function u_1' has at most one zero to the left of r_1. For, if it vanishes more than once on the left of r_1, then on the left r_1 there lies at least one zero of u_1, which, however, conflicts with the definition of u_1. The function u_1' has therefore precisely one zero, t_2, to the left of r_1.

We now consider a number $t \in (a, r_1)$ and an integral u of (q), whose derivative u' vanishes at t: $u'(t) = 0$. By a similar argument to that used above, and applying the ordering theorem (3) of § 2.3, we conclude that; if $t > t_2$ then the integral u has one zero to the left of t but if $t < t_2$ then u has no zero to the left of t. It follows that $t_2 = r_4$.

(b) Let the fundamental number r_4 be improper. In this case every integral of the differential equation (q) has at least one zero to the left of a zero of its derivative. If, therefore, the derivative u_1' of the fundamental integral u_1 has a zero to the left of r_1, then u_1 has also such a zero. But the fundamental integral u_1 has no zero to the left o r_1, consequently the derivative u_1' of u_1 has no zero to the left of r_1.

We continue to employ the symbol R_κ ($\kappa = 1, 2, 3, 4$) with the meaning used above (see § 2.4).

Let $t \in R_3$, and consider an integral u of (q) which vanishes at t; $u(t) = 0$. The derivative u' of u has a zero, x, to the left of t. Because $a = r_4$ we have $x \in R_4$. The integral u has therefore a zero t_1 lying to the left of x, which is obviously left 1-conjugate with t. Hence $R_3 \subset R_1$ and $r_3 \geqslant r_1$, whence (3.1) gives $r_3 = r_1$.

Let x be the first zero of the derivative u_1' of u_1 lying on the right of r_1. We consider a number $t \in j$ and an integral u of (q) whose derivative vanishes at t; $u'(t) = 0$. Since $t \in R_4$, u has a greatest zero t_1 lying to the left of t. It is now necessary to distinguish two cases, according as $t > x$ or $t < x$. In the first case the ordering theorem (4) implies that $t_1 > r_1$ and since $r_3 = r_1$ it follows that $t_1 \in R_3$. The function u' has therefore at least one zero to the left of t_1, which obviously is left 2-conjugate with t. Hence $t \in R_2$.

In the second case the ordering theorem (4) gives $t_1 < r_1$, from which, taking into account the fact that $r_3 = r_1$ it follows that $t_1 \notin R_3$. The function u' has therefore no zero lying to the left of t_1. Hence $t \notin R_2$.

Grouping these results together: the set R_2 coincides with the interval (x, b). From this it follows that $x = r_2$ and the proof is complete.

3.9 Singular bases

The concept of fundamental integrals is closely connected with that of singular bases of a differential equation (q).

Let (q_λ) be a differential equation with λ-conjugate numbers, $\lambda = 1, 2$, and let (u, v) be a basis of (q_λ). We call the basis (u, v) a *left* (*right*) *principal basis of the λ-th kind*, more briefly a *left* (*right*) *λ-principal basis*, if the first term u is a left (right) λ-fundamental integral of (q_λ). Further, we call the basis (u, v) a *principal basis of the λ-th kind*, more briefly a *λ-principal basis* if the first term u is a left or right λ-fundamental integral of (q_λ) and also v is a right or left λ-fundamental integral, respectively, of (q_λ). In our study of principal bases it is therefore important to notice whether u is a left and v a right λ-fundamental integral or conversely. Principal bases, and left or right principal bases, will be called *singular bases* of the differential equation (q_λ). In the case of a λ-general differential equation (q_λ) of finite type with λ-conjugate numbers, there are left and right principal bases as well as principal bases of the λ-th kind. In the case of λ-special differential equations (q_λ) there occur left and right principal bases of the λ-th kind, and in fact every left or right principal basis of the λ-th kind is the same as a right or left principal basis of the λ-th kind. In left (right) oscillatory differential equations (q_λ) we only have right (left) principal bases of the λ-th kind. In oscillatory differential equations (q_λ) there are no singular bases).

One can write down without difficulty the system consisting of all the left (right) λ-principal bases or all the λ-principal bases of the differential equation (q_λ). Let u be a left (right) λ-fundamental integral of the differential equation (q_λ). Then the functions ρu, with arbitrary constant $\rho \neq 0$, constitute all the left (right) λ-fundamental integrals of the differential equation (q_λ). The set of all left (right) λ-principal bases is simply $(\rho u, \bar{v})$, in which $\bar{v}$ is any integral of (q_λ) independent of u. If the integral v of (q_λ) is chosen to be independent of u, then $\bar{v}$ can be represented in the form $\bar{v} = \sigma v + \bar{\sigma} u$, with appropriate constants σ, $\bar{\sigma}$ $(\sigma \neq 0)$. Plainly, therefore, the left (right) λ-principal bases of the differential equation (q_λ) form a two-parameter system: $(\rho u, \sigma v + \bar{\sigma} u)$; $\rho\sigma \neq 0$. If the differential equation (q_λ) is of finite type and λ-general, then one can choose v as a right (left) fundamental integral of (q_λ). The functions σv, constructed with arbitrary constant $\sigma \neq 0$, then constitute all the right (left) λ-fundamental integrals of (q_λ); the situation can therefore be summarized as: the set of all λ-principal bases of the differential equation (q_λ) are of the form $(\rho u, \sigma v)$, $\rho\sigma \neq 0$ and thus constitute two two-parameter systems, in one of which u is a left and v a right λ-fundamental integral of (q_λ), and in the other conversely.

3.10 Differential equations (q) with 1-conjugate numbers

In this section we give a survey of differential equations (q) with 1-conjugate numbers, based on the above results. In order to simplify our explanation, we shall now leave out the attribute "1", using the terms conjugate numbers, fundamental sequences, principal bases, etc. instead of 1-conjugate numbers, 1-fundamental sequences, 1-principal bases and so on. Differential equations (q) with conjugate numbers are either of finite type (m) with $m \geqslant 2$ or of infinite type.

I. Let (q) be a differential equation of finite type (m), $m \geqslant 2$.

The differential equation (q) then possesses integrals with m zeros in the interval j but none with $m + 1$ zeros. In this case the ends a, b of j are obviously not cluster points of zeros of the individual integrals of (q); consequently both fundamental numbers r_1 and s_1 are proper, and the differential equation (q) possesses left and right fundamental integrals and also left and right fundamental sequences, each of which is composed of $m - 1$ terms. To simplify our notation we now denote these fundamental sequences by

$$a_1 < a_2 < \cdots < a_{m-1}; \qquad b_{-1} > b_{-2} > \cdots > b_{-m+1},$$

where $a_1 = r_1$; $b_{-1} = s_1$. In this, $a_1, a_2, \ldots, a_{m-1}$; $b_{-1}, b_{-2}, \ldots, b_{-m+1}$ are the singular numbers of (q).

Between the numbers $a_\mu, b_{-\mu}$ ($\mu = 0, 1, \ldots, m-1, a_0 = a, b_0 = b$) there hold the following inequalities

$$a < b_{-m+1} \leqslant a_1 < b_{-m+2} \leqslant \cdots \leqslant a_r < b_{-m+r+1}$$
$$\leqslant a_{r+1} < \cdots \leqslant a_{m-2} < b_{-1} \leqslant a_{m-1} < b. \tag{3.2}$$

In these relations, either the strict inequality holds throughout, in which case the differential equation (q) is general, or equality holds everywhere, in which case the differential equation is special.

(a) Let the differential equation (q) be general.

The differential equation (q) then possesses left and right fundamental integrals, a left and a right fundamental integral are independent, and the differential equation (q) admits of principal bases.

Every integral of (q) which vanishes in one of the intervals $(a_\mu, b_{-m+\mu+1})$ has in every such interval one and only one zero; it follows that such an integral has precisely m zeros in the interval j. All other integrals of (q) have precisely $m-1$ zeros in j.

(b) Let the differential equation (q) be special.

The differential equation (q) then possesses left and right fundamental integrals, and these coincide. The equation (q) admits of left and right principal bases; each left or right principal basis is the same as a right or left principal basis; each integral of (q) which is independent of a fundamental integral has precisely m zeros in j, while each fundamental integral has precisely $m-1$ zeros.

II. Now let (q) be a differential equation of infinite type.

In this case the differential equation (q) is left or right oscillatory or oscillatory.

If the differential equation (q) is left (right) oscillatory then its left (right) fundamental number $r_1(s_1)$ is improper, while the right (left) is proper. In this situation the differential equation (q) possesses right (left) fundamental integrals, and the right (left) fundamental sequence

$$b_{-1} > b_{-2} > \cdots \qquad (a_1 < a_2 < \cdots), \tag{3.3}$$

which is always infinite; then $b_{-1}, b_{-2}, \ldots$ $(a_1, a_2, \ldots)$ are the singular numbers of the differential equation (q). The equation possesses right (left) principal bases.

Every integral of (q) independent of a right (left) fundamental integral has precisely one zero in every interval $(b_{-\mu}, b_{-\mu-1})$, $((a_\mu, a_{\mu+1}))$ $\mu = 0, 1, \ldots; b_0 = b$ $(a_0 = a)$.

If the differential equation (q) is oscillatory, then both fundamental numbers r_1, s_1 of (q) are improper; consequently in this case there are no fundamental integrals and naturally no fundamental sequences or singular bases.

Surveying these results, we see that among all types of differential equations (q), those which are general and of finite type (m) with $m \geqslant 2$ are the only ones for which there exist independent fundamental integrals (naturally, one left and one right) and consequently also principal bases.

3.11 Differential equations (q) with conjugate numbers of all four kinds

In this paragraph we collect together those properties of differential equations (q) with conjugate numbers of all four kinds, which are of importance for our further study.

Let (q) be a differential equation with conjugate numbers of all four kinds. We assume that $q(t) < 0$ for $t \in j$. For each kind κ (= 1, 2, 3, 4) an open interval $i_{\kappa,\nu}$ or $j_{\kappa,\nu}$, $\nu = 1, 2, \ldots$, is formed by those numbers $t \in j$ for which the ν-th left or right κ-conjugate number exists. If the zeros of integrals of the differential equation (q) cluster towards the left or right end point a or b of the interval j, so that (q) is left or right oscillatory, or oscillatory, then $i_{\kappa,\nu} = j$ or $j_{\kappa,\nu} = j$ for $\kappa = 1, 2, 3, 4$ and for all $\nu = 1, 2, \ldots$. If on the other hand the zeros of integrals of the differential equation (q) do not cluster about the left end point a of j, so that the differential equation (q) is of finite type or right oscillatory, then for every $\kappa = 1, 2, 3, 4$ we have

$$i_{\kappa,\nu} = (r_{\kappa,\nu-1}, b); \qquad a \leqslant r_{\kappa,0} < r_{\kappa,1} < \cdots < b$$

in which the sequence $\{r_{\kappa,\nu-1}\}$, $\nu = 1, 2, \ldots$, is finite if the differential equation (q) is of finite type, while it is infinite in the case of a right oscillatory differential equation (q).

Similarly, if the zeros of integrals of (q) do not cluster about b, then (q) is of finite type or left oscillatory, so $j_{\kappa,\nu} = (a, s_{\kappa,\nu-1})$; $a < \cdots < s_{\kappa,1} < s_{\kappa,0} \leqslant b$ and the sequence $\{s_{\kappa,\nu-1}\}$ is finite or infinite according as (q) is finite or left oscillatory. If the differential equation (q) is oscillatory, then corresponding to every number $t \in j$ there is a left and a right κ-conjugate ν-th number for all $\kappa = 1, 2, 3, 4$ and all $\nu = 1, 2, \ldots$.

3.12 Bilinear relations between integrals of the differential equation (q)

In the transformation theory to be considered later certain bilinear relations between integrals of a differential equation (q) play an important role. We consider a differential equation (q).

The fundamental theorem is the following:

Theorem. For an arbitrary basis (u, v) of the differential equation (q) *the following relations hold at two different points $t, x \in j$*

$$\begin{aligned} &1.\ u(t)v(x) - u(x)v(t) = 0, \\ &2.\ u'(t)v'(x) - u'(x)v'(t) = 0, \\ &3.\ u(t)v'(x) - u'(x)v(t) = 0 \end{aligned}$$

if and only if the numbers t, x are connected, respectively, in the following ways:

1. *t, x are 1-conjugate,*
2. *t, x are 2-conjugate,*
3. *x is 3-conjugate with t, and consequently t is 4-conjugate with x.*

Proof. We shall confine ourselves to the proof of 1.

(a) Let the bilinear relation 1 hold between two given distinct numbers $t, x \in j$. Then the linear equations

$$c_1u(t) + c_2v(t) = 0, \qquad c_1u(x) + c_2v(x) = 0$$

are satisfied for appropriate values c_1, c_2 with $c_1^2 + c_2^2 \neq 0$ and consequently the numbers t, x are zeros of the integral $y = c_1u + c_2v$ of (q); hence the numbers t, x are 1-conjugate.
(b) Let the numbers t, x be 1-conjugate. Then $t \neq x$, and there is an integral $y = c_1u + c_2v$ of (q) which vanishes at t and x, with $c_1^2 + c_2^2 \neq 0$. From this follows the bilinear relationship 1.

This theorem can obviously be formulated briefly as: conjugate numbers of the various kinds represent zeros of basis functions of the differential equation (q).

The ratio u/v of the integrals u, v of a basis of the differential equation (q), or the ratio u'/v' of their derivatives, respectively, take the same value at two different points $t, x \in j$ (i.e.

$$\frac{u(t)}{v(t)} = \frac{u(x)}{v(x)} \quad \text{or} \quad \frac{u'(t)}{v'(t)} = \frac{u'(x)}{v'(x)}$$

respectively) if and only if the numbers t and x are 1-conjugate or 2-conjugate respectively. Moreover, the relationship

$$\frac{u(t)}{v(t)} = \frac{u'(x)}{v'(x)}$$

holds if and only if x is 3-conjugate with t and t is consequently 4-conjugate with x.

4 Centro-affine differential geometry of plane curves

Linear homogeneous differential equations of the second order stand in a close relationship with centro-affine differential geometry of plane curves. This geometry is the theory of those concepts and properties of plane curves which are invariant under transformations of the curve parameter and the plane linear homogeneous group of transformations. Linear homogeneous transformations are for brevity called "centro-affine transformations".

4.1 Representation of plane curves

A plane curve $\mathfrak{K}$ is the set of points which are given by the values of a vector function $x(t) = [u(t), v(t)]$ with two components $u(t)$, $v(t)$. We assume that the function x and consequently its components u, v are defined in an open interval j and $\in C_2$ or, if necessary, a higher class. The determinant $u^{(\mu)}v^{(\nu)} - u^{(\nu)}v^{(\mu)}$ formed from the derivatives $x^{(\mu)}$, $x^{(\nu)}$ we shall usually denote by $(x^{(\mu)}x^{(\nu)})$.

We shall refer to the function $x(t)$ as the *representation* of the curve $\mathfrak{K}$. Its components $u(t)$, $v(t)$ can be interpreted as parametric coordinates of the curve $\mathfrak{K}$ with respect to a coordinate system formed from the two vectors x_1, x_2 with origin O. If the function $x \in C_\mu$, then the curve $\mathfrak{K}$ is said to be also of the class C_μ. The independent variable t is called the parameter of $\mathfrak{K}$ in the representation $x(t)$.

We consider a plane curve $\mathfrak{K}$ with the representation $x(t)$. The curve $\mathfrak{K}$ admits of infinitely many representations, but in what follows we only consider those representations which arise from a single representation (for example, $x(t)$) by means of a transformation of parameter $T = T(t)$. We assume that the function $T(t)$ is defined in j, belongs to the class C_2 or if necessary to a higher class, and that its derivative T' is always non-zero. In this case the range of values of the function $T(t)$ is an open interval J, and in this interval there exists the inverse function, $t(T)$, to $T(t)$. Obviously $J = T(j)$, $j = t(J)$. We call any two numbers $t \in j$, $T \in J$ *homologous* if they are related to each other by the formulae $T = T(t)$, $t = t(T)$. If, now, we are given a representation $x(t)$ and a parameter transformation $T = T(t)$ we define the resulting representation $X(T) = [U(T), V(T)]$ of the curve $\mathfrak{K}$ by specifying that the value of the function X at each point $T \in J$ is precisely the value $x(t)$ of the function x at the homologous point $t \in j$, that is

$$X(T) = x(t) \tag{4.1}$$

for $T = T(t) \in J$, $t = t(T) \in j$. It is clear that each of the two representations x, X of the curve $\mathfrak{K}$ is uniquely determined by the other, and each representation of the curve

$\mathfrak{K}$ arises from the other by a parameter transformation. We stress that the values of two representations of the curve at homologous points give the same point of the curve.

The function $X(T)$ obviously belongs to the class C_2 or to a higher class. Denoting derivatives with respect to T by a dot, the following formulae hold at two homologous points $t \in j$, $T \in J$

$$\left.\begin{aligned} \dot{X}(T) &= x'(t)\dot{t}, & x'(t) &= \dot{X}(T)T', \\ \ddot{X}(T) &= x''(t)\dot{t}^2 + x'(t)\ddot{t}, & x''(t) &= \ddot{X}(T)T'^2 + \dot{X}(T)T'' \end{aligned}\right\} \tag{4.2}$$

and further

$$\left.\begin{aligned} (X\dot{X}) &= (xx')\dot{t}, & (xx') &= (X\dot{X})T', \\ (X\ddot{X}) &= (xx'')\dot{t}^2 + (xx')\ddot{t}, & (xx'') &= (X\ddot{X})T'^2 + (X\dot{X})T'', \\ (\dot{X}\ddot{X}) &= (x'x'')\dot{t}^3, & (x'x'') &= (\dot{X}\ddot{X})T'^3. \end{aligned}\right\} \tag{4.3}$$

If, therefore, at any point $t \in j$ we have $(xx') \neq 0$ or $(xx') = 0$, then at the homologous point $T \in J$ we have, respectively, $(X\dot{X}) \neq 0$ or $(X\dot{X}) = 0$. In the first case we have, moreover,

$$\operatorname{sgn} T' = \operatorname{sgn}(xx') \cdot \operatorname{sgn}(X\dot{X}) = \operatorname{sgn} \dot{t}.$$

The same is true of the functions $(x'x'')$, $(\dot{X}\ddot{X})$.

4.2 Centro-affine representatives of a plane curve and its representations

We now consider a linear homogeneous, (hence centro-affine) transformation of the plane

$$\left.\begin{aligned} \bar{x}_1 &= c_{11}x_1 + c_{12}x_2, \\ \bar{x}_2 &= c_{21}x_1 + c_{22}x_2, \end{aligned}\right\} \tag{4.4}$$

with the matrix $C = (c_{ik})$ $(|C| =)\, |c_{ik}| \neq 0$, $i, k = 1, 2$; such transformations obviously leave the point (0, 0), the so-called *centre* of the centro-affine plane, invariant. By the transformation (4) the function $x(t)$ goes over into the function $(Cx(t) =)\,\bar{x}(t)$ with the components $\bar{u}(t) = c_{11}u(t) + c_{12}v(t)$, $\bar{v}(t) = c_{21}u(t) + c_{22}v(t)$. We describe the curve $C\mathfrak{K} = \bar{\mathfrak{K}}$, determined by the function $\bar{x}(t)$, as a centro-affine *representative* or, more precisely, the *C-representative* of $\mathfrak{K}$; we also associate this term with the function $\bar{x}(t)$ itself and call $\bar{\mathfrak{K}}$ the C-representative of $x(t)$. Two points $x(t) \in \mathfrak{K}$, $\bar{x}(t) \in \bar{\mathfrak{K}}$, which are defined by the same value $t \in j$ of the parameter, we call (mutually) *associated*. In particular, the curve $\mathfrak{K}$ and its representation $x(t)$ is its own representative ($c_{11} = c_{22} = 1$, $c_{12} = c_{21} = 0$), while the curve $\mathfrak{K}$ and its representation $x(t)$ is a representative, more precisely the C^{-1}-representative, of $\bar{\mathfrak{K}}$ and $\bar{x}(t)$.

Let $X(T)$ be the representation of the curve $\mathfrak{K}$ arising from $x(t)$ by means of the parameter transformation $T = T(t)$. The C-representatives Cx, CX of the representations x, X of the curve $\mathfrak{K}$ obviously take the same value at two homologous points t, T. It follows that CX is the representation of the curve $C\mathfrak{K}$ arising from Cx by means of the parameter transformation $T = T(t)$. The values taken by two representations x, X of the curve $\mathfrak{K}$ at two homologous points, and the corresponding representatives Cx, CX at the same points, give associated points of the curves $\mathfrak{K}$ and $C\mathfrak{K}$. The determinants

$(\bar{x}\bar{x}')$, $(\bar{x}'\bar{x}'')$ constructed from a representative $\bar{x}(t) = Cx(t)$ of the function $x(t)$ differ from (xx'), (x', x'') by the same non-zero factor, namely the determinant $|C|$. It follows that if at a point $t \in j$ the relation $(xx') \neq 0$ or $(xx') = 0$ holds, then for every representative $\bar{x}$ of x at the same point t the analogous relation $(\bar{x}\bar{x}') \neq 0$ or $(\bar{x}\bar{x}') = 0$ holds. Moreover sgn $(\bar{x}\bar{x}) = \text{sgn}\,|C| \cdot \text{sgn}\,(xx')$. The same is true for the functions $(x'x'')$, $(\bar{x}'\bar{x}'')$.

4.3 Centro-affine invariants of plane curves

If we have a scalar function $f[x(t), x'(t), x''(t)]$ constructed from a representation $x(t)$ of $\mathfrak{K}$ and some of its derivatives such as $x'(t)$, $x''(t)$, then we call it a (centro-affine) *relative invariant* of the curve $\mathfrak{K}$, if for arbitrary choice of the representatives $\bar{x}(t)$, $\bar{X}(T)$ of two representations $x(t)$, $X(T)$ of the curve $\mathfrak{K}$ at two homologous points t, T, the values taken by this function are the same or the same with opposite sign, i.e.

$$f[\bar{x}(t), \bar{x}'(t), \bar{x}''(t)] = \pm f[\bar{X}(T), \dot{\bar{X}}(T), \ddot{\bar{X}}(T)].$$

We call the function f a (centro-affine) *absolute invariant*, if it takes the same value at two homologous points t, T for arbitrary choice of $\bar{x}(t)$, $\bar{X}(T)$. Obviously the absolute value of a relative invariant is an absolute invariant.

Among the simplest relative invariants of the curve $\mathfrak{K}$ are the functions mentioned above, namely sgn (xx'), sgn $(x'x'')$. Let us observe the geometrical significance of these, without going into details (for which see [80]). At a point $t \in j$, according as sgn $(xx') = \pm 1$, or sgn $(xx') = 0$, the centre of the centro-affine plane lies off or on the tangent to the curve $\mathfrak{K}$ at the point $x(t)$. According as sgn $(x'x'') = \pm 1$ or sgn $(x'x'') = 0$, the curve $\mathfrak{K}$ in the neighbourhood of the point $x(t)$ lies on one side of the tangent at the point $x(t)$, or the point $x(t)$ is a turning point of $\mathfrak{K}$, respectively.

4.4 Regular curves

We call the curve $\mathfrak{K}$ *regular*, if the two relationships sgn $(xx') = \pm 1$, sgn $(x'x'') = \pm 1$ hold everywhere in the interval j.

If therefore the curve $\mathfrak{K}$ is regular, then none of its tangents pass through the centre of the centro-affine plane, the curve is locally convex and has no turning points.

Now let $\mathfrak{K}$ be a regular curve with the representation $x(t)$, $t \in j$. We associate with this representation the functions

$$a(t) = \frac{(xx'')}{(xx')}, \quad b(t) = -\frac{(x'x'')}{(xx')} \tag{4.5}$$

which belong to the class C_0 or a higher class.

It is easy to verify that the function $x(t)$ satisfies the second order differential equation

$$x'' = a(t)x' + b(t)x, \tag{4.6}$$

constructed with the coefficients (5); that is to say, each of the components $u(t)$, $v(t)$ of the function $x(t)$ satisfies the scalar differential equation (6). Conversely, if the function $x(t)$ satisfies a differential equation of the form (6), then the coefficients $a(t)$, $b(t)$ are uniquely determined by $x(t)$ through the formulae (5). Further, the functions $\bar{a}(t)$, $\bar{b}(t)$ associated with a representative $\bar{x}(t)$ of $x(t)$ coincide with $a(t)$, $b(t)$; $\bar{a}(t) = a(t)$, $\bar{b}(t) = b(t)$; $t \in j$. Thus the functions $a(t)$, $b(t)$ are invariant no matter what choice is made of a representative of x, so every representative of the representation $x(t)$ provides a solution of the differential equation (6). We call the functions $a(t)$, $b(t)$ the *first* and *second centro-affine semi-invariants* of the representation $x(t)$.

Now let $x(t)$, $t \in j$ and $X(T)$, $T \in J$ be two representations of the curve $\mathfrak{K}$ and $a(t)$, $A(T)$ and $b(t)$, $B(T)$ be their first and second centro-affine semi-invariants. From (5) and (3) we find that the values of these functions at two homologous points $t \in j$, $T \in J$ are related as follows

$$\left.\begin{aligned} A(T) &= a(t)\dot{t} + \frac{\ddot{t}}{\dot{t}}, \quad B(T) = b(t)\dot{t}^2, \\ a(t) &= A(T)T' + \frac{T''}{T'}, \quad b(t) = B(T)T'^2. \end{aligned}\right\} \tag{4.7}$$

From these relations it follows that the two centro-affine semi-invariants $b(t)$, $B(T)$ of the representations $x(t)$, $X(T)$ at two homologous points $t \in j$, $T \in J$, have the same sign:

$$(\varepsilon =) \operatorname{sgn} b(t) = \operatorname{sgn} B(T).$$

We see that this sign ε is an absolute centro-affine invariant of the curve $\mathfrak{K}$. Since the curve $\mathfrak{K}$ is regular it follows from (5) that ε takes the same value for all representations of $\mathfrak{K}$ and its representatives for all values of the parameter. Regarding the geometrical significance of ε, we remark that according as $\varepsilon = -1$ or $= +1$, the centre of the centro-affine plane and the curve $\mathfrak{K}$ lie on the same or different sides of each tangent to the curve.

4.5 Centro-affine curvature

Let $t_0 \in j$, $T_0 \in J$ be two arbitrary homologous values. Then at two homologous points $t \in j$, $T \in J$ we have (from (7))

$$\operatorname{sgn}(xx') \int_{t_0}^{t} \sqrt{|b(\sigma)|}\, d\sigma = \operatorname{sgn}(X\dot{X}) \int_{T_0}^{T} \sqrt{|B(\sigma)|}\, d\sigma \tag{4.8}$$

and moreover

$$\frac{\operatorname{sgn}(xx')}{\sqrt{|b(t)|}}\left[a(t) - \frac{1}{2}\frac{b'(t)}{b(t)}\right] = \frac{\operatorname{sgn}(X\dot{X})}{\sqrt{|B(T)|}}\left[A(T) - \frac{1}{2}\frac{\dot{B}(T)}{B(T)}\right]. \tag{4.9}$$

Now we consider the following functions, defined in the interval j,

$$s(t) = \operatorname{sgn}(xx') \int_{t_0}^{t} \sqrt{|b(\sigma)|}\, d\sigma, \qquad k(t) = \frac{\operatorname{sgn}(xx')}{\sqrt{|b(t)|}}\left[a(t) - \frac{1}{2}\frac{b'(t)}{b(t)}\right]. \tag{4.10}$$

By (8), (9) it follows that each of the functions $s(t)$, $k(t)$ takes the same value at two homologous points t, T in an arbitrary representation $X(T)$ of $\mathfrak{K}$. For an arbitrary choice of a representative $\bar{x}(t) = Cx(t)$ of $x(t)$, it obviously takes the same value at every point t or takes the same value with opposite sign, according as sgn $|C| = 1$ or -1. It follows that the functions $s(t)$, $k(t)$ are relative invariants of the curve $\mathfrak{K}$. The function s is called the *centro-affine oriented arc-length* of the curve $\mathfrak{K}$; its value $s(t)$ gives the length of the centro-affine oriented arc from the point $x(t_0)$ to the point $x(t)$. The function k is called the *centro-affine curvature* of the curve $\mathfrak{K}$; its value $k(t)$ gives the centro-affine curvature of $\mathfrak{K}$ at the point $x(t)$.

From (5) we have

$$k(t) = \frac{\operatorname{sgn}(xx')}{2}\sqrt{\frac{|(xx')|}{|(x'x'')|}}\left[3\frac{(xx'')}{(xx')} - \frac{(x'x''')}{(x'x'')}\right]. \tag{4.11}$$

The function $s(t)$, $t \in j$, belongs to the class C_1 or to a higher class, and its derivative s' is always non-zero. It maps the interval j onto an interval i including zero, i.e. $0 \in i$.

Let $Y(s)$, $s \in i$ be the representation of the curve $\mathfrak{K}$ arising from the parametric transformation $s = s(t)$. Then at two homologous points $s \in i$, $t \in j$ we have

$$A(s) = k(t), \qquad B(s) = \varepsilon.$$

The value $A(s)$ $(= K(s))$ of the first semi-invariant A at each point $s \in i$ gives the centro-affine curvature of the curve $\mathfrak{K}$ at the point $Y(s)$, while the second semi-invariant B takes the constant value ε $(= \pm 1)$. The representation $Y(s)$ of $\mathfrak{K}$ satisfies the linear differential equation of the second order

$$\ddot{Y} = K(s)\dot{Y} + \varepsilon Y. \tag{4.12}$$

If we write $\dot{Y}(s) = Z(s)$, then this differential equation can be replaced by the system of (vector) differential equations of the first order

$$\left.\begin{aligned} \dot{Y} &= Z, \\ \dot{Z} &= \varepsilon Y + K(s)Z. \end{aligned}\right\} \tag{4.13}$$

These differential equations (13) are called the *Serret-Frenet formulae* of centro-affine plane curve theory, and the curves with representations $\dot{Y}(s)$ and $\ddot{Y}(s)$, $s \in i$, are designated the *tangent* and *curvature curves*, respectively, of $\mathfrak{K}$.

4.6 Application of the above theory to integral curves of the differential equation (q)

We consider a differential equation (q) in which we specify that $q \in C_1$. Let (u, v) be a basis of the differential equation (q) and $\mathfrak{K}$ the integral curve of (q) determined by this basis. The curve $\mathfrak{K}$ therefore admits of the representation $x(t) = [u(t), v(t)]$, $t \in j$. The various integral curves of (q) naturally constitute the representatives of the curve $\mathfrak{K}$. Obviously we have

$$(xx') = w\ (\neq 0); \qquad (xx'') = 0, \qquad (x'x'') = -qw, \qquad (x'x''') = -q'w.$$

The curve $\mathfrak{K}$ is then regular if and only if $q(t) \neq 0$ for $t \in j$.

Now we assume that the curve $\mathfrak{K}$ is regular, so that $q(t) \neq 0$ for $t \in j$. In the two cases $q < 0$ and $q > 0$ the centre of the centro-affine plane and the curve $\mathfrak{K}$ lie respectively on the same and on opposite sides of each tangent to the curve. The centro-affine oriented arc of the curve $\mathfrak{K}$ and its centro-affine curvature are given by the formulae

$$\left.\begin{aligned} s(t) &= \operatorname{sgn} w \cdot \int_{t_0}^{t} \sqrt{|q(\sigma)|}\, d\sigma, \\ k(t) &= \operatorname{sgn} w \left(\frac{1}{\sqrt{|q(t)|}}\right)', \end{aligned}\right\} \tag{4.14}$$

The carrier q can be expressed in terms of the functions s, k as follows: $q(t) = \operatorname{sgn} q(t_0) \cdot s'^2(t)$,

$$q(t) = \frac{q(t_0)}{\left[1 + \operatorname{sgn} w \cdot \sqrt{|q(t_0)|} \int_{t_0}^{t} k(\sigma)\, d\sigma\right]^2}. \tag{4.15}$$

The representation $Y(s)$ of the curve $\mathfrak{K}$ given by the parameter transformation $s = s(t)$ satisfies a linear differential equation of the second order of the form (12) in which $\varepsilon = \operatorname{sgn} q$.

Let $Q(s)$ be the function which the carrier $q(t)$ of (q) becomes on transforming to the variable s. The function Q is therefore defined by the relation $Q(s) = q(t)$ at every two homologous points $s \in i$, $t \in j$.

From (14) we have

$$\dot{Q}(s) = \operatorname{sgn} w \cdot \frac{q'(t)}{\sqrt{|q(t)|}},$$

and hence

$$K(s) = -\frac{1}{2} \cdot \frac{\dot{Q}(s)}{Q(s)}.$$

We see that if the centro-affine oriented arc of the curve $\mathfrak{K}$ is chosen as parameter for the representation of $\mathfrak{K}$, then the centro-affine curvature K of this curve is given by the formula

$$K(s) = \frac{d}{ds} \log \frac{1}{\sqrt{|Q(s)|}}. \tag{4.16}$$

From (16) it follows that, at every two homologous points $t \in j$, $s \in i$,

$$q(t) = q(t_0) \exp\left(-2 \int_0^s K(\sigma)\, d\sigma\right). \tag{4.17}$$

Phase theory of ordinary linear homogeneous differential equations of the second order

In this chapter we shall develop the topic of *phase theory* of the differential equations to be considered; this provides the appropriate methodological basis for the transformation theory which we shall build upon it. The reader may perhaps be surprised to discover the richness and breadth into which this phase theory, whose basis depends merely on the properties of a function $q(t)$, can be expanded.

5 Polar coordinates of bases

5.1 Introduction

In this paragraph we start our development with a few elementary facts.

Let (q) be a differential equation in the interval j. We consider a basis (u, v) of (q) and the corresponding integral curve $\mathfrak{K}$ with representation $x(t) = [u(t), v(t)]$, relative to a rectangular coordinate system formed from the vectors x_1, x_2 with origin O. As positive sense of variation of angle we choose that which is given by the rotation from x_2 to x_1.

Let $x_0 = x(t_0)$, $t_0 \in j$, be an arbitrary point of $\mathfrak{K}$ and r_0 (> 0), α_0 be its polar coordinates with respect to the pole O and the polar axis OX_2. Then r_0 represents the modulus of the vector x_0, while the number α_0, which we specify to lie in the interval $[0, 2\pi)$, is the value of the angle (x_2, x_0). Obviously we have $u(t_0) = r_0 \sin \alpha_0$, $v(t_0) = r_0 \cos \alpha_0$. Now we define, in the interval j, the function $r(t)$ by means of the formula

$$r(t) = \sqrt{u^2(t) + v^2(t)}. \tag{5.1}$$

Moreover we let $\alpha(t)$ be the (unique) continuous function, defined in the interval j, which takes the value α_0 at the point t_0 and in the interval j satisfies the equation*

$$\tan \alpha(t) = \frac{u(t)}{v(t)} \tag{5.2}$$

except, of course, at the zeros of v.

Then we have, for all $t \in j$,

$$u(t) = r(t) \sin \alpha(t), \qquad v(t) = r(t) \cos \alpha(t). \tag{5.3}$$

Besides the function α, there are obviously other continuous functions in the interval j which satisfy the equation (2) everywhere in j apart from the zeros of v. Every such function has the form $\alpha_n = \alpha + n\pi$, n being an integer; consequently their totality forms a countable system. Each member is uniquely determined by its initial value $\alpha_n(t_0) = \alpha_0 + n\pi$. For n even the functions r, α_n satisfy the formulae (3), while for n odd the right side of (3) gives the basis $(-u, -v)$, which is proportional to (u, v).

The functions r and α_n constitute the polar coordinates of the basis (u, v). In special cases these ideas can be taken over to the ordered pair (u', v') formed from the derivatives u', v' of the functions u, v.

* This change to polar coordinates is commonly known as the Prüfer substitution (Trans.)

Now we wish to consider the information given here in full detail. We denote the Wronskian of the basis (u, v) by w.

5.2 Amplitudes

The functions r, s defined by the formulae

$$r = \sqrt{u^2 + v^2}, \qquad s = \sqrt{u'^2 + v'^2} \tag{5.4}$$

in the interval j will be called the *first* and *second* amplitudes of the basis (u, v) respectively. These functions are obviously always positive and belong to the classes C_2, C_1 respectively.

Clearly, the amplitudes of the inverse basis (v, u) are also r and s, we can thus refer more briefly to the amplitudes of the (independent) integrals u, v or, if we are concerned with their values, to the amplitudes of the point $x(t) = [u(t), v(t)]$.

We now show that the amplitudes r, s satisfy the following non-linear differential equations of the second order

$$\left.\begin{aligned} r'' &= qr + \frac{w^2}{r^3}, \\ s'' &= qs + \frac{w^2q^2}{s^3} + \frac{q'}{q}s', \end{aligned}\right\} \tag{5.5}$$

the first in the interval j, and the second in every sub-interval $i \subset j$ in which the function q is differentiable and non-zero.

Starting from the formulae

$$r^2 = u^2 + v^2, \qquad rr' = uu' + vv', \qquad w = uv' - u'v, \qquad s^2 = u'^2 + v'^2 \tag{5.6}$$

there follows the relation

$$r^2(s^2 - r'^2) = (uv' - u'v)^2 = w^2,$$

and the further relation

$$s^2 - r'^2 = \frac{w^2}{r^2} \tag{5.7}$$

The second formula (6) leads to the equation

$$rr'' = s^2 - r'^2 + qr^2,$$

and from the last two relations there follows the first of the differential equations (5). To obtain the second, we differentiate the equation (7), use the first equation (5) and so obtain the relation

$$ss' = qrr'. \tag{5.8}$$

Then, assuming the differentiability of q, we get

$$ss'' + s'^2 = qs^2 + q'rr' + q^2r^2,$$

then (8) gives

$$q(ss'' + s'^2) = q^2s^2 + q'ss' + q^3r^2. \tag{5.9}$$

Eliminating r, r' between the formulae (7), (8), (9) we obtain the second equation (5), for $q \neq 0$.

5.3 First phases of a basis

By a *first phase of the basis* (u, v) we mean any function α, continuous in the interval j, which satisfies in this interval the relationship

$$\tan \alpha(t) = \frac{u(t)}{v(t)} \tag{5.10}$$

except at the zeros of v. When convenient we refer to "phases" instead of "first phases" of the basis (u, v).

We note that there is precisely one countable system of phases of the basis (u, v). This system we call the *first phase system of the basis* (u, v), or more briefly the phase system of the basis (u, v). The individual phases of this system differ from each other by integral multiples of π.

Let (α) denote the phase system of (u, v). Let us choose an arbitrary phase $\alpha \in (\alpha)$, then the phase system (α) is composed of the set of functions

$$\alpha_\nu(t) = \alpha(t) + \nu\pi \qquad (\nu = 0, \pm 1, \pm 2, \ldots; \quad \alpha_0 = \alpha) \tag{5.11}$$

and these can clearly be ordered as follows:

$$\cdots < \alpha_{-2} < \alpha_{-1} < \alpha_0 < \alpha_1 < \alpha_2 < \cdots. \tag{5.12}$$

The value of each phase $\alpha_\nu \in (\alpha)$ at a zero of u or v is respectively an even or odd multiple of $\frac{1}{2}\pi$; conversely every point in j, at which a phase $\alpha_\nu \in (\alpha)$ takes the value of an even or odd multiple of $\frac{1}{2}\pi$, is a zero of u or v respectively.

If the integral u does not vanish in j, then there is precisely one phase $\alpha_\nu \in (\alpha)$, whose values lie entirely between 0 and π. If, however, u possesses zeros in j, then corresponding to each of these zeros there is precisely one phase in (α) which vanishes there.

From the first formula (2.1) we deduce that each phase $\alpha_\nu \in (\alpha)$ increases or decreases in j, according as $-w > 0$ or $-w < 0$. The integrals u, v are expressed in terms of the amplitude r, and an arbitrary phase $\alpha_\nu \in (\alpha)$ of the basis (u, v), as follows:

$$u(t) = \varepsilon_\nu r(t) \cdot \sin \alpha_\nu(t), \qquad v(t) = \varepsilon_\nu r(t) \cdot \cos \alpha_\nu(t) \qquad (t \in j); \tag{5.13}$$

in which ε_ν, the so-called *signature of the phase* α_ν, takes the value $+1$ or -1. The phase α_ν is called *proper* or *improper* (with respect to the basis (u, v), according as $\varepsilon_\nu = 1$ or $\varepsilon_\nu = -1$.

Two phases $\alpha_\nu, \alpha_\mu \in (\alpha)$, for which the difference $\nu - \mu$ is even, are both proper or both improper, while if $\nu - \mu$ is odd, then one of them is proper and the other improper. If α_ν, α_μ are both proper or both improper they are said to be *of the same kind*;

otherwise we describe them as *of different kind.* Clearly, in the ordering (12) of the phase system (α) the individual phases are alternately proper and improper: the successor of a proper phase is improper and conversely.

Let t_0 and n be arbitrary numbers, with $t_0 \in j$ and n an integer; then there is precisely one proper and one improper phase whose values at the point t_0 lie in the interval $[2n\pi, (2n + 2)\pi)$. Every proper (improper) phase with respect to the basis (u, v) is improper (proper) with respect to the basis $(-u, -v)$.

The geometrical significance of first phases of the basis (u, v) is as follows:—

Let α be a first phase of the basis (u, v), and let $\mathrm{W}\alpha(t)$ be the (unique) number in the interval $[0, 2\pi)$ which is congruent to $\alpha(t)$ modulo 2π; that is to say $\alpha(t) = \mathrm{W}\alpha(t) + 2\pi n$, n $(= n(t))$ integral, $0 \leqslant \mathrm{W}\alpha(t) < 2\pi$, $t \in j$.

We consider the integral curve $\mathfrak{K}$ with the vector representation $x(t) = [u(t), v(t)]$. Then $\mathrm{W}\alpha(t)$ is the angle formed between the vector $x(t)$ or the vector $-x(t)$ and the co-ordinate vector x_2, according as α is proper or improper. In other words $\alpha(t)$ is congruent modulo 2π to that angle in the range 0 to 2π which lies between $x(t)$ or $-x(t)$ and the co-ordinate vector x_2.

5.4 Boundedness of a first phase

Let α be a phase of the basis (u, v); the range of α in the interval j obviously forms an open interval.

We have the following theorem:

Theorem. The phase α is bounded in the interval j if and only if the differential equation (q) *is of finite type.*

Proof. Let J be the range of α in the interval j; in this interval j there hold formulae of the type (13).

(a) Assume that α is bounded in j, then the interval J contains only a finite number, say m $(\geqslant 0)$, of distinct multiples of the number π. From that it follows, using (13), that the integral u vanishes precisely m times in the interval j. Now let $\bar{u}$ be an arbitrary integral of (q). If $\bar{u}$ is linearly dependent upon u, then $\bar{u}$ has precisely m zeros in the interval j, (the same zeros as u). If, however, the integrals u, $\bar{u}$ are linearly independent, then $\bar{u}$ has $m - 1$ or m or $m + 1$ zeros in j, for between every two neighbouring zeros of u there is precisely one zero of $\bar{u}$ and conversely (§ 2.3). Thus the differential equation (q) is of finite type m or $m + 1$.

(b) Let the differential equation (q) be of finite type m $(\geqslant 0)$. Then the integral u has at most m zeros in the interval j. Consequently, from (13), the interval J contains at most m distinct integral multiples of π, so the phase α is bounded, and the proof is complete.

By similar reasoning we can obtain the following result:

If the phase α is increasing (decreasing), then it is unbounded below and bounded above in the interval j if and only if the differential equation (q) is left (right) oscillatory; similarly α is bounded below and unbounded above in the interval j if and only if the differential equation (q) is right (left) oscillatory while α is unbounded both below and above in the interval j if and only if the differential equation (q) is oscillatory.

The theorem of § 3.12 can be supplemented by the following remark:

The values $\alpha(t)$, $\alpha(x)$ of the phase α at two distinct points $t, x \in j$ differ by an integral multiple of π if and only if the numbers t, x are 1-conjugate.

5.5 Continuity property of a first phase

The phase α belongs to the class C_3. To show this we take an arbitrary number $x \in j$, not a zero of v, and form, in the interval j, the function

$$\bar{\alpha}(t) = \alpha(x) + \int_x^t \frac{-w}{r^2}\, d\sigma.$$

Obviously $\bar{\alpha} \in C_3$.

Now we consider the function $F(t) = \alpha(t) - \bar{\alpha}(t)$. From the definitions of α and $\bar{\alpha}$ it follows that

1. $F(x) = 0$,
2. $F \in C_0$,
3. In every interval $i \subset j$, in which v does not vanish, the derivative of F vanishes identically so that F takes a constant value.

(a) Let there be no zero of v in the interval j, then from 1, 3 we have $F(t) \equiv 0$ and consequently $\alpha(t) = \bar{\alpha}(t) \in C_3$.

(b) Let $t_0 \in j$ be a zero of v. This is an isolated zero of v, consequently there are maximal open intervals $i_1 \subset j$ and $i_2 \subset j$, with right and left end point t_0 respectively, in which v is non-zero. From 3° it follows that $F(t) = k_1$ for $t \in i_1$, $F(t) = k_2$ for $t \in i_2$, k_1, k_2 being constants, and from 2° we obtain $k_1 = F(t_0) = k_2$. There exists therefore a constant k such that $F(t) = k$ for all $t \in j$. Now 1° shows that $k = 0$; it follows that $\alpha(t) = \bar{\alpha}(t) \in C_3$, and the proof is complete.

It is easy to establish the validity of the following formulae in the interval j:

$$\alpha' = \frac{-w}{r^2}, \qquad \alpha'' = 2w\frac{rr'}{r^4}, \qquad \alpha''' = 2w\left(\frac{q}{r^2} - 3\frac{s^2}{r^4} + 4\frac{w^2}{r^6}\right); \qquad (5.14)$$

in which naturally r, s denote the first and second amplitudes of the basis (u, v). We easily deduce that in the interval j,

$$\alpha' \neq 0, \qquad (5.15)$$

and the phase α satisfies the non-linear differential equation of the third order

$$-\{\alpha, t\} - \alpha'^2(t) = q(t) \qquad (5.16)$$

where the symbol $\{\alpha, t\}$ denotes the Schwarzian derivative of α at the point t $(\in j)$ (§ 1.7).

Then a brief calculation gives the relation

$$\{\alpha, t\} + \alpha'^2(t) = \{\tan \alpha, t\}; \qquad (5.17)$$

in this relationship, if the function on the right has any singularities we assign to it at such points the corresponding values taken by the left hand side. Hence the non-linear differential equation (16) can be more briefly written as

$$-\{\tan \alpha, t\} = q(t). \qquad (5.18)$$

We see that *the phase* α *serves to determine the carrier* q *of the differential equation* (q) *uniquely, through the formulae* (16) *and* (18).

Finally, we note the important first formula (14) which expresses the relation between the phase α and the first amplitude of the basis (u, v).

5.6 First phases of the differential equation (q)

By a *first phase of the differential equation* (q) we mean a first phase of an arbitrary basis of the differential equation (q). Obviously, the above results are valid for all first phases of the differential equation (q) and for the corresponding bases (u, v).

5.7 Phase functions

In the course of our study we shall frequently encounter functions known as "phase functions". By a *phase function* we mean a function α defined in an open interval j with the following properties:

1. $\alpha \in C_1$;
2. $\alpha' \neq 0$ for all $t \in j$.

The following theorem can be established without difficulty:

Each phase function $\alpha \in C_3$ represents, in its interval of definition j, a first phase of the differential equation (q) constructed by formula (16), and the functions

$$u = |\alpha'|^{-\frac{1}{2}} \sin \alpha, \qquad v = |\alpha'|^{-\frac{1}{2}} \cos \alpha$$

are independent integrals of this differential equation (q), α being a first phase of the basis (u, v).

5.8 Second phases of a basis

We now wish to define second phases of the basis (u, v) in a manner analogous to that for first phases. In order to achieve this we assume that the zeros of the first derivative v' of v, in so far as they exist, are isolated. We shall always make this assumption in what follows, when we are concerned with second phases of a basis (u, v) of the differential equation (q). It holds for instance if the carrier q is non-zero in j (§ 2.1).

By a *second phase of the basis* (u, v) we mean any continuous function β in the interval j, which satisfies the relation

$$\tan \beta(t) = \frac{u'(t)}{v'(t)} \tag{5.19}$$

at every point of this interval with the exception of the zeros of v'.

The second phases of the basis (u, v) have, in general, similar properties to those of first phases; we shall therefore only recount them briefly.

The countable system of second phases associated with the basis (u, v) we shall call the *second phase system of the basis* (u, v) and denote it by (β).

If we choose a second phase $\beta \in (\beta)$, then the system (β) comprises the functions

$$\beta_\nu(t) = \beta(t) + \nu\pi \qquad (\nu = 0, \pm 1, \pm 2, \ldots; \beta_0 = \beta), \tag{5.20}$$

and they can clearly be ordered as follows:

$$\cdots < \beta_{-2} < \beta_{-1} < \beta_0 < \beta_1 < \beta_2 < \cdots. \tag{5.21}$$

From the second formula (2.1) we deduce that each second phase $\beta_\nu \in (\beta)$ is increasing or decreasing in j according as $wq > 0$ or < 0.

The derivatives u', v' of the integrals u, v may be expressed in terms of the second amplitude s and an arbitrary second phase $\beta_\nu \in (\beta)$ of the basis (u, v) as follows:

$$u'(t) = \varepsilon'_\nu s(t) \cdot \sin \beta_\nu(t), \qquad v'(t) = \varepsilon'_\nu s(t) \cdot \cos \beta_\nu(t) \qquad (t \in j), \tag{5.22}$$

in which ε'_ν, the so-called *signature of the second phase* β_ν, takes the value $+1$ or -1. The second phase β_ν is called *proper* or *improper* (with respect to the basis (u, v)) according as $\varepsilon'_\nu = 1$ or $\varepsilon'_\nu = -1$. By means of the ordering (21) of the second phase system (β) the individual second phases are alternately proper and improper: the successor of a proper second phase is improper and conversely. Every proper (improper) second phase of a basis (u, v) is improper (proper) with respect to the basis $(-u, -v)$.

The geometrical significance of second phases of the basis (u, v) is as follows:

Let β be a second phase of the basis (u, v). Moreover let $W\beta(t)$ be that value lying in the interval $[0, 2\pi)$ which is congruent to $\beta(t)$ modulo 2π: that is $\beta(t) = W\beta(t) + 2\pi n$, n $(= n(t))$ integral, $0 \leqslant W\beta(t) < 2\pi$; $t \in j$.

We consider the integral curve $\mathfrak{K}$ with the vectorial representation $x(t) = [u(t), v(t)]$. Then $x'(t) = [u'(t), v'(t)]$ is the tangent vector to the curve κ at the point $P[u(t), v(t)]$, and $W\beta(t)$ is the angle between the tangent vector $x'(t)$ or the opposite vector $-x'(t)$ and the co-ordinate vector x_2, according as β is proper or improper. In other words $\beta(t)$ is congruent modulo 2π to that angle in the range $[0, 2\pi)$ between the vector $x'(t)$ or $-x'(t)$ and the coordinate vector x_2.

5.9 Boundedness of a second phase

Let β be a second phase of the basis (u, v). If the carrier q of (q) is non-zero in the interval j, then the second phase β is bounded or not according to the type of the differential equation (q) in the interval j, and similar statements can be made as for a first phase (§ 5.4). The theorem of § 3.12 can be extended as follows: the values $\beta(t)$,

$\beta(x)$ of the phase β at two distinct points t, $x \in j$ differ by an integral multiple of π if and only if the numbers t, x are 2-conjugate.

5.10 Continuity property of a second phase

The second phase β belongs to the class C_1. The proof of this statement follows the lines of that in § 5.5 for the first phase. In this case we consider the function

$$\bar{\beta}(t) = \beta(x) + \int_x^t \frac{wq}{s^2}\, d\sigma.$$

Assuming the existence of the appropriate derivatives of q, β belongs to a higher class than C_1 and we have the formulae

$$\left.\begin{aligned} \beta' &= w\frac{q}{s^2}, \qquad \beta'' = w\left(\frac{q'}{s^2} - 2\frac{qs'}{s^3}\right), \\ \beta''' &= w\left(\frac{q''}{s^2} - 2\frac{2q's' + qs''}{s^3} + 6\frac{qs'^2}{s^4}\right). \end{aligned}\right\} \tag{5.23}$$

The first formula shows that the zeros of the function β' coincide with those of the carrier q.

5.11 Connection of a second phase with the associated differential equation

Now we assume that the carrier q of (q) does not vanish in j, and that $q \in C_2$. Then the functions

$$u_1 = \frac{u'}{\sqrt{|q|}}, \qquad v_1 = \frac{v'}{\sqrt{|q|}}$$

form a basis of the differential equation ($\hat{\text{q}}_1$), the associated differential equation of (q) (§ 1.9). From the relation $u'/v' = u_1/v_1$ we see that the second phase system of the basis (u, v) coincides with the first phase system of the basis (u_1, v_1). There follows the relationship, valid for $t \in j$,

$$-\tan\{\beta, t\} = \hat{q}_1(t). \tag{5.24}$$

Moreover

The differential equation (q) *and the associated differential equation* ($\hat{\text{q}}_1$) *have the same oscillatory character; that is, both are simultaneously of finite type or are oscillatory of the same kind.*

For, let β be an increasing second phase of the basis (u, v) and consequently also an increasing first phase α_1 of (u_1, v_1): $\beta = \alpha_1$. If the differential equation (q) is of finite type, then (by § 5.9) the function β i.e. α_1 is bounded; § 5.4 then shows that the differential equation ($\hat{\text{q}}_1$) is of finite type. If the differential equation (q) is of infinite type and is left or right oscillatory or oscillatory, then by § 5.9 the function β is respectively unbounded below and bounded above, or bounded below and unbounded above, or unbounded on both sides, and the phase α_1 naturally has the same properties. Then from § 5.4 we conclude that the differential equation ($\hat{q}_1$) is of infinite type, and respectively left oscillatory, or right oscillatory, or oscillatory.

By a similar argument, a given oscillatory character of ($\hat{q}_1$) implies the same character of (q).

5.12 Second phases of the differential equation (q)

By a *second phase of the differential equation* (q) we mean a second phase of any basis of (q). Obviously the results obtained in §§ 5.8–5.10 are valid for any second phase of the differential equation (q) and the corresponding basis (u, v). We observe that the problem of determining the differential equation (q) with a non-vanishing carrier $q \in C_2$, when given one of its second phases β is equivalent to the problem of integrating the non-linear second-order differential equation

$$X'' = -\{\tan \beta, t\} \cdot X + \frac{\varepsilon}{X} \qquad (\varepsilon = \pm 1).$$

This may be seen when we write the formula (1.18) in the following way:

$$-\{\tan \beta, t\} \frac{1}{\sqrt{|q(t)|}} = \frac{\operatorname{sgn} q(t)}{\dfrac{1}{\sqrt{|q(t)|}}} + \left(\frac{1}{\sqrt{|q(t)|}}\right)''.$$

5.13 Integrals of the differential equation (q) and their derivatives expressed in polar coordinates

Let (u, v) be a basis of the differential equation (q), r, s be its amplitudes and α, β a first and second phase of the basis (u, v). We have already seen ((13), (22)) that the integrals u, v and their derivatives u', v' can be represented in the interval j by the formulae

$$\left.\begin{aligned} u(t) &= \varepsilon r(t) \cdot \sin \alpha(t), & v(t) &= \varepsilon r(t) \cdot \cos \alpha(t),\\ u'(t) &= \varepsilon' s(t) \cdot \sin \beta(t), & v'(t) &= \varepsilon' s(t) \cdot \cos \beta(t) \end{aligned}\right\}$$
$$(\varepsilon, \varepsilon' = \pm 1) \qquad (5.25)$$

in which the values of $\varepsilon, \varepsilon'$ depend on the choice of the phases α, β.

Making use of (14), (23) we can also write

$$\left.\begin{aligned} u(t) &= \varepsilon \sqrt{|w|} \frac{\sin \alpha(t)}{\sqrt{|\alpha'(t)|}}, & v(t) &= \varepsilon \sqrt{|w|} \frac{\cos \alpha(t)}{\sqrt{|\alpha'(t)|}},\\ \sqrt{|\beta'(t)|}\, u'(t) &= \varepsilon' \sqrt{|wq(t)|} \sin \beta(t), & \sqrt{|\beta'(t)|}\, v'(t) &= \varepsilon' \sqrt{|wq(t)|} \cos \beta(t). \end{aligned}\right\} \qquad (5.26)$$

Consequently we have, for the general integral y of the differential equation (q) and its derivative y', the expressions

$$\left.\begin{aligned} y(t) &= k_1 \frac{\sin[\alpha(t) + k_2]}{\sqrt{|\alpha'(t)|}}, \\ \sqrt{|\beta'(t)|}\, y'(t) &= \pm k_1 \sqrt{|q(t)|}\, \sin[\beta(t) + k_2], \end{aligned}\right\} \tag{5.27}$$

in which k_1, k_2 are arbitrary constants. In the second formula (27) we take the sign + or − according as the signatures $\varepsilon, \varepsilon'$ of the phases α, β are the same or different.

If $k_2 = n\pi + k_2'$, $0 \leqslant k_2' < \pi$, n being an integer, then the value of the right hand side of (27) is not changed if we replace k_1 by $k_1' = (-1)^n k_1$ and k_2 by k_2'. Consequently, in the formula (27) we can assume without loss of generality that $0 \leqslant k_2 < \pi$.

5.14 Ordering relations between first and second phases of the same basis

We consider an arbitrary basis (u, v) of the differential equation (q) with the Wronskian w $(= uv' - u'v)$. Let $\alpha \in (\alpha)$, $\beta \in (\beta)$ be respectively a first and second phase of the basis (u, v) and $\varepsilon, \varepsilon'$ be the corresponding signatures.

From the definition of w and the formula (25) there follows (for $t \in j$) the relation

$$r \cdot s \cdot \sin(\beta - \alpha) = \varepsilon\varepsilon'(-w). \tag{5.28}$$

Since the right side of this equation is everywhere non-zero, there is an integer n such that the difference $\beta - \alpha$ lies between $n\pi$ and $(n + 1)\pi$ $\forall\, t \in j$,

$$n\pi < \beta - \alpha < (n + 1)\pi. \tag{5.29}$$

We set $\alpha_0 = \alpha + n\pi$, $\beta_0 = \beta$ and define the phases $\alpha_\nu \in (\alpha)$, $\beta_\nu \in (\beta)$ as in the formulae (11), (20). It is clear that the system formed from all first and second phases of the basis (u, v), which we call the *mixed phase system* of the basis (u, v), can be ordered in the following way:

$$\cdots < \alpha_{-1} < \beta_{-1} < \alpha_0 < \beta_0 < \alpha_1 < \beta_1 < \cdots. \tag{5.30}$$

This ordering is obviously such that in the interval j two neighbouring phases α_ν, β_ν or $\beta_\nu, \alpha_{\nu+1}$ satisfy respectively the relations

$$0 < \beta_\nu - \alpha_\nu < \pi, \qquad -\pi < \beta_\nu - \alpha_{\nu+1} < 0.$$

It follows from (28) that, respectively,

$$\operatorname{sgn} \varepsilon_\nu \varepsilon_\nu'(-w) = 1, \qquad \operatorname{sgn} \varepsilon_\nu' \varepsilon_{\nu+1}(-w) = -1.$$

We have, clearly, to consider two cases, according as $-w > 0$ or $-w < 0$; correspondingly, the first phase α_ν is increasing or decreasing.

In the case $-w > 0$ we have

$$\operatorname{sgn} \varepsilon_\nu \varepsilon_\nu' = 1; \qquad \operatorname{sgn} \varepsilon_\nu' \varepsilon_{\nu+1} = -1.$$

Then the ordering (30) of the mixed phase system of the basis (u, v) has the effect that each proper (improper) first phase α_ν is followed by a proper (improper) second phase

β_ν, while after each proper (improper) second phase β_ν there follows an improper (proper) first phase $\alpha_{\nu+1}$.

In the case $-w < 0$ we have

$$\operatorname{sgn} \varepsilon_\nu \varepsilon'_\nu = -1, \qquad \operatorname{sgn} \varepsilon'_\nu \varepsilon_{\nu+1} = 1,$$

and then each proper (improper) first phase α_ν is followed by an improper (proper) second phase β_ν, while each proper (improper) second phase β_ν is followed by a proper (improper) first phase $\alpha_{\nu+1}$.

This provides a relationship between first and second phases of the same basis of the differential equation (q).

5.15 Some consequences

We now wish to develop some further relations. First, from (28), (14) and (23) there follows:

$$\frac{\alpha'\beta'}{\sin^2(\beta - \alpha)} = -q. \tag{5.31}$$

Moreover from the first formulae (14), (23) we obtain

$$\frac{\beta' \cdot s^2}{\alpha' \cdot r^2} = -q. \tag{5.32}$$

If at a point $t \in j$, we have $-q > 0$, or $-q < 0$, then the functions α', β' have the same or opposite signs at this point respectively. Hence if the function q is everywhere non-zero in the interval j, the phases α, β have the property that in the case $-q > 0$ both phases α, β increase or decrease together, while in the case $-q < 0$ one increases while the other decreases.

Consequently if the function q is everywhere non-zero in the interval j, then the mixed phase system (30) behaves as follows: in the case $-q > 0$ both phases α_ν, β_ν increase or decrease together, while in the case $-q < 0$ one of the phases α_ν, β_ν increases while the other decreases.

We can supplement the theorem of § 3.12 as follows: the values $\alpha(t)$, $\beta(x)$ of the phases α, β at two points t, $x \in j$ differ by an integral multiple of π if and only if x is 3-conjugate with t and consequently t is 4-conjugate with x.

5.16 Explicit connection between first and second phases of the same basis

By §§ 5.5, 5.10, arbitrary phases α, β of the basis (u, v) are such that, in the interval j

$$\alpha \in C_3, \qquad \beta \in C_1, \qquad \alpha' \neq 0. \tag{5.33}$$

Now we prove the following theorem:

Theorem. Two functions α, β *in an open interval* j, *with the properties* (33), *represent a first and second phase of a basis* (u, v) *of a differential equation* (q) *if and only if*

$$\beta = \alpha + \operatorname{Arccot} \frac{1}{2}\left(\frac{1}{\alpha'}\right)'. \tag{5.34}$$

Here Arccot denotes a particular or an arbitrary branch of the function.
If the relation (34) *is satisfied then the functions defined in* j *by the following relations*

$$u = |\alpha'|^{-\frac{1}{2}} \sin \alpha, \qquad v = |\alpha'|^{-\frac{1}{2}} \cos \alpha \tag{5.35}$$

have the desired property and the corresponding carrier q *is determined by the formula* (16).

Proof. (a) Let (u, v) be a basis of a differential equation (q) and α, β a first and second phase respectively of (u, v). The functions α, β have the properties (33) and they satisfy formulae such as (10), (19), (29). From the relation

$$\cot(\beta - \alpha) = -\frac{\sin\alpha \sin\beta + \cos\alpha\cos\beta}{\sin\alpha\cos\beta - \sin\beta\cos\alpha} = -\frac{uu' + vv'}{uv' - u'v} = -\frac{1}{w}rr',$$

and the first formula (14), there follows the relation (34).

(b) Let α, β be arbitrary functions in an open interval j with the properties (33), (34). In j we define the functions q, u, v by means of formulae (16) and (35). Obviously the functions $u, v \in C_2$ and at every point $t \in j$ there hold the formulae

$$\left.\begin{aligned} u' &= \varepsilon|\alpha'|^{\frac{1}{2}}\left[\cos\alpha + \frac{1}{2}\left(\frac{1}{\alpha'}\right)'\sin\alpha\right], \\ v' &= \varepsilon|\alpha'|^{\frac{1}{2}}\left[-\sin\alpha + \frac{1}{2}\left(\frac{1}{\alpha'}\right)'\cos\alpha\right] \end{aligned} \quad (\varepsilon = \operatorname{sgn}\alpha').\right\} \tag{5.36}$$

Moreover,

$$\begin{aligned} u'' &= [-\{\alpha, t\} - \alpha'^2]|\alpha'|^{-\frac{1}{2}}\sin\alpha, \\ v'' &= [-\{\alpha, t\} - \alpha'^2]|\alpha'|^{-\frac{1}{2}}\cos\alpha. \end{aligned}$$

The functions u, v thus form a basis of the differential equation (q) with the Wronskian $w = -\varepsilon$.

From (35), $u/v = \tan\alpha$ throughout the interval j, except for zeros of v, while (36) and (34) give

$$\frac{u'}{v'} = \frac{\cos\alpha + \frac{1}{2}\left(\frac{1}{\alpha'}\right)'\sin\alpha}{-\sin\alpha + \frac{1}{2}\left(\frac{1}{\alpha'}\right)'\cos\alpha} = \frac{\cos\alpha + \sin\alpha\cdot\cos(\beta-\alpha)}{-\sin\alpha + \cos\alpha\cdot\cot(\beta-\alpha)} = \tan\beta$$

except at the zeros of v'.

Clearly, α is a first and β a second phase of the basis (u, v) and the proof is complete.

Finally we remark that arbitrary phases α, β of the basis (u, v) in the interval j are related by means of the following "bracket formula"

$$\{\tan \alpha, t\} - \{\tan \beta, t\} + \left\{\int_{t_0}^{t} q\, d\sigma, t\right\} = 0 \quad (t_0 \in j), \tag{5.37}$$

this relation being obtained from (18), (24) and (1.20).

5.17 Phases of different bases of the differential equation (q)

We now wish to study relationships between the first and second phases of two different bases of the differential equation (q).

Let (u, v), $(\bar{u}, \bar{v})$ be bases of the differential equation (q) and w, $\bar{w}$ their Wronskians; moreover let α, $\bar{\alpha}$ and β, $\bar{\beta}$ be first and second phases of these bases and c_{11}, c_{12}, c_{21}, c_{22} be constants such that the determinant $\Delta = |c_{ik}|$ is not zero. Then we have the following theorem

Theorem. If

$$\left.\begin{aligned} \bar{u} &= c_{11}u + c_{12}v, \\ \bar{v} &= c_{21}u + c_{22}v \end{aligned}\right\} \tag{5.38}$$

then

$$\tan \bar{\alpha} = \frac{c_{11} \tan \alpha + c_{12}}{c_{21} \tan \alpha + c_{22}}, \qquad \tan \bar{\beta} = \frac{c_{11} \tan \beta + c_{12}}{c_{21} \tan \beta + c_{22}}. \tag{5.39}$$

Conversely, from the first relation (39), *or from the second relation if $q \neq 0$ in j, it follows that*

$$\left.\begin{aligned} \bar{u} &= \pm \sqrt{\frac{\bar{w}}{w\Delta}}\,(c_{11}u + c_{12}v), \\ \bar{v} &= \pm \sqrt{\frac{\bar{w}}{w\Delta}}\,(c_{21}u + c_{22}v). \end{aligned}\right\} \tag{5.40}$$

The relations (39) are understood to hold throughout j except for singular points of the functions involved.

Proof. The first part of the theorem is obviously valid. We assume therefore that the first relation (39) holds; then $\bar{u}/\bar{v} = (c_{11}u + c_{12}v)/(c_{21}u + c_{22}v)$ and moreover, on taking account of (2.1) we have also $\bar{w}/\bar{v}^2 = w\Delta/(c_{21}u + c_{22}v)^2$. Then (40) follows immediately.

Now we assume that the second relation (39) holds, and that $q \neq 0$ for all $t \in j$. We then have $\bar{u}'/\bar{v}' = (c_{11}u' + c_{12}v')/(c_{21}u' + c_{22}v')$, then (2.1) and the hypothesis $q \neq 0$ yield the relation $\bar{w}/\bar{v}'^2 = w\Delta/(c_{21}u' + c_{22}v')^2$. From this we obtain the formulae (40).

We now have the following corollaries:

1. The first phase systems of two proportional bases of the differential equation (q) coincide; the second phase systems also coincide. Conversely, if two bases of (q) have

a common first or second phase, and q is non-zero in the interval j, then these bases are proportional.

2. If $\alpha(\beta)$ is a first (second) phase of the basis (u, v), then $\frac{1}{2}\pi - \alpha$ $(\frac{1}{2}\pi - \beta)$ is a first (second) phase for the inverse basis (v, u) of (u, v).

That is to say, one obtains the elements of the first or second phase system of the inverse basis (v, u) by multiplying the corresponding elements of the basis (u, v) by -1 and increasing by $\frac{1}{2}\pi$.

3. If $\alpha(\beta)$ is a first (second) phase of the basis (u, v) and λ is arbitrary then $\alpha + \lambda$, $(\beta + \lambda)$ is a first (second) phase of the basis (u, v) transformed by the orthogonal substitution

$$\begin{pmatrix} \cos\lambda & \sin\lambda \\ -\sin\lambda & \cos\lambda \end{pmatrix} \tag{5.41}$$

that is to say the basis

$$\bar{u} = u \cdot \cos\lambda + v \cdot \sin\lambda,$$
$$\bar{v} = -u \cdot \sin\lambda + v \cdot \cos\lambda.$$

When we have a first (second) phase α (β) of the differential equation (q) then the system formed from this by means of the formula $\bar{\alpha} = \alpha + \lambda$ $(\bar{\beta} = \beta + \lambda)$ with arbitrary λ, we call the *complete phase system* of the phase $\alpha(\beta)$. For this we use the notation $[\alpha]$, $([\beta])$. Obviously, given any number $t_0 \in j$ there is precisely one first (second) phase of the differential equation (q) in the system $[\alpha]$ $([\beta])$, which vanishes at the point t_0.

4. Two first or second phases $\alpha, \bar{\alpha}$ or $\beta, \bar{\beta}$ of the differential equation (q) are connected by means of the formulae (39). If conversely $\alpha(\beta)$ is a first (second) phase of (q) and there holds a formula similar to (39) for a function $\bar{\alpha}$ $(\bar{\beta})$ defined in the interval j, then $\bar{\alpha}$ $(\bar{\beta})$ is also a first (second) phase of (q).

5.18 Calculation of the integrals $\int_{x_0}^{x_1} g(\sigma)d\sigma$, $\int_{x_0}^{x_1} h(\sigma)d\sigma$ in the neighbourhood of singular points

As an application of the concept of phases, and of their properties obtained so far, we show how to evaluate the integrals considered in § 2.4.

We revert to the situation described there, using the same notation. In particular y denotes an integral of the differential equation (q) with a zero c; j_{-1} or j_0 denotes a left or right neighbourhood of c, in which the integral y does not vanish, and $g(\sigma)$ for $\sigma \in j_{-1}$ or $\sigma \in j_0$ denotes the function

$$g(\sigma) = \frac{1}{y^2(\sigma)} - \frac{1}{y'^2(c)} \cdot \frac{1}{(\sigma - c)^2}$$

We know that for $x_0 \in j_{-1}$, $x_1 \in j_0$ the integral $\int_{x_0}^{x_1} g(\sigma)\,d\sigma$ exists. We now wish to determine its value.

Let $t \in (x_0, c)$ be arbitrary. Then we have

$$\int_{x_0}^{t} g(\sigma)\, d\sigma = \int_{x_0}^{t} \frac{d\sigma}{y^2(\sigma)} + \frac{1}{y'^2(c)}\left[\frac{1}{t-c} - \frac{1}{x_0 - c}\right].$$

Let α be the first phase, vanishing at c, of that basis (u, v) of the differential equation (q) determined by the initial values $u(c) = 0$, $u'(c) = 1$; $v(c) = 1$, $v'(c) = 0$. Then we have

$$\alpha(c) = 0, \qquad \alpha'(c) = 1, \qquad \alpha''(c) = 0,$$

and the first formula (27) gives

$$y(\sigma) = y'(c)\,\frac{\sin\alpha(\sigma)}{\sqrt{\alpha'(\sigma)}}.$$

Thus

$$y'^2(c)\int_{x_0}^{t}\frac{d\sigma}{y^2(\sigma)} = \int_{x_0}^{t}\frac{\alpha'(\sigma)\, d\sigma}{\sin^2\alpha(\sigma)} = \int_{\alpha(x_0)}^{\alpha(t)}\frac{d\sigma}{\sin^2\sigma} = -\cot\alpha(t) + \cot\alpha(x_0),$$

and consequently

$$y'^2(c)\int_{x_0}^{t} g(\sigma)\, d\sigma = -\cot\alpha(t) + \frac{1}{t-c} + \cot\alpha(x_0) - \frac{1}{x_0 - c}.$$

On account of the fact that

$$\lim_{t\to c-}\left[-\cot\alpha(t) + \frac{1}{t-c}\right] = 0$$

we have

$$y'^2(c)\int_{x_0}^{c} g(\sigma)\, d\sigma = \cot\alpha(x_0) - \frac{1}{x_0 - c}. \tag{5.42}$$

Similarly we obtain, for arbitrary $x_1 \in j_0$

$$y'^2(c)\int_{c}^{x_1} g(\sigma)\, d\sigma = -\cot\alpha(x_1) + \frac{1}{x_1 - c}. \tag{5 43}$$

Clearly therefore, for arbitrary numbers $x_0 \in j_{-1}$, $x_1 \in j_0$ we have

$$\int_{x_0}^{x_1}\left[\frac{1}{y^2(\sigma)} - \frac{1}{y'^2(c)}\frac{1}{(\sigma - c)^2}\right] d\sigma$$
$$= \frac{1}{y'^2(c)}\left[-\cot\alpha(x_1) + \cot\alpha(x_0) + \frac{1}{c - x_0} + \frac{1}{x_1 - c}\right]. \tag{5 44}$$

If the numbers x_0, x_1 are 1-conjugate, then the quantities $\alpha(x_0)$, $\alpha(x_1)$ differ by an integral multiple of π (§ 5.4) In this case we have

$$\int_{x_0}^{x_1}\left[\frac{1}{y^2(\sigma)} - \frac{1}{y'^2(c)}\,\frac{1}{(\sigma - c)^2}\right] d\sigma = \frac{1}{y'^2(c)}\left[\frac{1}{c - x_0} + \frac{1}{x_1 - c}\right]. \tag{5.45}$$

If we apply these results to the calculation of the integral $\int_{x_0}^{x_m} g_m(\sigma)\,d\sigma$ considered in § 2.4, then we first obtain the formula

$$\int_{x_0}^{x_m} g_m(\sigma)\,d\sigma = \sum_{\nu=1}^{m} \frac{1}{y'^2(c_\nu)}[-\cot\alpha_\nu(x_\nu) + \cot\alpha_\nu(x_{\nu-1})] + \sum_{\nu=1}^{m} \frac{1}{y'^2(c_\nu)}\left[\frac{1}{c_\nu - x_0} + \frac{1}{x_m - c_\nu}\right]; \tag{5.46}$$

in which α_ν is naturally the first phase, vanishing at the point c_ν, of that basis (u, v) of the differential equation (q) determined by the initial values $u(c_\nu) = 0$, $u'(c_\nu) = 1$; $v(c_\nu) = 1$, $v'(c_\nu) = 0$; $\nu = 1, \ldots, m$.

If, in particular, the numbers x_0, x_m are 1-conjugate, then

$$\int_{x_0}^{x_m} g_m(\sigma)\,d\sigma = \sum_{\nu=1}^{m} \frac{1}{y'^2(c_\nu)}\left[\frac{1}{c_\nu - x_0} + \frac{1}{x_m - c_\nu}\right]. \tag{5.47}$$

Similarly, we can express the values of the integrals $\int_{x_0}^{x_1} h(\sigma)\,d\sigma$, $\int_{x_0}^{x_m} h_m(\sigma)\,d\sigma$ considered in § 2.5, in terms of appropriate second phases β, β_ν of the differential equation (q), as follows:

$$\int_{x_0}^{x_1}\left[\frac{q(\sigma)}{y'^2(\sigma)} - \frac{1}{q(e)y^2(e)}\cdot\frac{1}{(\sigma - e)^2}\right]d\sigma = \frac{1}{q(e)y^2(e)}\left[-\cot\beta(x_1) + \cot\beta(x_0) + \frac{1}{e - x_0} + \frac{1}{x_1 - e}\right]; \tag{5.48}$$

$$\int_{x_0}^{x_m} h_m(\sigma)\,d\sigma = \sum_{\nu=1}^{m} \frac{1}{q(e_\nu)y^2(e_\nu)}[-\cot\beta_\nu(x_\nu) + \cot\beta_\nu(x_{\nu-1})] + \sum_{\nu=1}^{m} \frac{1}{q(e_\nu)y^2(e_\nu)}\left[\frac{1}{e_\nu - x_0} + \frac{1}{x_m - e_\nu}\right]. \tag{5.49}$$

If the numbers x_0, x_1 or x_0, x_m are 2-conjugate, then there hold the simpler formulae (§ 5.9).

$$\int_{x_0}^{x_1}\left[\frac{q(\sigma)}{y'^2(\sigma)} - \frac{1}{q(e)y^2(e)}\cdot\frac{1}{(\sigma - e)^2}\right]d\sigma = \frac{1}{q(e)y^2(e)}\left[\frac{1}{e - x_0} + \frac{1}{x_1 - e}\right], \tag{5.50}$$

$$\int_{x_0}^{x_m} h_m(\sigma)\,d\sigma = \sum_{\nu=1}^{m} \frac{1}{q(e_\nu)y^2(e_\nu)}\left[\frac{1}{e_\nu - x_0} + \frac{1}{x_m - e_\nu}\right] \tag{5.51}$$

6 Polar functions

In this paragraph we shall be concerned with the so-called *polar functions* of a differential equation (q). A polar function is given by the difference $\beta - \alpha$ of two phases α, β belonging to a basis of the differential equation (q). Such functions are important principally because of their meaning in problems of a geometrical nature. The introduction of polar functions into the theory of the differential equation (q) provides a notable enrichment of the analysis and finds expression in many elegant relations.

In order to simplify our study, in this section we assume that $q(t) \neq 0$ for all $t \in j$.

6.1 The concept of polar functions

We consider a differential equation (q) ($q \neq 0$). Let (u, v) be a basis of this differential equation.

By a *polar function* of the basis (u, v) we mean the following function formed from a first and a second phase α, β of the basis u, v.

$$\theta = \beta - \alpha \qquad (t \in j). \tag{6.1}$$

We call the phases α, β the *components* of θ; more precisely, α is the *first* and β the *second* component of θ. There is therefore precisely one countable system (θ) of polar functions of the basis (u, v), whose elements differ by a multiple of π. If α_0, β_0 are two neighbouring elements in the ordered mixed phase system of (u, v) (5.30), so that $0 < \beta_0 - \alpha_0 < \pi$ or $-\pi < \beta_0 - \alpha_0 < 0$; and $\theta_0 = \beta_0 - \alpha_0$, then the system (θ) is composed of the functions

$$\theta_\nu(t) = \theta_0(t) + \nu\pi \qquad (\nu = 0, \pm 1, \pm 2, \ldots).$$

Every polar function $\theta \in (\theta)$ obviously has a countable infinity of first and second components: $\alpha = \alpha_0 + n\pi$, $\beta = \beta_0 + n\pi$ (n integral). We see that every first (second) phase α (β) of the basis (u, v) can be expressed by means of a first (second) component of θ, from which the second (first) component is uniquely determined.

For every first component α of θ we have, from (5.34), the relation

$$\theta = \text{Arccot}\, \frac{1}{2}\left(\frac{1}{\alpha'}\right)'; \tag{6.2}$$

in which Arccot denotes an appropriate branch of this function.

In view of this relation, we call the first component α the *generator* of θ and say that the polar function θ is *generated* by the first phase α.

Conversely, every branch of the above function constructed from an arbitrary first phase α of the basis (u, v) represents a polar function $\theta \in (\theta)$. Every polar function $\theta \in (\theta)$ obviously belongs to the class C_1, and according to (5.29) satisfies for $t \in j$ the relation $n\pi < \theta < (n + 1)\pi$, where n is an integer.

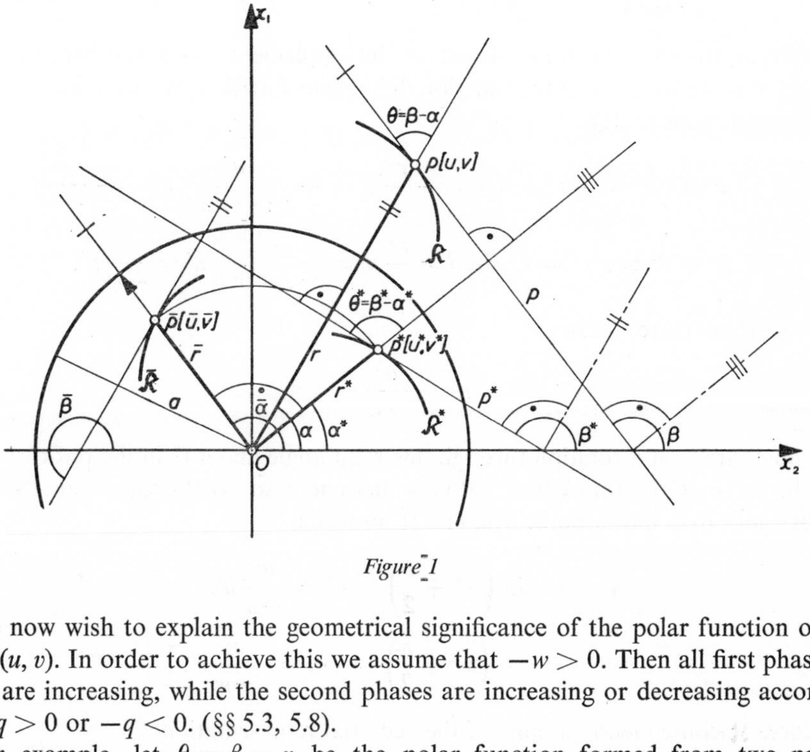

Figure 1

We now wish to explain the geometrical significance of the polar function of the basis (u, v). In order to achieve this we assume that $-w > 0$. Then all first phases of (u, v) are increasing, while the second phases are increasing or decreasing according as $-q > 0$ or $-q < 0$. (§§ 5.3, 5.8).

For example, let $\theta = \beta - \alpha$ be the polar function formed from two proper neighbouring phases α, β; that is to say such that $0 < \beta - \alpha < \pi$. We consider the integral curve $\mathfrak{K}$ of the differential equation (q) with the vector representation $x(t) = [u(t), v(t)]$. Then $x'(t) = [u'(t), v'(t)]$ is the tangent vector to the curve $\mathfrak{K}$ at the point $P[u(t), v(t)]$.

Let $W\alpha(t)$, $W\beta(t)$, $W\theta(t)$ be those numbers lying in the interval $[0, 2\pi)$ which are congruent to $\alpha(t)$, $\beta(t)$, $\theta(t)$ modulo 2π. We know that $W\alpha(t)$ is the angle formed by the vector $x(t)$ and the coordinate vector x_2; $W\beta(t)$ is the angle between the tangent vector $x'(t)$ and the coordinate vector x_2 (§§ 5.3, 5.8).

Moreover, we have $W\theta(t) = W\beta(t) - W\alpha(t)$ or $= 2\pi + [W\beta(t) - W\alpha(t)]$, according as $W\beta(t) > W\alpha(t)$ or $W\beta(t) < W\alpha(t)$. Consequently $W\theta(t)$ is the angle between the vectors $x(t)$, $x'(t)$.

We see that the value $\theta(t)$ gives modulo 2π the angle formed at the point $P(t)$ between the oriented straight line $OP(t)$ and the oriented tangent p to the curve at the point $P(t)$. The orientation has the same sense of that of the vectors $x(t)$, $x'(t)$ (see Fig. 1; the angles $W\alpha(t)$, $W\beta(t)$, $W\theta(t)$ are denoted by α, β, θ.)

We now wish to spend a moment in the consideration of this figure. It shows that under the inversion $\boldsymbol{K}_a$ with respect to a circle of radius a with centre O, the integral curve $\mathfrak{K}$ is transformed into another curve $\mathfrak{K}^*$. The pair P, p go over into another pair $P^*[u^*, v^*], p^*$, the numbers r; α, β, θ corresponding to numbers r^*; $\alpha^*, \beta^*, \theta^*$, by the relations

$$r^* = \frac{a^2}{r \cdot \sin\theta}; \qquad \alpha^* = \beta - \frac{\pi}{2}, \qquad \beta^* = \alpha + \frac{\pi}{2}, \qquad \theta^* = \pi - \theta;$$

in particular, the angle θ is transformed into its supplement $\pi - \theta$. It is because of this property that we have called the function θ the polar function. We have moreover the following relations (5.28)

$$u^* = r^* \sin\alpha^* = -r^* \cos\beta = -\frac{r^*}{s} v' = -\frac{a^2}{r \cdot s \cdot \sin\theta} v' = -\frac{a^2}{-w} v',$$

$$v^* = r^* \cos\alpha^* = r^* \sin\beta = \frac{r^*}{s} u' = \frac{a^2}{r \cdot s \cdot \sin\theta} u' = \frac{a^2}{-w} u',$$

and from these there follows

$$r^* = \frac{a^2}{-w} s.$$

If we now carry out a rotation through one right angle about O in the positive sense, then the curve $\mathfrak{K}^*$ is transformed into another curve $\bar{\mathfrak{K}}$. At the same time the point $P^* \in \mathfrak{K}^*$ goes over into a point $\bar{P}[\bar{u}, \bar{v}) \in \bar{\mathfrak{K}}$, in which

$$\bar{u} = r^* \sin\left(\alpha^* + \frac{\pi}{2}\right) = v^* = \frac{a^2}{-w} u',$$

$$\bar{v} = r^* \cos\left(\alpha^* + \frac{\pi}{2}\right) = -u^* = \frac{a^2}{-w} v'.$$

The curve $\bar{\mathfrak{K}}$ consequently admits of the vectorial representation

$$\bar{x}(t) = \left[\frac{a^2}{-w} u'(t), \frac{a^2}{-w} v'(t)\right]. \tag{6.3}$$

The tangent vector to the curve $\bar{\mathfrak{K}}$ at the point $\bar{P}(t)$ is obviously

$$\bar{x}'(t) = \left[\frac{a^2}{-w} q(t)u(t), \frac{a^2}{-w} q(t)v(t)\right]. \tag{6.4}$$

This is therefore opposite to or in the same sense as $x(t)$, according as $-q > 0$ or $-q < 0$. The angle formed by the tangent vector $\bar{x}'$ at the point $\bar{P}$ and the coordinate vector x_2 is $\bar{\beta} = \alpha \pm \pi$, in which we take the $+$ or $-$ sign according as $0 \leqslant \alpha < \pi$ or $\pi \leqslant \alpha < 2\pi$.

We have therefore

$$\bar{r} = \frac{a^2}{-w} s, \qquad \bar{\alpha} = \beta, \qquad \bar{\beta} = \alpha \pm \pi. \tag{6.5}$$

Let $\boldsymbol{R}$ be the transformation, arising from the inversion $\boldsymbol{K}_{\sqrt{-w}}$ $(a = \sqrt{-w})$ followed by a rotation about O in the positive sense through one right angle. We see that the integral curve $\mathfrak{K}$ is carried by this transformation into a curve $\overline{\mathfrak{K}}$. At the same time an arbitrary point $P \in \mathfrak{K}$ goes into a point $\bar{P} \in \overline{\mathfrak{K}}$, in such a manner that the radius vectors $x = \overrightarrow{OP}$, $\bar{x} = \overrightarrow{O\bar{P}}$ and the corresponding tangent vectors x', $\bar{x}'$, are related in the following way:

$$\bar{x} = x', \qquad \bar{x}' = q \cdot x, \tag{6.6}$$

while the corresponding amplitudes $\bar{r}$, s and angles α, β; $\bar{\alpha}$, $\bar{\beta}$ are transformed as follows

$$\bar{r} = s, \qquad \bar{\alpha} = \beta, \qquad \bar{\beta} = \alpha \pm \pi. \tag{6.7}$$

If the curve $\mathfrak{K}$ is so specialized that it goes into itself by the transformation $\boldsymbol{R}$, then it obviously has the following "ellipse property": the two straight lines passing through the point O and the corresponding points $P, \bar{P} \in \mathfrak{K}$ are parallel to the tangents to the curve $\mathfrak{K}$ at the points $\bar{P}$, P. Curves with this property are called *Radon curves*. They were first studied by J. Radon [Ber. Verh. Sächs. Akad. Leipzig, 68 (1916), 123–128]; we shall meet them later (§ 16.5) in a quite different connection.

6.2 General polar form of the carrier q

Let (u, v) be a basis of the differential equation (q) and θ a polar function of (u, v).

The function $\cot \theta$ constructed from the function θ is obviously the same for all polar functions of the basis (u, v), it is therefore determined uniquely by this basis. From (2) and (5.14) we have the formulae

$$-w \cdot \cot \theta = rr', \qquad 2 \cot \theta = \left(\frac{1}{\alpha'}\right)', \qquad \alpha' \cdot \cot \theta = (\log r)'; \tag{6.8}$$

in this r, α naturally represent the first amplitude and a first phase of the basis (u, v).

Let $t_0 \in j$ be an arbitrary number. The value which any function takes at the point t_0 we shall as a rule indicate by the suffix $_0$; for instance $r(t_0)$ will be denoted by r_0.

From (8) it follows that for $t \in j$

$$r^2 = r_0^2 - 2w \int_{t_0}^{t} \cot \theta(\tau)\, d\tau$$

and moreover, taking account of (5.14),

$$r^2 = r_0^2 \left(1 + 2\alpha_0' \int_{t_0}^{t} \cot \theta(\tau)\, d\tau\right). \tag{6.9}$$

This formula gives

$$1 + 2\alpha_0' \int_{t_0}^{t} \cot \theta(\tau)\, d\tau > 0.$$

Clearly, in the interval j the inequality

$$\int_{t_0}^{t} \cot \theta(\tau)\, d\tau \gtrless -\frac{1}{2\alpha_0'} \tag{6.10}$$

holds, according as $\alpha_0' > 0$ or $\alpha_0' < 0$. Consequently the function $\int_{t_0}^{t} \cot\theta(\tau)\,d\tau$ is bounded in j at least on one side by the number $-\frac{1}{2}(\alpha_0')^{-1}$, this bound being below or above according as $\alpha_0' > 0$ or $\alpha_0' < 0$. From (8) we have, for $t \in j$,

$$\alpha = \alpha_0 + \alpha_0' \int_{t_0}^{t} \frac{d\sigma}{1 + 2\alpha_0' \int_{t_0}^{\sigma} \cot\theta(\tau)\,d\tau} \tag{6.11}$$

and further, by (5.16)

$$q = -\frac{\alpha_0'}{\sin^2\theta} \cdot \frac{\alpha_0' + \theta'\left(1 + 2\alpha_0' \int_{t_0}^{t} \cot\theta(\tau)\,d\tau\right)}{\left(1 + 2\alpha_0' \int_{t_0}^{t} \cot\theta(\tau)\,d\tau\right)^2}. \tag{6.12}$$

We call the expression on the right of this formula the *general polar form of the carrier* q. We speak also of the general polar form of the *differential equation* (q), when the carrier q is put into the general polar form (12).

Consequently the inequality

$$\theta' \gtrless -\frac{\alpha_0'}{1 + 2\alpha_0' \int_{t_0}^{t} \cot\theta(\tau)\,d\tau} \tag{6.13}$$

holds in j, according as $-q\alpha_0' > 0$ or $-q\alpha_0' < 0$.

6.3 Determination of the carrier from a polar function

By a *polar function* of the carrier q or of the differential equation (q) we mean a polar function of any basis of the differential equation (q). Clearly, every polar function θ of the differential equation (q) has the following properties in the interval j:

1. $\theta \in C_1$:
2. $n\pi < \theta < (n + 1)\pi$ (n being an integer); (6.14)
3. the function $\int_{t_0}^{t} \cot\theta(\tau)\,d\tau$ ($t_0 \in j$ fixed) is bounded at least on one side.

The function $\theta(t) = c$ (= constant) in the interval $j = (-\infty, \infty)$ cannot, for instance, represent a polar function of the differential equation (q) if $n\pi < c < (n + 1)\pi$, $c \neq (n + \frac{1}{2})\pi$, n integral.

Now we wish to consider how far a given polar function determines the differential equation (q).

Let θ be a function, defined in j, with the above properties 1–3. We choose a number $t_0 \in j$. From property 2, the function $\int_{t_0}^{t} \cot\theta(\tau)\,d\tau$ exists in j, and from property 3 it is bounded at least on one side; for instance, let it be bounded below. Then we have, for a certain constant $A > 0$,

$$1 + 2A \int_{t_0}^{t} \cot \theta(\tau)\, d\tau > 0 \qquad (t \in j). \tag{6.15}$$

We now choose an arbitrary number α_0, and also a number α_0' such that $0 < \alpha_0' < A$; then there holds an inequality of the form (10) with the sign $>$. We shall now establish the following result:

Precisely one function q defined in the interval j can be constructed, so that the differential equation (q) *admits of the function θ as polar function with the initial values $\alpha(t_0) = \alpha_0$, $\alpha'(t_0) = \alpha_0'$ of its generator α.*

Proof. First we observe that there can exist at most one function q with the designated properties. For, given any function q of this kind the generator α of θ is uniquely determined by θ, α_0, α_0' by means of formula (11). From (5.16) there is precisely one carrier q which admits the function α as first phase.

We now show that there is at least one function q with the above properties. With this aim we construct the function $\alpha(t)$ $(t \in j)$ in accordance with formula (11). According to property 1, $\alpha \in C_3$, and by (15) we have $\alpha'(t) > 0$. Consequently α represents a phase function (§ 5.7) and by application of (5.16) a calculation shows that α is a first phase of the carrier q determined by a formula similar to (12). Moreover formula (11) yields, when differentiated, a relationship such as (2).

By § 5.15, the function

$$\beta = \alpha + \operatorname{Arccot} \frac{1}{2}\left(\frac{1}{\alpha'}\right)' \tag{6.16}$$

represents a second phase of each basis, determined from α, of the carrier q; in this formula Arccot is that branch of the function lying between $n\pi$ and $(n + 1)\pi$. From (2), (16) we have $\beta = \alpha + \theta$. Consequently θ is a polar function of the carrier q, and the proof is complete.

6.4 Radon functions

We now go back to the situation of § 6.1 and for convenience assume that $q(t) < 0$ for all $t \in j$. Then the functions α', β' always have the same sign in the interval j (§ 5.14), and in fact $\alpha' > 0$, $\beta' > 0$ or $\alpha' < 0$, $\beta' < 0$, according as $-w > 0$ or $-w < 0$.

By a *Radon function* of the basis (u, v) we mean a function

$$\zeta = \beta + \alpha \qquad (t \in j) \tag{6.17}$$

constructed from a first phase α and a second phase β of the basis (u, v). A Radon function ζ of the basis (u, v) is also called a *Radon parameter* of the basis (u, v). We make use of this designation in recognition of the elegant study made by J. Radon of the curves described above (§ 6.1), possessing the ellipse property, in which functions of this kind appear.

Obviously there is a countable system of Radon functions of the basis (u, v) and the individual functions of this system differ from each other by an integral multiple of π. Clearly $\zeta \in C_1$ and moreover $\zeta' > 0$ or $\zeta' < 0$ according as $-w > 0$ or $-w < 0$, for $t \in j$.

Two functions θ, ζ constructed from the same phases α, β of the basis (u, v)

$$\theta = \beta - \alpha, \qquad \zeta = \beta + \alpha \tag{6.18}$$

we call *associated*. The formulae (18) yield the results

$$\alpha = \frac{1}{2}(\zeta - \theta), \qquad \beta = \frac{1}{2}(\zeta + \theta). \tag{6.19}$$

Clearly, $\zeta - \theta \in C_3$. From $\alpha'\beta' > 0$, $\operatorname{sgn} \alpha' = \operatorname{sgn}(-w)$, we have

$$-\zeta' < \theta' < \zeta' \qquad \text{or} \qquad \zeta' < \theta' < -\zeta', \tag{6.20}$$

according as $-w > 0$ or $-w < 0$.

Now let us start from the formulae

$$\beta' \cot \theta = (\alpha' + \theta') \cot \theta = \alpha' \cot \theta + (\log|\sin \theta|)'.$$

By making use of (8) we obtain

$$\zeta' \cot \theta = (\log r^2 |\sin \theta|)',$$

and it follows that for $t \in j$

$$r^2 = r_0^2 \frac{\sin \theta_0}{\sin \theta} \exp \int_{t_0}^{t} \zeta'(\tau) \cot \theta(\tau)\, d\tau; \tag{6.21}$$

moreover, by (5.28),

$$s^2 = \alpha_0'^2 r_0^2 \frac{1}{\sin \theta_0 \cdot \sin \theta} \exp\left(-\int_{t_0}^{t} \zeta'(\tau) \cot \theta(\tau)\, d\tau\right). \tag{6.22}$$

Finally, formula (5.32) gives

$$q = -\frac{\alpha_0'^2}{\sin^2 \theta_0} \cdot \frac{\zeta' + \theta'}{\zeta' - \theta'} \exp\left(-2\int_{t_0}^{t} \zeta'(\tau) \cot \theta(\tau)\, d\tau\right). \tag{6.23}$$

The expression on the right of (23) is called the *Radon polar form* of the carrier q. We speak of the *Radon polar form of the differential equation* (q), if the carrier q is put in the Radon polar form.

6.5 Normalized polar functions

In the following study we wish to relate the values of a polar function to its components or to the Radon parameter, regarding the latter as an independent variable. A polar function transformed in this way we call a normalized polar function. We assume $q(t) \neq 0$ for all $t \in j$.

Let $\theta = \beta - \alpha$ be a polar function of the basis (u, v) of the differential equation (q). We choose an arbitrary number $t_0 \in j$ and denote, as before, by the suffix $_0$ the values taken by the functions concerned at the point t_0. The function θ naturally has the properties (14) above.

6.6 Normalized polar functions of the first kind

First we shall represent the polar function θ as a function of the independent variable α.

Since the function $\alpha(t)$, in the interval j, is strictly increasing ($\alpha' > 0$) or decreasing ($\alpha' < 0$), the inverse function α^{-1} exists. This is defined in the range J_1 of the function α in the interval j. Consequently J_1 is an open interval, and $\alpha_0 \in J_1$.

Now we observe that if the differential equation (q) is of finite type, then the interval J_1 is bounded. If (q) is left (right) oscillatory, and $\alpha' > 0$ ($\alpha' < 0$), then the interval J_1 is of the form $(-\infty, A)$, A being finite; if on the other hand the differential equation (q) is left (right) oscillatory and $\alpha' < 0$ ($\alpha' > 0$), then the interval J_1 is of the type (A, ∞), A being finite. Finally if the differential equation (q) is oscillatory then the interval J_1 is unbounded on both sides.

The function α^{-1} obviously belongs to the class C_3. Moreover it forms a one-to-one mapping of the interval J_1 on j; in particular we have $\alpha^{-1}(\alpha_0) = t_0$. From the definition of α^{-1} it follows that for $\alpha \in J_1$, $\alpha^{-1}(\alpha) = t \in j$ is that number for which $\alpha(t) = \alpha$. We again use the term *homologous* to describe two such numbers $t = \alpha^{-1}(\alpha) \in j$ and $\alpha = \alpha(t) \in J_1$.

Now we define, in the interval J_1, the function $h(\alpha)$ or more briefly $h\alpha$, by assigning to it at each point $\alpha \in J_1$ the value of the function θ at the homologous point $t \in j$, that is

$$h(\alpha) = \theta\alpha^{-1}(\alpha) = \theta(t). \tag{6.24}$$

Consequently $h\alpha$ is the polar function θ regarded as a function of the independent variable α. We call h a *normalized polar function of the first kind of the basis* (u, v), more briefly a *first normalized* (or 1-*normalized*) *polar function of the basis* (u, v)

Obviously the function h in the interval J_1 has the following properties:

1. $h \in C_1$;
2. $n\pi < h < (n + 1)\pi$, n being an integer.

We emphasize that if the differential equation (q) is oscillatory then the interval of definition J_1 of the first normalized polar function h is the interval $(-\infty, \infty)$.

From (8) and (5.14) we have the following formulae, valid in the interval j

$$r(t) = r_0 \exp \int_{\alpha_0}^{\alpha} \cot h(\rho)\, d\rho, \tag{6.25}$$

$$\alpha'(t) = \alpha_0' \exp\left(-2 \int_{\alpha_0}^{\alpha} \cot h(\rho)\, d\rho\right). \tag{6.26}$$

When we have a function of α ($\in J_1$) we shall denote its derivative with respect to α by means of an oblique dash ($^\backprime$).

From (26) it follows that

$$t^{\backprime}(\alpha) = \frac{1}{\alpha_0'} \exp 2 \int_{\alpha_0}^{\alpha} \cot h(\rho)\, d\rho, \tag{6.27}$$

and this relationship gives

$$t = t_0 + \frac{1}{\alpha_0'} \int_{\alpha_0}^{\alpha} \left(\exp 2 \int_{\alpha_0}^{\sigma} \cot h(\rho)\, d\rho\right) d\sigma. \tag{6.28}$$

Obviously this expression represents the inverse function $t = \alpha^{-1}(\alpha)$ of the first phase $\alpha(t)$.

From (5.16) and (26) we obtain the following formula relating any two homologous points $t \in j$, $\alpha \in J_1$:

$$q(t) = -\alpha_0'^2 \frac{1 + h^{\backprime}(\alpha)}{\sin^2 h(\alpha)} \exp\left(-4 \int_{\alpha_0}^{\alpha} \cot h(\rho)\, d\rho\right). \tag{6.29}$$

The expression on the right of this formula is called *the first polar form of the carrier q*. We speak of the *first polar form of the differential equation* (q), when the carrier q is in its first polar form.

Obviously, in the interval J_1 we have the inequality

$$h^{\backprime}(\alpha) \gtrless -1,$$

according as $-q > 0$ or $-q < 0$.

We observe that the formula (25) represents the equation of the integral curve $x(t) = [u(t), v(t)]$ in polar coordinates.

6.7 Determination of the carrier from a first normalized polar function

By a *normalized polar function of the first kind of the carrier q* or of the *differential equation* (q), which we shall more briefly call the first normalized polar function of the carrier q or of the differential equation (q), we mean a first normalized polar function of any basis of the differential equation (q).

Every first normalized polar function h of the differential equation (q) has the following properties in its interval of definition J_1.

1. $h \in C_1$;
2. $n\pi < h < (n + 1)\pi$ (n integral)
3. $h^{\backprime} > -1$ or $h^{\backprime} < -1$. (6.30)

We now study the question of how far the differential equation (q) is determined from a first normalized polar function.

Let h be a function defined in an open interval J_1 with the above properties (30). We choose a number t_0 and further numbers $\alpha_0 \in J_1$, $\alpha_0' \neq 0$. Now we have the following theorem.

Theorem. Precisely one function q, defined and never zero in an open interval $j(t_0 \in j)$, *can be constructed so that the differential equation* (q) *admits of the function h as first normalized polar function, and this polar function is generated by a first phase* α *of the differential equation* (q) *with initial values* $\alpha(t_0) = \alpha_0$, $\alpha'(t_0) = \alpha_0'$.

We shall only sketch the proof, which is essentially similar to that of § 6.3.

If there exist a differential equation (q) and a first phase $\alpha(t)$ as described in the theorem, then the latter is uniquely determined as the inverse of the function $t(\alpha)$ defined by formula (28). There can therefore be at most one differential equation (q) as specified by the theorem. Now we define the function $t(\alpha)$ in terms of the function h and the numbers α_0, α_0' by the formula (28). This functions maps the interval J_1 onto an open interval j, and $t_0 \in j$. Let $\alpha(t)$ be the inverse function to $t(\alpha)$ defined in the interval j; this is obviously a phase function. Let $q(t)$ be the carrier of the differential equation (q) (defined in the sense of (5.16)), for which the function $\alpha(t)$ represents a first phase. Then there is an analogous formula to (29), and by (30), 3 it follows that $q(t) \neq 0$ for all $t \in j$. Further, let θ be the polar function of the differential equation (q), which is generated by the first phase α and lies between $n\pi$ and $(n + 1)\pi$. By (26) we have, at two homologous points $t \in j$, $\alpha \in J_1$,

$$\cot \theta(t) = \frac{1}{2}\left(\frac{1}{\alpha'}\right)' = \frac{1}{2}\frac{1}{\alpha_0'}\left[\exp\left(2\int_{\alpha_0}^{\alpha} \cot h(\rho)\, d\rho\right)\right]^{\cdot} \cdot \alpha_0' \exp\left(-2\int_{\alpha_0}^{\alpha} \cot h(\rho)\, d\rho\right)$$
$$= \cot h(\alpha).$$

Consequently $\theta(t) = h(\alpha)$ and the proof is complete.

6.8 Normalized polar functions of the second kind

In the second place we shall regard the polar function θ as a function of the independent variable β.

Following the lines of our study in § 6.6 we consider the function β^{-1} inverse to the function β; this exists in an open interval J_2, and we have $\beta_0 \in J_2$. With regard to the character of the interval J_2, analogous remarks can be made to those in § 6.6 about the interval J_1.

Now we define, in the interval J_2, the function $-k(\beta)$, or more shortly $-k\beta$, in such a way that to every point $\beta \in J_2$ there corresponds the value of the function θ at the homologous point $\beta^{-1}(\beta) = t \in j$, that is:

$$-k(\beta) = \theta\beta^{-1}(\beta) = \theta(t). \tag{6.31}$$

Consequently $-k\beta$ is the polar function θ regarded as a function of the independent variable β. We call $-k$ *a normalized polar function of the second kind of the basis* (u, v) or more shortly *a second normalized* (or *2-normalized*) *polar function of the basis* (u, v). Obviously the function $-k$ has the following properties in the interval J_2:

1. $-k \in C_1$;
2. $n\pi < -k < (n + 1)\pi$, n integral.

We emphasize that if the differential equation (q) is oscillatory then the definition interval J_2 of the second normalized polar function $-k$ is the interval $(-\infty, \infty)$. The derivative of a function of β $(\in J_2)$ with respect to β we shall similarly denote by an oblique dash $\diagdown$. Now, from (1), (31) we have the following relation holding at two homologous points $t \in j$, $\beta \in J_2$

$$\alpha(t) = \beta + k(\beta), \tag{6.32}$$

and consequently also

$$\alpha'(t) = [1 + k^{\diagdown}(\beta)]\beta'(t). \tag{6.33}$$

Moreover, we obviously have

$$\int_{t_0}^{t} \alpha'(\tau) \cot \theta(\tau)\, d\tau = \int_{t_0}^{t} [1 + k^{\diagdown}\beta(\tau)] \cot \theta(\tau) \cdot \beta'(\tau)\, d\tau = -\int_{\beta_0}^{\beta} [1 + k^{\diagdown}(\rho)] \cot k(\rho)\, d\rho$$

and hence, by (8), (33),

$$t^{\diagdown}(\beta) = \frac{1}{\alpha_0'} [1 + k^{\diagdown}(\beta)] \exp\left(-2\int_{\beta_0}^{\beta} [1 + k^{\diagdown}(\rho)] \cot k(\rho)\, d\rho\right). \tag{6.34}$$

This relationship gives

$$t = t_0 + \frac{1}{\alpha_0'} \int_{\beta_0}^{\beta} [1 + k^{\diagdown}(\sigma)] \exp\left(-2\int_{\beta_0}^{\sigma} [1 + k^{\diagdown}(\rho)] \cot k(\rho)\, d\rho\right) d\sigma. \tag{6.35}$$

Obviously this formula represents the inverse function $t = \beta^{-1}(\beta)$ to the second phase $\beta(t)$.

From (5.31) and (33), (34) we obtain the following formula, valid at two homologous points $t \in j$, $\beta \in J_2$

$$q(t) = -\frac{\alpha_0'^2}{\sin^2 k(\beta)} \cdot \frac{\exp 4 \int_{\beta_0}^{\beta} [1 + k^{\diagdown}(\rho)] \cot k(\rho)\, d\rho}{1 + k^{\diagdown}(\beta)}. \tag{6.36}$$

The expression on the right of this formula is called the *second polar form of the carrier q*. We speak of the *second polar form of the differential equation* (q), when the carrier q is in the second polar form.

In the interval J_2 there hold therefore the inequalities

$$-k^{\diagdown}(\beta) \lessgtr 1,$$

according as $-q > 0$ or $-q < 0$.

6.9 Determination of the carrier from a second normalized polar function

By a normalized polar function of the second kind *of the carrier q* or of the *differential equation* (q), (more shortly, a second normalized polar function of the carrier q or the differential equation (q)) we mean a normalized polar function of the second kind of any basis of the differential equation (q). Every second normalized polar function

$-k$ of the differential equation (q) has the following properties in its interval of definition J_2

1. $-k \in C_1$;
2. $n\pi < -k < (n+1)\pi$ (n integral)
3. $-k^{\backslash} < 1$ or >1. (6.37)

As in the case of first normalized polar functions (§ 6.7), one can consider the problem of determining the differential equation (q) from its second normalized polar functions. There is, indeed a relevant theorem, analogous to that of § 6.7 but we shall not formulate it since no essentially new ideas are involved and there are no applications of it in our subsequent studies.

6.10 Normalized polar functions of the third kind

In the third place we go back to the situation examined in § 6.4 and set ourselves the object of representing the polar function $\theta = \beta - \alpha$ as a function of the Radon parameter $\zeta = \beta + \alpha$. We know that $\zeta \in C_1$ and that $\zeta' > 0$ or $\zeta' < 0$ according as $-w > 0$ or $-w < 0$.

Consider the inverse function ζ^{-1} of ζ. This is defined in an open interval J_3, and we have $\zeta_0 \in J_3$. With regard to the nature of the interval J_3, similar remarks apply to those on the interval J_1 in § 6.6.

In the interval J_3 we define the function $p(\zeta)$, more shortly $p\zeta$, by associating with every point $\zeta \in J_3$ the value of the function θ at the point homologous to ζ, namely $\zeta^{-1}(\zeta) = t \in j$;

$$p(\zeta) = \theta\zeta^{-1}(\zeta) = \theta(t). \tag{6.38}$$

Consequently $p\zeta$ is the polar function θ regarded as a function of the independent variable ζ. We call p a *normalized polar function of the third kind of the basis* (u, v) or more briefly a *third normalized* (or 3-*normalized*) *polar function of the basis* (u, v).

The third normalized polar function p clearly has, in the interval J_3, the following properties:

1. $p \in C_1$;
2. $n\pi < p < (n+1)\pi$, n integral.

The derivative of a function of $\zeta(\in J_3)$ with respect to ζ will be denoted by a $^{\backslash}$.

We have, in the interval j,

$$\zeta = \theta + 2\alpha$$

and consequently, from (21) and (5.14)

$$\zeta' = \theta' + 2\frac{\alpha_0'}{\sin\theta_0}\sin\theta\cdot\exp\left(-\int_{t_0}^{t}\zeta'(\tau)\cot\theta(\tau)\,d\tau\right).$$

This formula, together with (38), shows that any two homologous points $t = \zeta^{-1}\zeta \in j$, $\zeta = \zeta(t) \in J_3$ are so related that

$$\zeta'(t) = 2\frac{\alpha_0'}{\sin p_0}\frac{\sin p(\zeta)}{1 - p^{\backslash}(\zeta)}\exp\left(-\int_{\zeta_0}^{\zeta}\cot p(\rho)\,d\rho\right). \tag{6.39}$$

From (39) we find that

$$t = t_0 + \frac{1}{2}\frac{\sin p_0}{\alpha_0'}\int_{\zeta_0}^{\zeta}\frac{1 - p^{\backslash}(\sigma)}{\sin p(\sigma)}\left(\exp\int_{\zeta_0}^{\sigma}\cot p(\rho)\,d\rho\right)d\sigma; \qquad (6.40)$$

this formula gives the inverse function $\zeta^{-1}(\zeta)$ to the Radon parameter $\zeta(t)$. From (23) we obtain the following result, valid for two homologous points $t \in j$, $\zeta \in J_3$

$$q(t) = -\frac{\alpha_0'^2}{\sin^2 p_0}\frac{1 + p^{\backslash}(\zeta)}{1 - p^{\backslash}(\zeta)}\exp\left(-2\int_{\zeta_0}^{\zeta}\cot p(\rho)\,d\rho\right). \qquad (6.41)$$

The expression on the right of this formula is called the *third* or *Radon polar form* of the carrier q. We speak of the *third* or *Radon polar form of the differential equation* (q) when the carrier q is in this form.

In the interval J_3 we have the inequalities

$$-1 < p^{\backslash}(\zeta) < 1.$$

6.11 Determination of the carrier from a third normalized polar function

By a normalized polar function of the third kind of the carrier q or of the differential equation (q) we mean a normalized polar function of the third kind of any basis of (q). More briefly we call this a third normalized polar function of the carrier q or of the differential equation (q).

Every third-normalized polar function p of the differential equation (q) obviously has the following properties in its definition interval J_3:

1. $p \in C_1$;
2. $n\pi < p < (n + 1)\pi$ (n integral);
3. $-1 < p^{\backslash} < 1$. (6.42)

With regard to the problem of determining the differential equation (q) from its third-normalized polar functions, similar remarks hold as in the case of second-normalized polar functions (§ 6.9).

6.12 Some applications of polar function

First let us apply the above results to answering the following question:

What are the possible carriers q determined by a *constant* polar function with the value $(2n + 1)\pi/2$, n being an integer?

Each of the formulae (12), (23), (29), (36), (41) gives the answer

$$q(t) = -\alpha_0'^2.$$

The differential equations (q) with constant negative carriers q are therefore the only ones which admit of a constant polar function with a value congruent to $\frac{\pi}{2}$ (mod π). For such an equation $(-k^2)$ whose carrier $-k^2$ is a negative constant, there are precisely two first phases of $(-k^2)$ passing through any point (t_0, α_0); these have the directions $\alpha_0' = k$ and $\alpha_0' = -k$ in such a way that the polar functions generated by them have a constant value congruent to $\frac{\pi}{2}$ (mod π).

Because of their geometrical significance (§ 6.1), polar functions, and particularly normalized polar functions, of differential equations (q) occur in the study of curves with special properties of a centro-affine nature. We develop the comment in the following study.

Let $(P =)\ P(t)$, $(\bar{P} =)\bar{P}(\bar{t})$, $t \neq \bar{t}$, be arbitrary points of the integral curve $\mathfrak{K}$ with the vectorial representation $x = [u, v]$, and $p, \bar{p}$ be the corresponding tangents to the curve. The straight lines OP, $O\bar{P}$ we shall denote by $g, \bar{g}$. Moreover, let $\theta = \beta - \alpha$ be a polar function of the basis (u, v). We know (§ 6.1) that the values $\theta(t)$, $\theta(\bar{t})$ represent the angles (mod 2π) between the corresponding directed straight lines g, p and $\bar{g}, \bar{p}$.

If the points $P, \bar{P}$ lie on the same straight line g, (i.e. $\bar{g} = g$) and so are points of intersection of the integral curve $\mathfrak{K}$ with g, then the values $(\alpha =)\ \alpha(t)$ $(\bar{\alpha} =)\ \alpha(\bar{t})$ differ by an integral multiple of π, $\bar{\alpha} = \alpha + n\pi$ (n integral), and conversely.

If the tangents $p, \bar{p}$ are parallel then the values $(\beta =)\ \beta(t)$, $(\bar{\beta} =)\ \beta(\bar{t})$ differ by an integral multiple of π, $\bar{\beta} = \beta + n\pi$ (n integral) and conversely. If the straight lines $\bar{g}, p$ or $g, \bar{p}$ are parallel then $\bar{\alpha} = \beta + m\pi$ or $\bar{\beta} = \alpha + n\pi$ (m, n integral), and conversely.

If the first normalized polar function h of the basis (u, v) is periodic with period π, then we have the following situation: through the point O there pass straight lines which cut the curve $\mathfrak{K}$ in at least two points; moreover the tangents at all points of intersection of such a line with the curve $\mathfrak{K}$ are parallel to each other, and conversely.

If the second normalized polar function $-k$ of the basis (u, v) is periodic with period π then there are tangents to the curve $\mathfrak{K}$ to which parallel tangents exist; moreover the points of contact with the curve $\mathfrak{K}$ of all parallel tangents lie on a straight line passing through O, and conversely.

The third normalized polar function p of the basis (u, v) needs rather fuller consideration. We assume that the function p is defined in an interval J_3 of length $> \pi$ and satisfies the functional equation

$$p(\zeta) + p(\zeta + \pi) = \pi. \tag{6.43}$$

Naturally we are only considering points ζ, $\zeta + \pi$ such that both of them lie in J_3. At points ζ and $\bar{\zeta} = \zeta + \pi$ the components of p have certain values α, β and $\bar{\alpha}, \bar{\beta}$ and obviously

$$\bar{\beta} + \bar{\alpha} = \beta + \alpha + \pi. \tag{6.44}$$

Moreover it follows from (43) that

$$\bar{\beta} - \bar{\alpha} = -\beta + \alpha + \pi. \tag{6.45}$$

From these two equations we obtain

$$\bar{\alpha} = \beta; \qquad \bar{\beta} = \alpha + \pi. \tag{6.46}$$

Conversely (46) yields the relations (44), (45).

If the third normalized polar function p of the basis (u, v) satisfies the functional equation (43) then we have the following situation: there are tangents p to the curve $\mathfrak{K}$ such that the straight line $\bar{g}$ passing through the point O and parallel to p cuts the curve $\mathfrak{K}$ at least once; moreover the tangents to $\mathfrak{K}$ at its points of intersection with $\bar{g}$ are parallel to the straight line passing through O and the point of contact of p. This is the "ellipse property".

7 Local and boundary properties of phases

In this paragraph we discuss some further properties of phases of a differential equation (q). Mainly we shall be concerned with first phases, and consequently we shall refer to these simply as phases; when a second phase is being considered, this will be explicitly stated and in such a case we shall assume that the function q does not vanish in the interval j.

7.1 Unique determination of a phase from the Cauchy initial conditions

Let us consider a differential equation (q). We have the following result.

Theorem. Let $t_0 \in j$; X_0, $X_0' \neq 0$, X_0'' be arbitrary numbers. Then there exists precisely one phase α of the differential equation (q) *which satisfies at the point t_0 the Cauchy initial conditions:*

$$\alpha(t_0) = X_0, \qquad \alpha'(t_0) = X_0', \qquad \alpha''(t_0) = X_0''. \tag{7.1}$$

This phase α is included in the phase system of the basis (u, v) of the differential equation (q)*:*

$$\left.\begin{aligned} u(t) &= \left(X_0' \cos X_0 - \frac{1}{2}\frac{X_0''}{X_0'} \sin X_0\right) u_0(t) + \sin X_0 \cdot v_0(t), \\ v(t) &= -\left(X_0' \sin X_0 + \frac{1}{2}\frac{X_0''}{X_0'} \cos X_0\right) u_0(t) + \cos X_0 \cdot v_0(t), \end{aligned}\right\} \tag{7.2}$$

in which u_0, v_0 are those integrals of (q) *determined by the initial values*

$$u_0(t_0) = 0, \qquad u_0'(t_0) = 1; \qquad v_0(t_0) = 1, \qquad v_0'(t_0) = 0.$$

Proof. It is sufficiently general to carry out the proof for the case $X_0 = 0$ and then transform the basis found by means of the orthogonal substitution of (5.41), taking the value $\lambda = X_0$.

We therefore assume that there is a phase α of the differential equation (q) with the initial values 0, X_0' $(\neq 0)$, X_0'' and let

$$\begin{aligned} u(t) &= c_{11}u_0(t) + c_{12}v_0(t), \\ v(t) &= c_{21}u_0(t) + c_{22}v_0(t) \end{aligned}$$

be the first and second elements of a corresponding basis; u_0, v_0 have the significance given in the statement of the theorem, while naturally c_{11}, c_{12}, c_{21}, c_{22} represent appropriate constants. By an easy calculation, the following formulae are seen to hold at the point t_0:

$$r^2 = c_{12}^2 + c_{22}^2; \qquad rr' = c_{11}c_{12} + c_{21}c_{22}; \qquad -w = c_{11}c_{22} - c_{12}c_{21}$$
$$(r^2 = u^2 + v^2;\ w = uv' - u'v).$$

Now, on our assumptions, the functions α, α', α'' take the values 0, X_0', X_0'' at the point t_0. It follows, on making use of (5.14), that

$$c_{12} = 0, \qquad X_0' = \frac{c_{11}}{c_{22}}, \qquad X_0'' = -2\frac{c_{11}}{c_{22}}\frac{c_{21}}{c_{22}}.$$

Obviously $c_{22} \neq 0$. We can take $c_{22} = 1$ because if we were to multiply the integrals u, v by $1/c_{22}$ we would merely obtain a basis proportional to (u, v) and consequently derive the same phase system (§ 5.17). Then we have

$$u(t) = X_0'u_0(t), \qquad v(t) = -\frac{1}{2}\frac{X_0''}{X_0'}u_0(t) + v_0(t). \tag{7.3}$$

There is therefore at most one phase α with the initial values 0, X_0', X_0'' and this must be included in the phase system determined by the formula (3). Now it is easy to calculate that that phase α which is included in this phase system and vanishes at the point t_0 does indeed satisfy the given initial conditions. This completes the proof.

It should be noticed that the above theorem implies the following formula

$$\tan \alpha(t) = \frac{\sin X_0 + \left(X_0' \cos X_0 - \dfrac{1}{2}\dfrac{X_0''}{X_0'} \sin X_0\right) \tan \alpha_0(t)}{\cos X_0 - \left(X_0' \sin X_0 + \dfrac{1}{2}\dfrac{X_0''}{X_0'} \cos X_0\right) \tan \alpha_0(t)} \tag{7.4}$$

In this $\alpha_0(t)$ represents an arbitrary phase of the basis (u_0, v_0). This formula is valid for all values $t \in j$, with the exception of zeros of $\cot \alpha_0(t)$, $\cot \alpha(t)$, at which points it has no meaning.

7.2 Boundary values of phases

Let α be a phase of the differential equation (q). Since we know that α is an increasing or decreasing function in the interval j $(= (a, b))$, there exist finite or infinite limits

$$c = \lim_{t \to a+} \alpha(t), \qquad d = \lim_{t \to b-} \alpha(t). \tag{7.5}$$

We call these numbers c and d (which may possibly be infinite) the *left* and *right boundary values* of the phase α. Obviously the boundary value c (d) is finite if the phase α is bounded in a right (left) neighbourhood of the left (right) end point a (b) of j; moreover (§ 5.4) the left (right) boundary value c (d) of α is finite if the differential equation (q) is of finite type or right (left) oscillatory; it is infinite if (q) is left (right) oscillatory or oscillatory. From § 3.4 we conclude also that if the differential equation (q) possesses 1-conjugate numbers, then the boundary value c (d) of α is finite if and only if the left (right) 1-fundamental number $r_1(s_1)$ of (q) is proper. Obviously $c < d$ or $c > d$ according as the phase α is increasing or decreasing; $\operatorname{sgn}(d - c) = \operatorname{sgn} \alpha'$.

For any constant λ, obviously the phase $\alpha + \lambda$ of (q) has the boundary values $c + \lambda$, $d + \lambda$. In particular, the left (right) boundary values of the phases of the phase system of every basis of (q) differ from each other by an integral multiple of π.

The number $|c - d|$ we call the *oscillation* of the phase α in the interval j. Our notation is: $O(\alpha|j)$ or more briefly $O(\alpha)$. The oscillation $O(\alpha)$ is finite and positive, or is infinite, according as α is bounded or unbounded in j. All phases of the complete phase system $[\alpha]$ clearly have the same oscillation $O(\alpha)$ (§ 5.17).

Two phases α, $\bar{\alpha}$ of the differential equation (q) are linked, as we know, by the formula (5.39). Consequently if the boundary values c, d of the phase α differ by an integral multiple of π, then the same is true for the boundary values $\bar{c}$, $\bar{d}$ of $\bar{\alpha}$. If the oscillation $O(\alpha)$ of α is an integral multiple of π, then the same is true for the oscillation of every phase of the differential equation (q). Such a value of $O(\alpha)$ can naturally only occur when the differential equation (q) is of finite type (m), $m \geqslant 1$. In the case $m \geqslant 2$ it will be shown (§ 7.16) that the differential equation (q) is general or special according as

$$(m - 1)\pi < O(\alpha) < m\pi \qquad \text{or} \qquad O(\alpha) = m\pi$$

where $O(\alpha)$ is the oscillation of every phase α.

This result prompts the following definition: we call a differential equation (q) of type (1) *general* or *special* according as, for the oscillation $O(\alpha)$ of each of its phases α, we have $0 < O(\alpha) < \pi$ or $O(\alpha) = \pi$.

Then we have the following theorem.

Theorem. The differential equation (q) *of finite type* (m), $m \geqslant 1$, *is general or special according as, for the oscillation of each of its phases, we have:* $(m - 1)\pi < O(\alpha) < m\pi$ *or* $O(\alpha) = m\pi$.

7.3 Normalized boundary values of phases

We shall continue in this paragraph to make use of the above notation.

Let $\bar{a}$, $\bar{b}$ denote the numbers a, b or b, a according as $\operatorname{sgn} \alpha' > 0$ or < 0. Correspondingly, let $\bar{c}$, $\bar{d}$ denote the numbers c, d or d, c. Explicitly:

$$\left.\begin{aligned} \bar{a} &= \frac{1}{2}(1 + \varepsilon)\,a + \frac{1}{2}(1 - \varepsilon)b; \qquad & \bar{b} &= \frac{1}{2}(1 - \varepsilon)a + \frac{1}{2}(1 + \varepsilon)b; \\ \bar{c} &= \frac{1}{2}(1 + \varepsilon)c + \frac{1}{2}(1 - \varepsilon)d; \qquad & \bar{d} &= \frac{1}{2}(1 - \varepsilon)c + \frac{1}{2}(1 + \varepsilon)\,d \end{aligned}\right\} \tag{7.6}$$

$$(\varepsilon = \operatorname{sgn} \alpha').$$

We call $\bar{c}$, $\bar{d}$ *normalized boundary values* of the phase α. Clearly

$$\lim_{t \to \bar{a}} \alpha(t) = \bar{c}; \qquad \lim_{t \to \bar{b}} \alpha(t) = \bar{d}, \tag{7.7}$$

in which naturally we are considering the corresponding right or left limit, and moreover

$$\bar{c} < \bar{d}. \tag{7.8}$$

It is convenient to call the numbers $\bar{a}$, $\bar{b}$ the *normalized ends* of the interval j with respect to the phase α.

7.4 Singular phases

In this and the following §§ 7.5–7.16 we are concerned, as noted above, with *first* phases and with conjugate numbers, fundamental numbers, fundamental integrals, fundamental sequences and singular bases, always of the *first* kind.

By a *singular phase* of the differential equation (q) we mean a phase of a singular basis of (q), that is to say a basis (u, v) whose first term u is a left or right fundamental integral of (q) (§ 3.9).

In the transformation theory which we are going to consider, the phase concept is of fundamental importance. Consequently, it is convenient to utilize those phases which are most closely associated with the differential equation (q) under consideration. A particular claim to this position is possessed by the singular phases, since the corresponding (singular) bases are largely determined by the type and kind of a given differential equation (q).

Let (q) be a differential equation with conjugate numbers and proper fundamental numbers, and let the corresponding left (right) fundamental number $r_1(s_1)$ be proper. The differential equation (q) consequently admits of left (right) fundamental integrals, the left (right) fundamental sequence $r_1 = a_1 < a_2 < a_3 < \ldots$ $(s_1 = b_{-1} > b_{-2} > b_{-3} > \ldots)$ and naturally also left (right) principal bases.

The fundamental theorem is the following.

Theorem. The left (right) boundary values of the phases included in the phase system (α) *of a basis* (u, v) *of the differential equation* (q) *are integral multiples of* π *if and only if* (u, v) *is a left (right) principal basis.*

Proof. (a) Let us assume that the left (right) boundary values of phases of the basis (u, v) are integral multiples of π. Then precisely one phase α of (u, v) has zero as its left (right) boundary value. For brevity, we call this phase the left (right) *null phase* of (u, v).

We now assert that α takes, at the point $r_1(s_1)$, the value $\varepsilon\pi$ $(-\varepsilon\pi)$; that is, $\alpha(r_1) = \varepsilon\pi$ $(\alpha(s_1) = -\varepsilon\pi)$, where $\varepsilon = \operatorname{sgn} \alpha'$. If this is so, then $r_1(s_1)$ is a zero of u (§ 5.3) and consequently (u, v) represents a left (right) principal basis of (q).

For simplicity, we shall assume for definiteness that α is the left null phase of (u, v).

We know, (§ 5.13) that the function

$$y(t) = k_1 \frac{\sin(\alpha(t) + k_2)}{\sqrt{|\alpha'(t)|}}$$

constructed with arbitrary constants k_1, k_2 represents the general integral of (q). Since the differential equation (q) admits of conjugate numbers, $O(\alpha) > \pi$; moreover, $c = 0$ by hypothesis; it follows that α takes the value $\varepsilon\pi$ at some point $x \in j$. We thus have to show that every number $t_0 \in j$ such that $t_0 > x$ possesses left conjugate numbers, while no number $t_0 \in (a, x)$ has left conjugate numbers.

Let $t_0 \in j$ be arbitrary and $n \geqslant 0$ the integer determined by the inequalities

$$n\varepsilon\pi \lessgtr \alpha(t_0) \lesseqgtr (n+1)\varepsilon\pi.$$

The symbol $\gtrless$ denotes $<$ or $>$ according as $\varepsilon = 1$ or $\varepsilon = -1$.

We associate with the number t_0 the integral $y\ (= y_0)$ constructed with the constants

$$k_1 = 1, \qquad k_2 = (n+1)\varepsilon\pi - \alpha(t_0);$$

we then have $0 \lesseqgtr k_2 \lessgtr \varepsilon\pi$, and the integral y_0 vanishes at the point t_0.

Now let $t_0 > x$. If $k_2 = 0$, then by the relationship $\alpha(x) = \varepsilon\pi$, $y_0(x) = 0$, so x is a left conjugate number of t_0. In the case $0 \lessgtr k_2 \lessgtr \varepsilon\pi$ we have $\alpha(x) + k_2 \gtrless \varepsilon\pi$ and also, since $c = 0$, $\alpha(t) + k_2 \lessgtr \varepsilon\pi$ for appropriate numbers $t \in (a, x)$. The function $\alpha + k_2 - \varepsilon\pi$ has therefore a zero in the interval (a, x) which is obviously a zero of y_0 and consequently a left conjugate number of t_0.

Now let $t_0 < x$. Then we have $\alpha(t_0) + k_2 = \varepsilon\pi$. It follows that for $t \in (a, t_0)$, we have $0 \lessgtr \alpha(t) + k_2 \lessgtr \varepsilon\pi$ and $y_0(t) \neq 0$. There is therefore no left conjugate number of t_0.

(b) Now let (u, v) be a left (right) principal basis of (q). Consider a phase α of (u, v) and put $\varepsilon = \operatorname{sgn} \alpha'$, and let c (d) be the left (right) boundary value of α. Since u is a left (right) fundamental integral of (q), we have (§ 5.3) $\alpha(r_1) = n\varepsilon\pi$ ($\alpha(s_1) = -n\varepsilon\pi$); n integral, $n \geqslant 0$.

But $\bar{\alpha}(t) = \alpha(t) - c$ ($\bar{\alpha}(t) = \alpha(t) - d$) is the left (right) null phase of a basis $(\bar{u}, \bar{v})$ of (q). So, from (a),

$$\varepsilon\pi = \bar{\alpha}(r_1) = \alpha(r_1) - c = n\varepsilon\pi - c \qquad (-\varepsilon\pi = \bar{\alpha}(s_1) = \alpha(s_1) - d = -n\varepsilon\pi - d),$$

hence

$$c = (n-1)\varepsilon\pi, \qquad (d = -(n-1)\varepsilon\pi),$$

and the proof is complete.

7.5 Properties of singular phases

We now examine more closely the properties of singular phases. Let (u, v) be a left (right) principal basis of the differential equation (q) and (α) its first phase system. u is therefore a left (right) fundamental integral, and consequently one which vanishes at the points $a_1 < a_2 < \ldots$ ($b_{-1} > b_{-2} > \ldots$), while v is an integral of the differential equation (q) independent of u. In every interval $(a_\mu, a_{\mu+1})$ ($(b_{-\mu-1}, b_{-\mu})$) this integral has precisely one zero x_μ ($x_{-\mu}$); $\mu = 0, 1, \ldots$; $a_0 = a$($b_0 = b$):

$$(a =)\, a_0 < x_0 < a_1 < x_1 < a_2 < \cdots \quad ((b =)\, b_0 > x_0 > b_{-1} > x_{-1} > b_{-2} > \cdots). \tag{7.9}$$

The phases $\alpha \in (\alpha)$ increase or decrease according as the Wronskian w of (u, v) is negative or positive; $\operatorname{sgn} \alpha' = \operatorname{sgn}(-w)$. (5.14). We set $\varepsilon = \operatorname{sgn}(-w)$.

7.6 Boundary values of null phases

From the theorem of § 7.4, the phase system (α) contains precisely one left (right) null phase α_0; this takes the value $\varepsilon\pi$ ($-\varepsilon\pi$) at the point a_1 ($= r_1$) (b_{-1} ($= s_1$)) and consequently at the points (9) it takes the values

$$0, \frac{1}{2}\varepsilon\pi, \varepsilon\pi, \frac{3}{2}\varepsilon\pi, 2\varepsilon\pi, \ldots \quad \left(0, -\frac{1}{2}\varepsilon\pi, -\varepsilon\pi, -\frac{3}{2}\varepsilon\pi, -2\varepsilon\pi, \ldots\right); \tag{7.10}$$

$$\alpha_0(a_\mu) = \mu\varepsilon\pi, \quad \alpha_0(x_\mu) = \left(\mu + \frac{1}{2}\right)\varepsilon\pi \quad \left(\alpha_0(b_{-\mu}) = -\mu\varepsilon\pi, \alpha_0(x_{-\mu}) = -\left(\mu + \frac{1}{2}\right)\varepsilon\pi\right);$$

$\mu = 0, 1, \ldots$; naturally, $\alpha_0(a_0)$ ($\alpha_0(b_0)$) denotes the left (right) boundary value.

We now determine the right (left) boundary value d_0 (c_0) of the null phase α_0. This boundary value depends upon the type and kind of the differential equation (q).

I. First, let the differential equation (q) be of finite type (m), $m \geqslant 2$.

In this case, both fundamental numbers r_1, s_1 are proper, both fundamental sequences

$$(r_1 =)\, a_1 < a_2 < \cdots < a_{m-1}, \qquad (s_1 =)\, b_{-1} > b_{-2} > \cdots > b_{-m+1}$$

contain precisely $m - 1$ terms, and relationships hold such as those of (3.2).

From (10) we obtain

$$(m-1)\varepsilon\pi \lesseqgtr d_0 \lesseqgtr m\varepsilon\pi \qquad (-m\varepsilon\pi \lesseqgtr c_0 \lesseqgtr -(m-1)\varepsilon\pi). \tag{7.11}$$

Now the equality sign holds if and only if the differential equation (q) is special, for if $d_0 = m\varepsilon\pi$ ($c_0 = -m\varepsilon\pi$) then the theorem of § 7.4 shows that u is a right (left) fundamental integral of (q), and consequently vanishes at both the points r_1, s_1 (s_1, r_1). The fundamental numbers r_1, s_1 are therefore conjugate, which implies that the differential equation (q) is special.

Conversely, if the differential equation (q) is special, then the fundamental numbers r_1, s_1 are conjugate, and consequently the left (right) fundamental integral u is also a right (left) fundamental integral. From the theorem of § 7.4 we deduce that $d_0(c_0)$ is an integral multiple of π, and from (11) we obtain

$$d_0 = m\varepsilon\pi \qquad (c_0 = -m\varepsilon\pi).$$

(a) Let (q) be general. In this case the conditions (11) hold, but equality is not possible. The right (left) boundary value d_0 (c_0) of α_0 depends upon the second element v of the left (right) principal basis (u, v). Now we show that if v is a right (left) fundamental integral, so that (u, v) is a principal basis of the differential equation (q), then we have

$$d_0 = \left(m - \frac{1}{2}\right)\varepsilon\pi \qquad \left(c_0 = -\left(m - \frac{1}{2}\right)\varepsilon\pi\right). \tag{7.12}$$

For in this case we have $x_\mu = b_{-m+\mu+1}$ ($x_{-\mu} = a_{m-\mu-1}$), $\mu = 0, \ldots, m-2$ and from (10) the phase α_0 takes the value $(m - \frac{3}{2})\varepsilon\pi$ ($-(m - \frac{3}{2})\varepsilon\pi$) at the point $b_{-1}(a_1)$.

The function $\bar{\alpha}_0 = \alpha_0 - d_0$ ($\bar{\alpha}_0 = \alpha_0 - c_0$) is obviously a right (left) null phase of (q) and so (from (10)) takes the value $-\varepsilon\pi$ ($\varepsilon\pi$) at the point $b_{-1}(a_1)$. Hence

$$-\varepsilon\pi = \left(m - \frac{3}{2}\right)\varepsilon\pi - d_0 \qquad \left(\varepsilon\pi = -\left(m - \frac{3}{2}\right)\varepsilon\pi - c_0\right),$$

and the relation (12) follows immediately.

(b) Let (q) be special. In this case we have, as was shown above,

$$d_0 = m\varepsilon\pi \qquad (c_0 = -m\varepsilon\pi), \tag{7.13}$$

so that d_0 is independent of the choice of the second element v of the corresponding left principal (and also right principal) basis (u, v).

II. Secondly, let the differential equation (q) be of infinite type. If (q) is right (left) oscillatory, then it admits only of left (right) infinite fundamental sequences $a_1 < a_2 < \ldots$ ($b_{-1} > b_{-2} > \ldots$); then (9) is an infinite sequence.

In this case, obviously,

$$d_0 = \varepsilon\infty \qquad (c_0 = -\varepsilon\infty). \tag{7.14}$$

7.7 Boundary values of other phases

We now turn back to the phase system (α) of the left (right) principal basis (u, v) of (q), considered in § 7.5.

The system (α) is obviously formed from the phases

$$\alpha_\nu(t) = \alpha_0(t) - \nu\varepsilon\pi \qquad (\alpha_\nu(t) = \alpha_0(t) + \nu\varepsilon\pi)$$
$$(\nu = 0, \pm 1, \pm 2, \ldots).$$

At the points a_μ, x_μ ($b_{-\mu}$, $x_{-\mu}$) the phase α_ν takes the following values

$$\left.\begin{aligned} &\alpha_\nu(a_\mu) = (\mu - \nu)\varepsilon\pi, \qquad \alpha_\nu(x_\mu) = \left(\mu - \nu + \frac{1}{2}\right)\varepsilon\pi \\ &\left(\alpha_\nu(b_{-\mu}) = -(\mu - \nu)\varepsilon\pi, \qquad \alpha_\nu(x_{-\mu}) = -\left(\mu - \nu + \frac{1}{2}\right)\varepsilon\pi\right) \end{aligned}\right\} \quad 7.15)$$

$$(\mu = 0, 1, \ldots).$$

The left (right) boundary value c_ν (d_ν) of the phase α_ν is

$$c_\nu = -\nu\varepsilon\pi \qquad (d_\nu = \nu\varepsilon\pi). \tag{7.16}$$

The right (left) boundary value d_ν (c_ν) of the phase α_ν according to the type and kind of the differential equation (q) is as follows,

I. Finite type (m), $m \geqslant 2$.

(a) General (q):

$$(m - \nu - 1)\varepsilon\pi \lesseqgtr d_\nu \lesseqgtr (m - \nu)\varepsilon\pi \qquad (-(m - \nu)\varepsilon\pi \lesseqgtr c_\nu \lesseqgtr -(m - \nu - 1)\varepsilon\pi). \tag{7.17}$$

If, in particular, v is a right (left) fundamental integral and consequently (u, v) a principal basis of the differential equation (q) then we have, independently of our choice of the fundamental integral v,

$$d_\nu = \left(m - \nu - \frac{1}{2}\right)\varepsilon\pi \qquad \left(c_\nu = -\left(m - \nu - \frac{1}{2}\right)\varepsilon\pi\right). \tag{7.18}$$

(b) Special (q):

$$d_\nu = (m - \nu)\varepsilon\pi \qquad (c_\nu = -(m - \nu)\varepsilon\pi). \tag{7.19}$$

II. Infinite type.

(a) Right oscillatory (q):

$$d_\nu = \varepsilon\infty. \tag{7.20}$$

(b) Left oscillatory (q):

$$c_\nu = -\varepsilon\infty. \tag{7.21}$$

7.8 Normal phases

The null phases of a differential equation (q) are characterized by the fact that they are always non-zero in the interval j. We now consider, instead, phases which have one (and naturally only one) zero in j. A phase of the differential equation (q) which vanishes at a point of j will in what follows be called a *normal phase*.

We consider a differential equation (q). Let (u, v) be a basis of (q) and (α) the phase system of this basis (u, v).

We observe first, that the phase system (α) contains normal phases if and only if the integral u has zeros in the interval j. In this case every zero of the integral u is a zero of a normal phase of (α); conversely, the zero of every normal phase of (α) coincides with one of the zeros of the integral u. It follows that:

If the differential equation (q) is of finite type (m) $(m \geqslant 1)$ then the phase system (α) contains $m - 1$ or m normal phases, while if (q) is of infinite type then it contains a countable infinity of normal phases; the zeros of these phases coincide with those of the integral u. If, in particular, the differential equation (q) is oscillatory, then the system (α) has only normal phases.

7.9 Structure of the set of singular normal phases of a differential equation (q)

By a *singular normal phase* of a differential equation (q) we naturally mean (§ 7.4) a normal phase of a singular basis of (q). Singular normal phases thus occur in differential equations (q) of finite type (m), $m \geqslant 2$, and in left or right oscillatory differential equations (q) but only in these cases. Let (q) be a differential equation with conjugate numbers and the corresponding left (right) proper fundamental number $r_1(s_1)$.

We start from the situation considered in § 7.5, letting (u, v) be a left (right) principal basis of the differential equation (q) and (α) its first phase system.

The zeros of the integrals u coincide with the singular numbers $a_1 < a_2 < \ldots$ $b_{-1} > b_{-2} > \ldots$ of the differential equation (q); hence to every number $a_r(b_{-r})$ $(r = 1, 2, \ldots)$ there corresponds precisely one normal phase in the phase system (α) with the zero a_r (b_{-r}). From (15) we see that this is the phase given by

$$\alpha_r(t) = \alpha_0(t) - r\varepsilon\pi \quad (\alpha_r(t) = \alpha_0(t) + r\varepsilon\pi). \tag{7.22}$$

Conversely, every normal phase included in the phase system (α) is one of these phases $\alpha_1, \alpha_2, \ldots$

By a *phase bundle with the apex* a_r (b_{-r}) or more briefly, an a_r*-bundle* (b_{-r}*-bundle*) of the differential equation (q) we mean the subset comprising all normal phases of all left (right) principal bases of (q) which vanish at the point $a_r(b_{-r})$, $r = 1, 2, \ldots$. Obviously the set of all singular normal phases of the differential equation (q) is the union of the phase bundles with the apices $a_1, a_2, \ldots$ $(b_{-1}, b_{-2}, \ldots)$.

7.10 Structure of a phase bundle

We now examine the structure of the phase bundles; for brevity we shall confine ourselves to left principal bases. These occur, as we know, only in differential equations (q) of finite type (m), $m \geqslant 2$, and in right oscillatory differential equations (q). (The study of right principal bases is entirely analogous).

Let (u, v) be a left principal basis of (q) and a_r a term of the left fundamental sequence of (q). We know (§ 3.9) that the left principal bases of (q) form the three-parameter system $(\rho u, \sigma v + \bar{\sigma} u)$, $\rho\sigma \neq 0$. Now for every choice of values given to the parameters $\rho, \sigma, \bar{\sigma}$ the bases $(\rho u, \sigma v + \bar{\sigma} u)$ and $\left(\frac{\rho}{\sigma} u, v + \frac{\bar{\sigma}}{\sigma} u\right)$ are proportional, and therefore have the same phase system and consequently the same normal phase vanishing at the point a_r. So the phase bundle of the differential equation (q) with the apex a_r obviously comprises precisely those normal phases of the two parameter basis system $(\rho u, v + \sigma u)$, $\rho \neq 0$, which vanish at the point a_r. Naturally, for every basis of this two-parameter system there exists precisely one normal phase of (q) with the zero a_r.

The basis system $(\rho u, v + \sigma u)$ now separates into one-parameter systems each of which is determined by a fixed value σ_0 of σ. Such a one-parameter system thus comprises the left principal bases $(\rho u, v + \sigma_0 u)$, $\rho \neq 0$. Those normal phases of the left principal bases of this one-parameter system which are included in the a_r-bundle, form a one-parameter sub-system $P(a_r|\sigma_0)$ of the a_r-bundle. We call $P(a_r|\sigma_0)$ the *phase bunch with the apex* a_r or more briefly the a_r*-bunch* of the differential equation (q).

The phase bundle of the differential equation (q) with apex a_r consequently consists of a one-parameter system of a_r-bunches $P(a_r|\sigma)$, each of which is formed from those normal phases of the left principal bases $(\rho u, v + \sigma u)$, $0 \neq \rho$, arbitrary, σ fixed, which vanish at the point a_r. Thus the study of the structure of the phase bundles is reduced to that of the phase bunches.

7.11 Structure of a phase bunch

Now we wish to study the structure of the phase bunches. We consider a phase bunch $P(a_r|\sigma)$, with the apex a_r, of the differential equation (q). Without loss of generality we can take $\sigma = 0$ and also assume that w $(= uv' - u'v) < 0$ For brevity we shall write $P(a_r)$ instead of $P(a_r|0)$ The phase bunch $P(a_r)$ consists of the normal phases of the left principal bases $(\rho u, v)$, $\rho \neq 0$ which vanish at the point a_r.

For every value ρ $(\neq 0)$, ρu is a left fundamental integral, consequently vanishing at the points $a_1 < a_2 < \ldots$, while v represents an integral of the differential equation (q) independent of u. This integral v has precisely one zero x_μ, $\mu = 0, 1, \ldots$ in each interval $(a_\mu, a_{\mu+1})$, and we have ordering relationships similar to (9). For brevity we shall denote the intervals (a_μ, x_μ), $(x_\mu, a_{\mu+1})$ by j_μ and j'_μ respectively, i.e. $j_\mu = (a_\mu, x_\mu)$, $j_\mu' = (x_\mu, a_{\mu+1})$; if the differential equation (q) is of finite type (m) $(m \geqslant 2)$, then naturally we only have to consider the intervals $j_0, j'_0, j_1, j'_1, \ldots, j_{m-1}$, where x_{m-1} stands for the upper end point b of j.

For every number ρ $(\neq 0)$ we write $\alpha_{r,\rho}$ or more briefly α_ρ for the normal phase of the basis $(\rho u, v)$ contained in the phase bunch $P(a_r)$, and $c_{r,\rho}$ or $d_{r,\rho}$, more briefly c_ρ or d_ρ, for its left or right boundary value. Obviously we have $\tan \alpha_\rho = \rho u/v$ and consequently

$$\tan \alpha_\rho = \rho \tan \alpha_1. \tag{7.23}$$

The Wronskian of the basis $(\rho u, v)$ is obviously ρw. We deduce, using our assumption $w < 0$, that the phase α_ρ increases for positive values of ρ and decreases for negative values of ρ, i.e. $\operatorname{sgn}(\alpha'_\rho) = \operatorname{sgn} \rho$ $(= \varepsilon)$.

At the points

$$x_0, a_1, x_1, \ldots, x_{r-1}, a_r, x_r, \ldots$$

the phase α_ρ takes the following values, which are independent of $|\rho|$:

$$-\left(r - \frac{1}{2}\right)\varepsilon\pi, \quad -(r-1)\varepsilon\pi, \qquad -\left(r - \frac{3}{2}\right)\varepsilon\pi, \ldots, -\frac{1}{2}\varepsilon\pi, 0, \frac{1}{2}\varepsilon\pi, \ldots$$

and consequently

$$\alpha_\rho(a_{\mu+1}) = -(r - \mu - 1)\varepsilon\pi, \qquad \alpha_\rho(x_\mu) = -\left(r - \mu - \frac{1}{2}\right)\varepsilon\pi \tag{7.24}$$

$$(\mu = 0, 1, \ldots).$$

From (16) we have

$$c_\rho = -r\varepsilon\pi. \tag{7.25}$$

For the right boundary value d_ρ of α_ρ we obtain from (17), (18), (19), (20) the following inequalities, according to the type and kind of the differential equation (q):

I. Finite type (m), $(m \geqslant 2)$:

(a) General differential equation (q):

$$(m - r - 1)\varepsilon\pi \lesseqgtr d_\rho \lesseqgtr (m - r)\varepsilon\pi. \tag{7.26}$$

If in particular v is a right fundamental integral and consequently $(\rho u, v)$ a principal basis of the differential equation (q) then $x_\mu = b_{-m+\mu+1}$ $(\mu = 0, \ldots, m-2)$ and moreover

$$d_\rho = \left(m - r - \frac{1}{2}\right) \varepsilon\pi. \tag{7.27}$$

(b) Special differential equation (q):

$$d_\rho = (m - r)\varepsilon\pi. \tag{7.28}$$

II. Infinite type. Right oscillatory differential equation (q):

$$d_\rho = \varepsilon\infty. \tag{7.29}$$

The phase bunch $P(a_r)$ separates into two sub-bunches, of which one, $P_1(a_r)$, (say) comprises the increasing and the other, $P_{-1}(a_r)$, the decreasing phases. The individual phases in the respective sub-bunches $P_1(a_r)$ and $P_{-1}(a_r)$ take the same values at the points $a_{\mu+1}, x_\mu (\mu = 0, 1, \ldots)$, namely $-(r - \mu - 1)\pi$, $-(r - \mu - \frac{1}{2})\pi$ and $(r - \mu - 1)\pi$, $(r - \mu - \frac{1}{2})\pi$ respectively and have the same left boundary value $-r\pi$ or $r\pi$. Their right boundary values are generally dependent on the individual phases, but not in the following cases:

When the differential equation (q) is of:

I. Finite type (m), $m \geqslant 2$,
   (a) General, and v is a right fundamental integral of (q)
   (b) Special

II. Infinite type, right oscillatory.

In these cases the individual phases of $P_1(a_r)$ and $P_{-1}(a_r)$ also have the same right boundary value; in case I(a) they are $(m - r - \frac{1}{2})\pi$, $-(m - r - \frac{1}{2})\pi$, respectively; in case I(b) they are $(m - r)\pi$, $-(m - r)\pi$, and in case II they are $-\infty$, ∞.

Clearly this situation can be described as follows:

All the curves $[t, \alpha_\rho(t)]$, $t \in j$, determined by the phases α_ρ of the sub-bunch $P_\varepsilon(a_r)$ $(\varepsilon = \pm 1)$ go through the points

$$\left(x_\mu, -\left(r - \mu - \frac{1}{2}\right) \varepsilon\pi\right), \qquad (a_{\mu+1}, -(r - \mu - 1)\varepsilon\pi) \quad (\mu = 0, 1, \ldots) \tag{7.30}$$

and tend on the left to the point $(a, -r\varepsilon\pi)$. In the cases I(a), I(b), II they also tend on the right to a common point, in fact towards $(b, (m - r - \frac{1}{2})\varepsilon\pi)$, $(b, (m - r)\varepsilon\pi)$, $(b, \varepsilon\infty)$ respectively. Moreover all these curves lie in the region B_ε formed by the union of the open rectangular regions

$$\left.\begin{aligned} &j_\mu \times \left(-(r - \mu)\varepsilon\pi, -\left(r - \mu - \frac{1}{2}\right) \varepsilon\pi\right) \\ &j'_\mu \times \left(-\left(r - \mu - \frac{1}{2}\right) \varepsilon\pi, -(r - \mu - 1)\varepsilon\pi\right) \quad \mu) = 0, 1, \ldots). \end{aligned}\right\} \tag{7.31}$$

To complete the picture, we shall establish the fact that through every point of the region B_ε there passes precisely one curve $[t, \alpha_\rho(t)]$; in other words the curves considered, $[t, \alpha_\rho(t)]$ fill the region B_ε completely and simply.

For, let $P_0(t_0, X_0)$ be a point which we may take, for instance, to lie in the rectangular region $j_\mu \times (-(r-\mu)\varepsilon\pi), \ -(r-\mu-\frac{1}{2})\varepsilon\pi)$ so that $t_0 \in (a_\mu, x_\mu)$, $X_0 \in (-(r-\mu)\varepsilon\pi, -(r-\mu-\frac{1}{2})\varepsilon\pi.)$

If the curve $[t, \alpha_\rho(t)]$ determined by a phase $\alpha_\rho \in P_\varepsilon(a_r)$ passes through the point P_0, then we have $\alpha_\rho(t_0) = X_0$ and moreover $\rho = \tan X_0/\tan \alpha_1(t_0)$ $(= \rho_0)$. We have therefore at most to consider the phase $\alpha_{\rho_0}(t)$. Moreover, from (23) we have $\tan \alpha_{\rho_0}(t_0) = \tan X_0$; consequently, since both the numbers $\alpha_{\rho_0}(t_0)$, X_0 lie in the interval $(-(r-\mu)\varepsilon\pi, -(r-\mu-\frac{1}{2})\varepsilon\pi)$, we have $\alpha_{\rho_0}(t_0) = X_0$. This proves our assertion.

We also see that for any two phases $\alpha_\rho, \alpha_{\bar\rho} \in P_\varepsilon(a_r)$, there holds the following relation in the interval j

$$|\alpha_\rho - \alpha_{\bar\rho}| < \frac{\pi}{2}. \tag{7.32}$$

The situation in the case Ia is shown in figure 2.

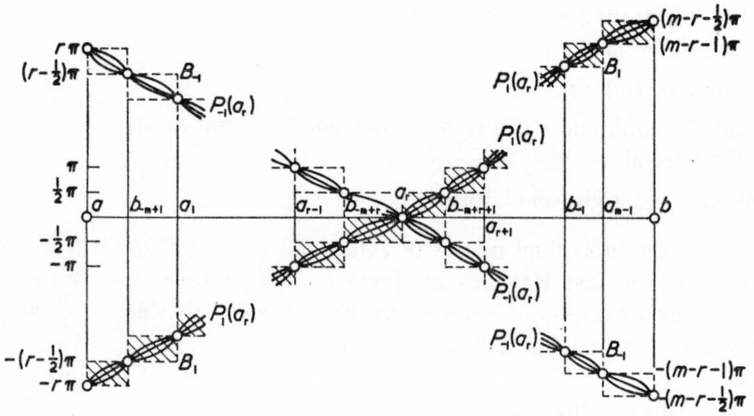

Figure 2

7.12 The mapping of ρ into the phase α_ρ

We now wish to study more closely the mapping $\mathbf{L}: \rho \to \alpha_\rho$. Let I_1, I_{-1} denote respectively the positive and negative real half-lines.

1. *The mapping* **L** *maps the interval* I_ε *on the sub-bunch* $P_\varepsilon(a_r)$ $(\varepsilon = \pm 1)$.
2. *The mapping* **L** *is simple.*

For, from $\rho, \bar\rho \in I_\varepsilon$, $\rho \neq \bar\rho$, it follows that $\alpha_\rho \neq \alpha_{\bar\rho}$. Conversely, if $\alpha_\rho \neq \alpha_{\bar\rho}$ for two phases $\alpha_\rho, \alpha_{\bar\rho} \in P_\varepsilon(a_r)$ and at the same time $\rho = \bar\rho$, then the formula (23) gives $\alpha_\rho = \alpha_{\bar\rho} + k\pi$, where k is an integer $\neq 0$. This is impossible, since both phases $\alpha_\rho, \alpha_{\bar\rho}$ vanish at the point a_r.

3. *For $\rho, \bar{\rho} \in I_\varepsilon$, $\rho < \bar{\rho}$ in every interval j_μ or j'_μ ($\mu = 0, 1, \ldots$) there holds the relation $\alpha_\rho < \alpha_{\bar{\rho}}$ or $\alpha_\rho > \alpha_{\bar{\rho}}$ respectively.*

For, obviously $\tan \alpha_1 > 0$ or $\tan \alpha_1 < 0$ in every interval j_μ or j'_μ respectively; then, taking account of (23), our assertion follows.

The sub-bunch $P_\varepsilon(a_\tau)$ admits of the following ordering relation, denoted by $\prec$: for $\alpha, \bar{\alpha} \in P_\varepsilon(a_\tau)$ we have $\alpha \prec \bar{\alpha}$ if and only if the inequality $\alpha < \bar{\alpha}$ or $\alpha > \bar{\alpha}$ holds in every interval j_μ or j'_μ, respectively.

*The mapping **L** is order-preserving with respect to this ordering.*

4. We define a metric in the set $P_\varepsilon(a_\tau)$, by means of the formula

$$d(\alpha_\rho, \alpha_{\bar{\rho}}) = \sup_{t \in j} |\alpha_\rho(t) - \alpha_{\bar{\rho}}(t)|.$$

In the interval I_ε we adopt the usual Euclidean metric.

We now show that:

*The mapping **L** is homeomorphic.*

Proof. (a) Let ρ, $\bar{\rho} \in I_\varepsilon$ be arbitrary numbers. At every point $t \in j$ other than $a_{\mu+1}, x_\mu$ ($\mu = 0, 1, \ldots$) we have the relation

$$\tan(\alpha_\rho - \alpha_{\bar{\rho}}) = \frac{\tan \alpha_\rho - \tan \alpha_{\bar{\rho}}}{1 + \tan \alpha_\rho \tan \alpha_{\bar{\rho}}} = \frac{\rho - \bar{\rho}}{\dfrac{v}{u} + \rho\bar{\rho}\dfrac{u}{v}}$$

whence, taking account of (32),

$$|\alpha_\rho - \alpha_{\bar{\rho}}| \leqslant \tan|\alpha_\rho - \alpha_{\bar{\rho}}| = \frac{|\rho - \bar{\rho}|}{\left|\dfrac{v}{u}\right| + \rho\bar{\rho}\left|\dfrac{u}{v}\right|} = \frac{|\rho - \bar{\rho}|}{\left(\sqrt{\left|\dfrac{v}{u}\right|} - \sqrt{\rho\bar{\rho}}\sqrt{\left|\dfrac{u}{v}\right|}\right)^2 + 2\sqrt{\rho\bar{\rho}}}.$$

From these relations it follows that

$$d(\alpha_\rho, \alpha_{\bar{\rho}}) \leqslant \frac{1}{2}\frac{|\rho - \bar{\rho}|}{\sqrt{\rho\bar{\rho}}}. \tag{7.33}$$

so that the mapping $\boldsymbol{L}$ is continuous at every point $\bar{\rho} \in I_\varepsilon$.

(b) Let α_ρ, $\alpha_{\bar{\rho}} \in P_\varepsilon(a_\tau)$ be arbitrary phases. Obviously, there exists a number $t_0 \in j$ such that $\tan \alpha_1(t_0) = \delta$ ($= \pm 1$). At this point t_0 we have, from (23),

$$\delta(\rho - \bar{\rho}) = \tan \alpha_\rho - \tan \alpha_{-} = (1 + \rho\bar{\rho}) \tan(\alpha_\rho - \alpha_{\bar{\rho}})$$

and further

$$\frac{|\rho - \bar{\rho}|}{1 + \rho\bar{\rho}} = \tan|\alpha_\rho - \alpha_{\bar{\rho}}| \leqslant \tan d(\alpha_\rho, \alpha_{\bar{\rho}}).$$

From these relationships it follows that

$$\frac{|\rho - \bar{\rho}|}{1 + \rho\bar{\rho}} \leqslant \tan d(\alpha_\rho, \alpha_{\bar{\rho}}); \tag{7.34}$$

hence the mapping L^{-1} is continuous at every point $\alpha_{\bar{p}} \in P_\varepsilon(a_r)$. This completes the proof.

7.13 Relations between zeros and boundary values of normal phases

In the course of our study of singular normal phases we encountered certain relations between the zeros and boundary values of these phases. For instance, Figure 2 shows that in the case we have considered every increasing or decreasing normal phase with the zero a_r possesses the boundary values $-r\pi$, $(m - r - \frac{1}{2})\pi$ or $r\pi$, $-(m - r - \frac{1}{2})\pi$. We now study in greater generality the relations between zeros and boundary values of normal phases.

Consider a differential equation (q) of finite type (m), $m \geqslant 1$, or a left or right oscillatory differential equation. The left and right 1-fundamental sequences of (q), if they exist, we denote by

$$(a <)\, r_1 = a_1 < a_2 < \cdots \qquad \text{and} \qquad (b >)\, s_1 = b_{-1} > b_{-2} > \cdots.$$

Let α be a normal phase of (q), $t_0 \in j$ its zero, and c, d its left and right boundary values. For simplicity, we put $\varepsilon = \operatorname{sgn} \alpha'$ and use the symbol $\lessgtr$ to mean $<$ when $\varepsilon = +1$, $>$ when $\varepsilon = -1$; similarly $\gtrless$ means $>$ when $\varepsilon = +1$, and $<$ when $\varepsilon = -1$.

(a) First we assume that the differential equation (q) is of finite type or is right oscillatory. In these cases the boundary value c is finite.

If the differential equation (q) is of type (1), and so without conjugate numbers of the first kind, then obviously we have the relations

$$-\pi\varepsilon \lessgtr c \lessgtr 0.$$

We now assume that the differential equation (q) admits of 1-conjugate numbers.

Let $r \geqslant 0$ be the integer defined by the stipulation $t_0 \in (a_r, a_{r+1}]$; $a_0 = a$. We consider the left null phase α_0 included in the system $[\alpha]$:

$$\alpha_0(t) = \alpha(t) - c.$$

We clearly have the relations

$$\alpha_0(a_r) \lessgtr \alpha_0(t_0) \underset{=}{\lessgtr} \alpha_0(a_{r+1}),$$

in which equality holds if and only if $t_0 = a_{r+1}$. It follows, when we take account of the monotonicity of α_0 and formula (10), that the two following relations hold simultaneously

$$a_r < t_0 \leqslant a_{r+1}; \qquad -(r+1)\pi\varepsilon \underset{=}{\lessgtr} c \lessgtr -r\pi\varepsilon.$$

(b) Secondly, we assume that the differential equation (q) is of finite type or left oscillatory. In these cases the boundary value d is finite.

As above, we find that if the differential equation (q) is of type (1) then we have

$$0 \lessgtr d \lessgtr \pi\varepsilon.$$

If the differential equation (q) admits of 1-conjugate numbers, then there hold simultaneously the formulae

$$b_{-s-1} \leqslant t_0 \lessgtr b_{-s}; \qquad s\pi\varepsilon \lessgtr d \underset{=}{\lessgtr} (s+1)\pi\varepsilon \qquad (s \geqslant 0;\ b_0 = b),$$

in which both equality signs must be taken at the same time.

These results may be summed up:

Theorem. Between the zero $t_0 \in j$ *and the boundary values* c, d *of a normal phase* α *of the differential equation* (q), *there hold the relations set out below, corresponding to the following table of type and kind of the differential equation* (q):

I. *finite type* (m), $m \geqslant 1$; (a) *general* (b) *special;*
II. *infinite type;* (a) *right oscillatory* (b) *left oscillatory* (c) *oscillatory;*

I. *The boundary values* c, d *are finite.*

(a) *There hold the relations* $(m-1)\pi\varepsilon \lessgtr d - c \lessgtr m\pi\varepsilon$, *and when*

$$m = 1:\ t_0 \in j,\ -\pi\varepsilon \lessgtr c \lessgtr 0 \lessgtr d \lessgtr \pi\varepsilon;$$
$$m \geqslant 2:\quad a_r < t_0 \leqslant b_{-m+r+1};\quad -(r+1)\pi\varepsilon \lessgtr c \lessgtr -r\pi\varepsilon;$$
$$(m-r-1)\pi\varepsilon \underset{=}{\lessgtr} d \lessgtr (m-r)\pi\varepsilon$$

or

$$b_{-m+r+1} < t_0 \leqq a_{r+1};\quad -(r+1)\pi\varepsilon \underset{=}{\lessgtr} c \lessgtr -r\pi\varepsilon;$$
$$(m-r-2)\pi\varepsilon \lessgtr d \lessgtr (m-r-1)\pi\varepsilon.$$

(b) *We have* $d - c = m\pi\varepsilon$ *and*

$$a_r < t_0 \leqslant a_{r+1};\quad -(r+1)\pi\varepsilon \underset{=}{\lessgtr} c \lessgtr -r\pi\varepsilon;\quad d = c + m\pi\varepsilon.$$

II. *At least one of the boundary values* c, d *is infinite.*

(a) $a_r < t_0 \leqslant a_{r+1};\quad -(r+1)\pi\varepsilon \underset{=}{\lessgtr} c \lessgtr -r\pi\varepsilon;\quad d = \varepsilon\infty.$

(b) $b_{-r-1} \leqslant t_0 < b_{-r};\quad r\pi\varepsilon \lessgtr d \underset{=}{\lessgtr} (r+1)\pi\varepsilon;\quad c = -\varepsilon\infty.$

(c) $t_0 \in j;\quad c = -\varepsilon\infty,\quad d = \varepsilon\infty.$

7.14 Boundary characteristics of normal phases

Let α be a normal phase of the differential equation (q), $t_0 \in j$ its zero and c, d its left and right boundary values. The ordered triad $(t_0; c, d)$ we shall call the *boundary characteristic* of the normal phase α; we shall use for it the notation $\chi(\alpha)$, or more briefly χ. The elements c, d are the *essential terms* of $\chi(\alpha)$. It is convenient also to speak of the values c, d when they denote the symbols $+\infty$ or $-\infty$.

From the above results, we see that there exist certain relationships between the terms of $\chi(\alpha)$, according to the type of the differential equation (q). In particular, for all types of the differential equation (q), we have $\alpha(t_0) = 0$, $\operatorname{sgn} \alpha' = -\operatorname{sgn} c = \operatorname{sgn} d$.

We now wish to study how far the normal phase α is determined by these relations.

By a *characteristic triple* of the differential equation (q) we mean an ordered triad $(\bar{t}_0; \bar{c}, \bar{d})$ whose terms satisfy the appropriate conditions I(a)–II(c) of § 7.13, according to the type and kind of the differential equation (q). Naturally, $t_0 \in j$ and $\bar{c}, \bar{d}$ may denote the symbols $\pm\infty$; also $\varepsilon = \operatorname{sgn}(\bar{d} - \bar{c})$.

Obviously, the boundary characteristic $\chi(\alpha)$ represents a characteristic triple of (q). The question which we now examine is: given any characteristic triple χ of the differential equation (q), do there correspond normal phases with the boundary characteristic χ? If so, how may these normal phases be determined?

7.15 Determination of normal phases with given characteristic triple

Let $\chi = (t_0; c, d)$ be a characteristic triple of the differential equation (q). We assume that there does indeed exist a normal phase α of (q) with the boundary characteristic χ. Let (u_0, v_0) be the basis of the differential equation (q) determined by the initial values

$$u_0(t_0) = 0, \quad u_0'(t_0) = -\operatorname{sgn} c; \quad v_0(t_0) = 1, \quad v_0'(t_0) = 0$$

and $\bar{\alpha}_0$ be the phase of (u_0, v_0) which vanishes at the point t_0: $\bar{\alpha}_0(t_0) = 0$. Obviously, $\bar{\alpha}_0$ is a normal phase of (q) and we have $\operatorname{sgn} \bar{\alpha}_0' = -\operatorname{sgn} c$. We denote by $\chi(\bar{\alpha}_0) = (t_0; \bar{c}_0, \bar{d}_0)$ the boundary characteristic of $\bar{\alpha}_0$. We have $-\operatorname{sgn} \bar{c}_0 = \operatorname{sgn} \bar{\alpha}_0' = -\operatorname{sgn} c$ and consequently $\operatorname{sgn} \bar{c}_0 = \operatorname{sgn} c$, $\operatorname{sgn} \bar{d}_0 = \operatorname{sgn} d$.

The functions α, $\bar{\alpha}_0$ are, however, phases of the same differential equation (q); it follows that if one of the numbers $c, \bar{c}_0$ or $d, \bar{d}_0$ is finite, so is the other. Further, the following formula holds in the interval j, except at the singular points of $\tan \alpha(t)$, $\tan \bar{\alpha}_0(t)$;

$$\tan \alpha(t) = \frac{c_{11} \tan \bar{\alpha}_0(t) + c_{12}}{c_{21} \tan \bar{\alpha}_0(t) + c_{22}},$$

with appropriate constants $c_{11}, c_{12}, c_{21}, c_{22}$. (See § 5.17, corollary 4).

Now $c_{12} = 0$, since both phases α, $\bar{\alpha}_0$ vanish at the point t_0. Obviously $c_{22} \neq 0$, and we can assume that $c_{22} = 1$, because the numerator and denominator may be divided by c_{22}. So we have

$$\tan \alpha(t) = \frac{c_{11} \tan \bar{\alpha}_0(t)}{c_{21} \tan \bar{\alpha}_0(t) + 1}.$$

This formula can be put into the following form, valid for all $t \in j$

$$\sin \bar{\alpha}_0(t) \cdot [c_{11} \cos \alpha(t) - c_{21} \sin \alpha(t)] = \cos \bar{\alpha}_0(t) \sin \alpha(t).$$

If the numbers $c, \bar{c}_0$ or $d, \bar{d}_0$ are finite then this relation gives, respectively,

$$\left.\begin{aligned} \sin \bar{c}_0 \cdot [c_{11} \cos c - c_{21} \sin c] &= \cos \bar{c}_0 \cdot \sin c, \\ \sin \bar{d}_0 \cdot [c_{11} \cos d - c_{21} \sin d] &= \cos \bar{d}_0 \cdot \sin d. \end{aligned}\right\} \tag{7.35}$$

We observe that the constants c_{11}, c_{21}, satisfy one or both of the linear equations (35) according to the type and kind of the differential equation (q) and are more or less determined by the boundary values c, $\bar{c}_0$ or d, $\bar{d}_0$. We now wish to study this situation in the various cases.

We use the notation of § 7.13. In particular, we put $\varepsilon = \operatorname{sgn} \bar{\alpha}'_0 = \operatorname{sgn} \alpha'$ and denote the 1-fundamental sequences, if they exist, by $(a <)\, r_1 = a_1 < a_2 < \cdots$, $(b >)\, s_1 = b_{-1} > b_{-2} > \cdots$.

I. Finite type (m), $m \geqslant 1$

$$c,\ d \text{ finite},\ (m-1)\pi\varepsilon \lessgtr d - c \underset{=}{\lessgtr} m\pi\varepsilon.$$

(a) General case:

$$(m-1)\pi\varepsilon \lessgtr d - c \lessgtr m\pi\varepsilon. \tag{7.36}$$

1. $m = 1$: $\sin c \sin d \neq 0$, $t_0 \in j$, $\sin \bar{c}_0 \sin \bar{d}_0 \neq 0$.

2. $m \geqslant 2$:

α) $\sin c \sin d \neq 0$, $b_{-m+r+1} \neq t_0 \neq a_{r+1}$, $\sin \bar{c}_0 \sin \bar{d}_0 \neq 0$,

β) $\sin c \neq 0$, $\sin d = 0$; $t_0 = b_{-m+r+1} < a_{r+1}$; $\sin \bar{c}_0 \neq 0$, $\sin \bar{d}_0 = 0$,

γ) $\sin c = 0$, $\sin d \neq 0$; $b_{-m+r+1} < t_0 = a_{r+1}$; $\sin \bar{c}_0 = 0$, $\sin \bar{d}_0 \neq 0$.

In cases 1 and 2α) the equations (35) can be written as follows

$$\left.\begin{aligned} c_{11} \cot c - c_{21} &= \cot \bar{c}_0, \\ c_{11} \cot d - c_{21} &= \cot \bar{d}_0. \end{aligned}\right\} \tag{7.37}$$

From (36) we have $\cot c - \cot d \neq 0$. It is clear that the constants c_{11}, c_{21} are uniquely determined.

In the respective cases 2β) (2γ)) the first (second) equation (35) can be replaced by the first (second) equation (37) while the second (first) equation (35) is satisfied identically. Clearly, of the two constants c_{11}, c_{21}, one is undetermined.

(b) Special case:

$$d - c = m\pi\varepsilon = \bar{d}_0 - \bar{c}_0.$$

1. $m = 1$: $\sin c \sin d \neq 0$, $t_0 \in j$, $\sin \bar{c}_0 \sin \bar{d}_0 \neq 0$.

2. $m \geqslant 2$:

α) $\sin c \sin d \neq 0$; $a_r < t_0 < a_{r+1}$, $\sin \bar{c}_0 \sin \bar{d}_0 \neq 0$.

β) $\sin c = \sin d = 0$; $t_0 = a_{r+1}$, $\sin \bar{c}_0 = \sin \bar{d}_0 = 0$.

In the cases 1 and 2α) the equations (35) are linearly dependent and can be replaced by one of the equations (37). Clearly, of the constants c_{11}, c_{21}, one is undetermined. In the case 2β) the equations (35) are satisfied identically: the constants c_{11}, c_{21} are arbitrary.

II. Infinite type

At least one of the boundary values c, d is infinite.

(a) Right oscillatory differential equation:

$$-(r+1)\pi\varepsilon \underset{=}{\lessgtr} c \lessgtr -r\pi\varepsilon; \quad d = \varepsilon\infty.$$

1. $\sin c \neq 0$, $a_r < t_0 < a_{r+1}$; $\sin \bar{c}_0 \neq 0$.

2. $\sin c = 0$, $t_0 = a_{r+1}$; $\sin \bar{c}_0 = 0$.

Clearly, in case 1 one of the constants c_{11}, c_{21} is undetermined, in case 2 both are arbitrary.

(b) Left oscillatory differential equation:

$$s\pi\varepsilon \lesseqgtr d \lesseqqgtr (s+1)\pi\varepsilon; \quad c = -\varepsilon\infty.$$

1. $\sin d \neq 0$, $b_{-s-1} < t_0 < b_{-s}$; $\sin \bar{d}_0 \neq 0$.
2. $\sin d = 0$, $t_0 = b_{-s-1}$; $\sin \bar{d}_0 = 0$.

Clearly, in case 1 one of the constants c_{11}, c_{21} is undetermined, in case 2 both are arbitrary.

(c) Oscillatory differential equation:

$$t_0 \in j; \quad c = -\varepsilon\infty, \quad d = \varepsilon\infty.$$

In this case both the constants c_{11}, c_{21} are arbitrary.

These results may be collected together as follows:

Theorem. For every characteristic triple $\chi = (t_0; c, d)$ *of the differential equation* (q) *there exist corresponding normal phases of* (q) *with the boundary characteristic* χ. *According to the type and kind of the differential equation* (q) *and according as to whether* t_0 *is singular or not, there is precisely one normal phase or there is precisely one system of normal phases (containing one or two parameters) of the differential equation* (q) *with the boundary characteristic* χ. *More precisely:*

There is precisely one normal phase of the differential equation (q) *with boundary characteristic* χ, *if* (q) *is a general differential equation either of type* (1), *or of type* (*m*), $m \geqslant 2$, t_0 *not being singular.*

There are precisely ∞^1 *normal phases of the differential equation* (q) *with the boundary characteristic* χ, *if* (q) *is a differential equation of type* (*m*), $m \geqslant 2$, *and* t_0 *is singular, or if the differential equation* (q) *is special of type* (1) *or of type* (*m*), $m \geqslant 2$, t_0 *not being singular; this is also true if the differential* (q) *is left or right oscillatory and* t_0 *is not singular.*

There are precisely ∞^2 *normal phases of the differential equation* (q) *with boundary characteristic* χ, *if* (q) *is special of type* (*m*), $m \geqslant 2$, *and* t_0 *is singular; this is also true if the differential equation* (q) *is left or right oscillatory and the number* t_0 *is singular; finally this is also true if the differential equation* (q) *is oscillatory.*

7.16 Determination of the type and kind of the differential equation (q) by means of the boundary values of a phase of (q)

We now show that, given the boundary values of a phase of the differential equation (q) the type and kind of (q) is uniquely determined.

Let (q) be a differential equation, α a phase of (q) and c, d the left and right boundary values of α.

Theorem. Let the numbers c, d be finite; then the differential equation (q) *is of finite type* (*m*), $m \geqslant 1$, *and is general with*† $m = [|d - c|/\pi] + 1$ *or special with* $m = |d - c|/\pi$ *according as* $|d - c|$ *cannot or can be divided by* π. *If c is finite and d infinite, then the differential equation* (q) *is right oscillatory; if c is infinite and d finite then* (q) *is left oscillatory. If c, d are both infinite then* (q) *is oscillatory.*

Proof. We first assume the numbers *c*, *d* are finite. Then, from the theorem of § 7.13, the differential equation (q) is of finite type (*m*), $m \geqslant 1$.

We choose λ so that $c^* = c + \lambda$, $d^* = d + \lambda$ have different signs. Then c^*, d^* are the boundary values of the normal phase $\alpha^* = \alpha + \lambda$ of (q); let t_0 be the zero of α^*.

If the differential equation (q) is general, then from the theorem of § 7.13 we have

$$(m - 1)\pi\varepsilon \lesseqgtr d^* - c^* \lesseqgtr m\pi\varepsilon \quad (\varepsilon = \operatorname{sgn}(d^* - c^*));$$

from which it follows that $|d - c|$ is not divisible by π and *m* has the value $[|d - c|/\pi] + 1$.

If the differential equation is special, then from the same theorem we have

$$d^* - c^* = m\pi\varepsilon;$$

in this case $|d - c|$ is divisible by π and *m* has the value $|d - c|/\pi$.

Secondly we assume that at least one of the numbers *c*, *d* is infinite. In this case our assertion follows immediately from the theorem of § 7.13.

In particular we have: a differential equation (q) of finite type (*m*), $m \geqslant 2$, is general or special according as the oscillation $O(\alpha)$ of each of its phases α satisfies the relations

$$(m - 1)\pi < O(\alpha) < m\pi \qquad \text{or} \qquad O(\alpha) = m\pi.$$

From the definition of § 7.2, this statement is also true for $m = 1$.

7.17 Properties of second phases

In the study of local and boundary properties of second phases of the differential equation (q), we make use of similar ideas and methods to those employed with respect to the first phases. We shall therefore abbreviate the discussion and bring out only a few of the relevant concepts and results. We assume from now on that the carrier *q* of the differential equation (q) is always non-zero in the interval *j*, and satisfies further properties according to circumstances. In particular, we shall recall that in the case $q \in C_2$ the associated differential equation ($\hat{q}_1$) of (q) exists and the first phases of this differential equation ($\hat{q}_1$) represent the second phases of (q) (§ 5.11).

(a) Theorem on the unique determination of a second phase from the Cauchy initial conditions.

Let $t_0 \in j$; Z_0, $Z_0' \neq 0$, Z_0'' *be arbitrary numbers. We assume the existence of* $q'(t_0)$. *There exists precisely one second phase* β *of the differential equation* (q) *which satisfies at the point* t_0 *the Cauchy initial conditions*:

$$\beta(t_0) = Z_0, \quad \beta'(t_0) = Z_0', \quad \beta''(t_0) = Z_0''. \tag{7.38}$$

† [*x*] denotes here, of course, the greatest integer not exceeding *x*. (Trans.)

This phase β is included in the second phase system of the basis (u, v) of the differential equation (q):

$$\left.\begin{aligned} u(t) &= \sin Z_0 \cdot u_0(t) + \frac{1}{q(t_0)}\left[Z_0' \cos Z_0 + \frac{1}{2}\left(\frac{q'(t_0)}{q(t_0)} - \frac{Z_0''}{Z_0'}\right) \sin Z_0\right] v_0(t), \\ v(t) &= \cos Z_0 \cdot u_0(t) + \frac{1}{q(t_0)}\left[-Z_0' \sin Z_0 + \frac{1}{2}\left(\frac{q'(t_0)}{q(t_0)} - \frac{Z_0''}{Z_0'}\right) \cos Z_0\right] v_0(t), \end{aligned}\right\} \tag{7.39}$$

in which u_0, v_0 are integrals of (q) *determined by the initial values*

$$u_0(t_0) = 0, \quad u_0'(t_0) = 1; \quad v_0(t_0) = 1, \quad v_0'(t_0) = 0,$$

(b) Let β be a second phase of the differential equation (q).

Assuming, as above, that $q(t) \neq 0$, $t \in j$, β is an increasing or decreasing function in the interval j $(= (a, b))$. The finite or infinite limiting values

$$c' = \lim_{t \to a+} \beta(t) \qquad \text{and} \qquad d' = \lim_{t \to b-} \beta(t)$$

are called respectively the *left* and *right boundary values of* β.

The circumstances under which these boundary values are finite or infinite are analogous to those for the first phases. In particular, we have the following fact: let the differential equation (q) possess 2-conjugate numbers; then the boundary value c' (d') of β is finite if and only if the left (right) 2-fundamental number $r_2(s_2)$ of (q) is proper.

The number $|c' - d'|$ is known as the *oscillation* of the phase β in the interval j. The notation used is $O(\beta|j)$, more briefly $O(\beta)$.

(c) The left (right) boundary values of the second phases of a left (right) principal basis of the second kind of the differential equation (q) are integral multiples of the number π.

The right (left) boundary values of the second phases of a 2-principal basis, whose first term is a left (right) 2-fundamental integral, are odd multiples of $\frac{1}{2}\pi$.

(d) We define second normal phases of the differential equation (q) and their boundary characteristics in a similar manner to those of first phases. With regard to the structure of the set of singular second normal phases, and the determination of second normal phases by their boundary characteristics, there are analogous results to those for first phases (§§ 7.9–7.15).

7.18 Relations between the boundary values of a first and second phase of the same basis

Let (u, v) be a basis of the differential equation (q) and α, β be first and second phases of (u, v). We choose the phases so that in the interval j the relations

$$0 < \beta(t) - \alpha(t) < \pi \tag{7.40}$$

hold; that is to say, we are dealing with two neighbouring phases of the mixed phase system of (u, v) (§ 5.14).

We know that the functions $y(t)$, $y'(t)$ constructed with arbitrary constants k_1, k_2

$$\left.\begin{aligned} y(t) &= k_1 \frac{\sin[\alpha(t) + k_2]}{\sqrt{|\alpha'(t)|}}, \\ y'(t) &= \pm k_1 \sqrt{|q(t)|}\, \frac{\sin[\beta(t) + k_2]}{\sqrt{|\beta'(t)|}} \end{aligned}\right\} \qquad (7.41)$$

represent the general integral of (q) and its derivative.

We now assume that the function q is negative throughout the interval j. Then the phases α, β either both increase or both decrease, i.e. $\operatorname{sgn}\alpha' \operatorname{sgn}\beta' = 1$.

Let c, c' and d, d' be the left and right boundary values of α, β. We know that the two numbers c, c' and also d, d' are simultaneously finite or infinite.

We consider the first case and assume from here onwards that the boundary values c, c' or d, d' are finite. It then follows from (40) that

$$0 \leqq c' - c \leqq \pi, \quad \text{or} \quad 0 \leqq d' - d \leqq \pi.$$

We now show that:

The relation $c' = c$ ($d' = d$) holds if and only if the left (right) 3-fundamental number r_3 (s_3) of (q) *is improper. In this case, we have therefore the situation described in* § 3.8. $a = r_3$, $r_4 = r_2 < r_1$ ($b = s_3$, $s_4 = s_2 > s_1$).

The relation $c' = c + \pi$ ($d' = d + \pi$) holds if and only if the left (right) 4-fundamental number r_4 (s_4) of (q) *is improper. In this case, we have $a = r_4$, $r_3 = r_1 < r_2$ ($b = s_4$, $s_3 = s_1 > s_2$).*

Proof. We restrict ourselves to proving the statements regarding the left boundary values c, c'.

First we note that each of the relations $c' = c$, $c' = c + \pi$ is invariant with respect to a transformation $\bar{\alpha}(t) = \alpha(t) + \lambda$, $\bar{\beta}(t) = \beta(t) + \lambda$ of the phases α, β, where λ is arbitrary. One can, in particular, take $\lambda = -c$, so without loss of generality we may assume $c = 0$. We shall also assume, for definiteness, that $\operatorname{sgn}\alpha' = \operatorname{sgn}\beta' = 1$.

(a) Let $c' = c = 0$. Let us select a number $x \in j$; we thus have to show that there exists a number which is left 3-conjugate with x.

From the fact that $c = 0$ and $\operatorname{sgn}\alpha' = 1$ we have $\alpha(x) > 0$. In the formulae (41) let us choose the constants k_1, k_2 as $k_1 = 1$, $k_2 = -\alpha(x)$. Then we have an integral y of (q) which vanishes at the point x, and its derivative y'.

From (40), we have the inequality

$$\beta(x) - \alpha(x) > 0. \qquad (7.42)$$

Since $c' = 0$ and $\operatorname{sgn}\beta' = 1$, the inequality $\beta(t) < \frac{1}{2}\alpha(x)$ holds for every $t \in j$ in a right neighbourhood of a, and consequently

$$\beta(t) - \alpha(x) < 0. \qquad (7.43)$$

From (42) and (43) we conclude that the left side of (43), regarded as a function of t, takes the value 0 at some point $t_0 \in (a, x)$. The number t_0 is obviously a zero of y' and consequently is a left 3-conjugate number of x.

(b) Let $a = r_3$. Then every number $x \in j$ possesses a left 3-conjugate number. We have to show that $c' = 0$.

We assume that $c' > 0$ and choose a number $x \in j$ satisfying the inequality $\alpha(x) < \frac{1}{2}c'$; this is possible since $c = 0$ and $\operatorname{sgn} \alpha' = 1$. Next, in the formulae (41), we choose the constants k_1, k_2 as $k_1 = 1, k_2 = -\alpha(x)$. Then we have an integral y of (q) which vanishes at the point x, and its derivative y'.

Now from the definition of c' and the fact that $\operatorname{sgn} \beta' = 1$ it follows that at every point $t \in (a, x)$

$$0 < \frac{1}{2}c' = c' - \frac{1}{2}c' < \beta(t) - \alpha(x) < \beta(x) - \alpha(x) < \pi,$$

and therefore $0 < \beta(t) - \alpha(x) < \pi$.

Obviously, the derivative y' of y has no zero to the left of x; consequently x has no left 3-conjugate number, which is a contradiction.

(c) Let $c' = \pi$. Let us choose a number $x \in j$; we then have to show that x has a left 4-conjugate number. Since $c' = \pi$ and $\operatorname{sgn} \beta' = 1$ we have $\beta(x) > \pi$.

In the formulae (41) we choose the constants k_1, k_2 as $k_1 = 1, k_2 = -\beta(x)$. Then we have an integral y of (q) whose derivative y' vanishes at the point x.

From (40) we have the inequality

$$\alpha(x) - \beta(x) > -\pi. \tag{7.44}$$

Since $c = 0$ and $\operatorname{sgn} \alpha' = 1$ the inequality $\alpha(t) < \frac{1}{2}(\beta(x) - \pi)$ holds for all t in a left neighbourhood of a, and consequently we also have

$$\alpha(t) - \beta(x) < -\pi. \tag{7.45}$$

From (44) and (45) we conclude that: the left side of (45), regarded as a function of t, takes the value $-\pi$ at a point $t_0 \in (a, x)$. This number t_0 is obviously a zero of y and so represents a left 4-conjugate number of x.

(d) Let $a = r_4$. Then to every number $x \in j$ there corresponds a left 4-conjugate number of x. We have to show that $c' = \pi$.

We assume that $c' < \pi$ and choose a number $x \in j$ satisfying the inequality $c' < \beta(x) < \pi$; since $c' < \pi$ and $\operatorname{sgn} \beta' = 1$, such a choice is possible.

Next, in the formulae (41) we choose the constants k_1, k_2 as $k_1 = 1, k_2 = -\beta(x)$. Then we have an integral y of (q) whose derivative y' vanishes at the point x.

Now, since $c = 0$ and $\operatorname{sgn} \alpha' = 1$, at every point $t \in (a, x)$ we have

$$-\pi < \alpha(t) - \pi < \alpha(t) - \beta(x) < \alpha(x) - \beta(x) < 0,$$

hence $-\pi < \alpha(t) - \beta(x) < 0$.

The integral y has therefore no zero to the left of x. Consequently there is no left 4-conjugate number of x, which is a contradiction.

8 Elementary phases

In this section we shall consider phases with certain special properties, known as elementary phases, which we shall often meet in the course of our researches. In order to set out this study as simply as possible, we shall introduce the concept of an elementary phase in a somewhat narrow sense, which however will suffice for our purposes. In this connection we assume that the length of the definition interval $j = (a, b)$ associated with the differential equation (q) is always greater than π. Naturally, we only have to introduce this assumption when j is finite, as it is automatically true for an unbounded interval j. Moreover, whenever a *second* phase is involved we always assume that the corresponding carrier q is negative in the whole interval j.

We consider a differential equation (q).

8.1 Introduction

A first or second phase $\gamma(t)$ of (q) will be called *elementary*, if for any two values t, $t + \pi$ lying in the interval j there holds the relation

$$\gamma(t + \pi) = \gamma(t) + \varepsilon\pi \qquad (\varepsilon = \operatorname{sgn} \gamma'). \tag{8.1}$$

We sometimes speak of elementary phases of the *carrier* q.

For instance, both the first and second phases of the carrier $q(t) = -1$, namely $\alpha(t) = t$ and $\beta(t) = \frac{1}{2}\pi + t$, are elementary.

Let $\gamma(t)$ be an elementary first or second phase of (q). Clearly, the phase γ may be represented in the form

$$\gamma(t) = \varepsilon t + G(t) \qquad (\varepsilon = \operatorname{sgn} \gamma'),$$

where $G(t)$, $t \in j$, is a function with the following properties:

1. G is periodic with period π,
2. $G \in C_3$ or $G \in C_1$ according as γ is a first or second phase,
3. $\operatorname{sgn} [\varepsilon + G'(t)] = \varepsilon$ for all $t \in j$.

Further we obtain from equation (1) and the monotonicity of γ the following result: the values $\gamma(t)$, $\gamma(t + \pi)$ are either both integral multiples of π or neither of them is such a multiple. In the first case, between the numbers t, $t + \pi$ there is no point at which the function γ takes the value of an integral multiple of π, while in the second case there is precisely one such point.

Further properties are:

Every phase of the complete phase system $[\gamma]$, hence every phase of the form $\gamma(t) + \lambda$, λ arbitrary, is also elementary.

The derivative γ' is a periodic function with period π.

8.2 Properties of equations with elementary phases

Now we obtain the following theorem:

Theorem. All first phases of the differential equation (q) *are elementary, if any one first phase of* (q) *possesses this property. The same statement holds also for second phases.*

Proof. Let us assume, for instance, that the first phase α_0 of (q) is elementary. Let α be an arbitrary first phase of (q). Then for every $t \in j$, with the exception of the singular points of the functions $\tan \alpha_0(t)$, $\tan \alpha(t)$, there holds a formula corresponding to (5.39), where α and α_0 are to be read in place of $\bar{\alpha}$ and α respectively, and the c_{ij} are appropriate constants. If we evaluate each side of this formula at two points t, $t + \pi \in j$, we find that $\tan \alpha(t + \pi) = \tan \alpha(t)$ and consequently $\alpha(t + \pi) = \alpha(t) + n\pi$, $n \neq 0$ being an integer. We have to show that $|n| = 1$.

To establish this, we choose two arbitrary values x, $x + \pi \in j$ and consider the following (first) phases of (q):

$$\bar{\alpha}_0(t) = \alpha_0(t) - \alpha_0(x), \quad \bar{\alpha}(t) = \alpha(t) - \alpha(x).$$

We first note that the phases $\bar{\alpha}_0$, $\bar{\alpha}$ have the common zero x; it follows that the two integrals

$$y(t) = \sin \bar{\alpha}_0(t)/\sqrt{|\bar{\alpha}_0'(t)|}, \quad \bar{y}(t) = \sin \bar{\alpha}(t)/\sqrt{|\bar{\alpha}'(t)|}$$

of the differential equation (q) both vanish at the point x and consequently have all their zeros in common.

Moreover, the phase $\bar{\alpha}_0$ is obviously elementary; we therefore have

$$\bar{\alpha}_0(x) = 0, \quad \bar{\alpha}_0(x + \pi) = \pi \cdot \operatorname{sgn} \bar{\alpha}_0'; \quad y(t) \neq 0 \quad \text{for} \quad t \in (x, x + \pi).$$

For the phase $\bar{\alpha}$, there hold the relations

$$\bar{\alpha}(x) = 0, \quad \bar{\alpha}(x + \pi) = |n|\pi \cdot \operatorname{sgn} \bar{\alpha}'.$$

If $|n| \neq 1$, then the function $\bar{\alpha}$ takes the value $\pi \operatorname{sgn} \bar{\alpha}'$ at some point $t_0 \in (x, x + \pi)$. In this case, t_0 is a zero of $\bar{y}$ and consequently also a zero of y. This, however, is impossible since it follows from the above that the integral y does not vanish in the interval $(x, x + \pi)$. We therefore have $|n| = 1$ and the proof is complete.

According to this result a differential equation (q) either has all its first (second) phases elementary or none of them are elementary. A differential equation (q) whose first (second) phases are elementary we shall call a *differential equation with elementary first* (*second*) *phases*, and apply the same terminology also to the corresponding carriers. It is convenient also to speak of differential equations and carriers as elementary with respect to their first (second) phases.

Now let (q) be a differential equation with elementary first (second) phases. The question arises: what properties have the second (first) phases of (q)?

In order to answer this, consider a first and a second phase α and β of the same basis of (q). Between the phases α, β there holds therefore a relationship similar to (5.34), and it follows that for any two values $t, t + \pi \in j$

$$\beta(t+\pi) - \beta(t) - \varepsilon\pi = \alpha(t+\pi) - \alpha(t) - \varepsilon\pi$$
$$+ \operatorname{Arccot}\left[\frac{1}{2}(1/\alpha'(t+\pi))'\right] - \operatorname{Arccot}\left[\frac{1}{2}(1/\alpha'(t))'\right] \tag{8.2}$$
$$(\varepsilon = \operatorname{sgn} \alpha' = \operatorname{sgn} \beta').$$

If the phase α is elementary, then α' is a periodic function with period π and so, consequently, is Arccot $[\frac{1}{2}(1/\alpha')']$. Formula (2) then shows that the phase β is also elementary.

If, conversely, the phase β is elementary, then the left side of (2) is identically zero and it is clear that the phase α is elementary if and only if the function $(1/\alpha')'$ has period π.

To sum up:

In a differential equation (q) *with elementary first phases, the second phases are also elementary.*

In a differential equation (q) *with elementary second phases the first phases α are also elementary if and only if the functions* $(1/\alpha')'$ *formed from them have period* π.

8.3 Properties of integrals, and their derivatives, of differential equations (q) with elementary phases

Integrals of differential equations (q) with elementary first or second phases, and their derivatives, are distinguished by particular properties which we now examine. Our investigation will be mainly concerned with differential equations (q) possessing elementary first phases; with regard to equations with elementary second phases we shall content ourselves with an indication of the results.

We show that

If the differential equation (q) *is elementary with respect to its first phases, then the values taken by every integral y of* (q) *at two arbitrary points $t, t + \pi \in j$ are equal and opposite in sign, that is*

$$y(t+\pi) = -y(t). \tag{8.3}$$

To show this, we assume that the first phases of (q) are elementary. Consider an integral y and a first phase α of (q). Then for an appropriate choice of the constants k_1, k_2 there holds a formula like the first of the formulae (5.27). Since the phase α is elementary, α' has period π, so, at two arbitrary points $t, t + \pi \in j$ the relation (3) holds.

Moreover

Theorem. The differential equation (q) *is elementary with respect to its first phases if and only if any two numbers $t, t + \pi \in j$ are neighbouring 1-conjugate numbers.*

Proof. (a) Let the differential equation (q) be elementary with respect to its first phases.

Let $t, t + \pi \in j$ be arbitrary numbers; let us choose a first phase α and an integral y of (q) which vanishes at t. Then we have a formula such as (5.27), in which $\alpha(t) + k_2 = n\pi$, n integral. Since the phase α is elementary, we have $\alpha(t + \pi) + k_2 = (n + \varepsilon)\pi$, $\varepsilon = \operatorname{sgn} \alpha'$, and between the numbers $t, t + \pi$ there is no point at which the value of the function $\alpha + k_2$ is an integral multiple of π. Consequently, $t + \pi$ is the first zero of y following t.

(b) Let two arbitrary numbers $t, t + \pi \in j$ be neighbouring 1- conjugate numbers.

We consider a first phase α of (q). Let $t, t + \pi \in j$ be arbitrary numbers, and let y be an integral of (q) which vanishes at t. Then we have a formula such as (5.27) in which $\alpha(t) + k_2 = n\pi$, where n is integral. According to our assumption, $t + \pi$ is a zero of y, and indeed the first zero of y following t; it follows that $\alpha(t + \pi) + k_2 = (n + \varepsilon)\pi$, $\varepsilon = \operatorname{sgn} \alpha'$. We have therefore $\alpha(t + \pi) = \alpha(t) + \varepsilon\pi$, which establishes our result.

The theorem which has just been proved can obviously be formulated as follows:

The differential equation (q) is elementary with respect to its first phases if and only if the zeros of all its integrals are situated at a distance π apart; that is, if two neighbouring zeros always have the same separation π.

The following property of a differential equation (q) with elementary first phases, in the case $b - a > 2\pi$, is worth mentioning: the integrals of such an equation are periodic functions with the fundamental period 2π.

For, by the relation (3), the integrals of (q) have period 2π, and for the fundamental period $p > 0$ of an integral y of (q) we have $0 < p \leqslant 2\pi$. If $p < 2\pi$, then for appropriate values $t, t + p \in j$ we have the inequality $y(t)\, y(t + p) < 0$, which conflicts with the definition of p.

Analogous properties are possessed by differential equations (q) with negative carriers and elementary second phases:

If the differential equation (q) *is elementary with respect to its second phases, then for the values of the derivative y' of every integral y of* (q) *at two arbitrary points $t, t + \pi$ there holds the relationship*

$$\frac{y'(t + \pi)}{\sqrt{|q(t + \pi)|}} = -\frac{y'(t)}{\sqrt{|q(t)|}}. \tag{8.4}$$

The differential equation (q) *is elementary with respect to its second phases if and only if every two numbers $t, t + \pi \in j$ are neighbouring 2-conjugate numbers.*

The differential equation (q) is elementary with respect to its second phases if and only if the zeros of the derivatives of all its integrals are separated by a distance π.

For a differential equation (q) with elementary second phases, in the case when $b - a > 2\pi$, the function $y'(t)/\sqrt{|q(t)|}$ constructed from an arbitrary integral y of the differential equation (q) is periodic with fundamental period 2π.

We know that for a differential equation (q) with elementary first phases, the second phases are also elementary (§ 8.2). From that, and the above results, we deduce that:

For a differential equation (q) *with a negative carrier q and elementary first phases, the successive zeros of any integral of* (q) *are separated by a distance π; so also are the successive turning points of such an integral.*

8.4 Determination of all carriers with elementary first phases

Our knowledge of elementary phases now makes it possible to determine explicitly all carriers with elementary first phases. For brevity, in §§ 8.4–8.7, we shall speak of phases and elementary carriers instead of first phases and carriers with elementary first phases.

From § 8.1, every elementary phase α of a differential equation (q) may be expressed in the form

$$\alpha(t) = \varepsilon t + A(t) \quad (\varepsilon = \operatorname{sgn} \alpha') \tag{8.5}$$

involving a function $A(t)$, $t \in j$, with the following properties:

1. A has period π,
2. $A \in C_3$,
3. $\operatorname{sgn} [\varepsilon + A'(t)] = \varepsilon$ for all $t \in j$.

From § 5.5, we know that the carrier q is determined uniquely by the phase α, being given precisely by the formula (5.16), so that

$$q(t) = -\frac{1}{2}\frac{A'''(t)}{\varepsilon + A'(t)} + \frac{3}{4}\frac{A''^2(t)}{(\varepsilon + A'(t))^2} - (\varepsilon + A'(t))^2. \tag{8.6}$$

If, conversely, we choose $\varepsilon = +1$ or $\varepsilon = -1$ and an arbitrary function $A(t)$, $t \in j$, with the above properties 1–3, then the function α defined by (5) represents a phase function (§ 5.7) with the property that $\alpha(t + \pi) = \alpha(t) + \varepsilon\pi$. Consequently, this function is an elementary phase of the carrier q determined by (6).

In this way we have determined all elementary carriers in the interval j:

All elementary carriers q in the interval j are given by the formula (6)*; ε denotes* $+1$ *or* -1, *and A represents an arbitrary function in the interval j with the above properties* 1–3.

8.5 Equations with elementary phases, defined over $(-\infty, \infty)$

With subsequent applications in mind, we now consider differential equations (q), with elementary phases, on the interval j, $j = (-\infty, \infty)$. From (5) we see that each phase of such a differential equation (q) is unbounded above and below. It follows (§ 5.4) that all differential equations (q) with elementary phases in the interval $j = (-\infty, \infty)$ are oscillatory. The corresponding carriers q are naturally given by formula (6).

An example, which we shall need, of a system of elementary carriers q in the interval $j = (-\infty, \infty)$ is given by the following formula, due to F. Neuman [53] (c is an arbitrary constant):

$$q(t|c) = \frac{\sin 4(t - c) + \frac{1}{3}\sin^4 (t - c)}{\left(1 - \frac{1}{3}\sin 2(t - c)\cdot\sin^2 (t - c)\right)^2} - 1. \tag{8.7}$$

This system is obtained from formula (6) by the following choice of the function $A(t)$:

$$A(t) = \begin{cases} \arctan\left(\frac{1}{6}\cos 2(t-c) - \cot(t-c)\right) - t + \nu\pi & \text{for} \quad t \in (c + \nu\pi, c + (\nu+1)\pi), \\ -\frac{\pi}{2} - c \quad \text{for} \quad t = \nu\pi; \quad (\nu = 0, \pm 1, \pm 2, \ldots). \end{cases}$$

Here the symbol arctan denotes that branch of the function which lies in the interval $(-\frac{1}{2}\pi, \frac{1}{2}\pi)$. Later we shall obtain this system of carriers by another method (§ 15.8).

It is easy to show that any two elementary carriers $q(t|c_1)$, $q(t|c_2)$ of the system (7) obtained from different values $c_1, c_2 \in [0, \frac{1}{4}\pi)$ represent different functions. It follows that:

The system of elementary carriers (7) *has the power of the continuum,* $\aleph$.

8.6 Power of the set of elementary carriers

We now seek to determine the power of the set of all elementary carriers in the interval $j = (-\infty, \infty)$. Let E denote this set. Since the elements of E can be obtained from an arbitrary π-periodic function A of class C_3 by means of the formula (6), it is reasonable to suppose that the power of E is $\aleph$. This can be proved formally as follows:

The system of elementary carriers given by the formula (7) represents a subset of E. Since this possesses the power $\aleph$, we have card $E \geqslant \aleph$. Moreover, the elements of E are continuous functions in the interval j, so that card $E \leqslant \aleph$; consequently:

The power of the set of all elementary carriers in the interval $j = (-\infty, \infty)$ *is precisely the power of the continuum,* $\aleph$.

8.7 Generalization of the concept of elementary phases

We now close this section with the following remarks. Consider a differential equation $(\bar{q})$ in the interval $\bar{j} = (\bar{a}, \bar{b})$. Let $c > 0$, $k > 0$ be arbitrary numbers and $\bar{b} - \bar{a} > c$.

A first or second phase $\bar{\gamma}(t)$ of $(\bar{q})$ will be called *quasi-elementary* if for every two values t, $t + c$ lying in the interval $\bar{j}$ there holds the relation

$$\bar{\gamma}(t + c) = \bar{\gamma}(t) + \varepsilon k \quad (\varepsilon = \operatorname{sgn} \bar{\gamma}').$$

It is easy to verify that the function

$$\gamma(t) = \frac{\pi}{k}\,\bar{\gamma}\left(\frac{c}{\pi}\,t\right)$$

defined by means of a quasi-elementary phase $\bar{\gamma}$ of $(\bar{q})$ in the interval (a, b), $a = (c/\pi)\bar{a}$, $b = (c/\pi)\bar{b}$ satisfies the relationship

$$\gamma(t + \pi) = \gamma(t) + \varepsilon\pi \quad (\varepsilon = \operatorname{sgn} \gamma').$$

If $\bar{\gamma}$ $(= \bar{\alpha})$ is a first quasi-elementary phase of $(\bar{q})$, then the function γ $(= \alpha) \in C_3$, and taking account of the formula (1.17), we have

$$\{\alpha, t\} = \frac{c^2}{\pi^2}\left\{\bar{\alpha}, \frac{c}{\pi}t\right\}.$$

It follows that

$$(q(t) =) -\{\alpha, t\} - \alpha'^2(t) = \frac{c^2}{\pi^2}\left[-\left\{\bar{\alpha}, \frac{c}{\pi}t\right\} - \bar{\alpha}'^2\left(\frac{c}{\pi}t\right)\right]$$
$$+ c^2\left(\frac{1}{\pi^2} - \frac{1}{k^2}\right)\bar{\alpha}'^2\left(\frac{c}{\pi}t\right) = \frac{c^2}{\pi^2}\bar{q}\left(\frac{c}{\pi}t\right) + c^2\left(\frac{1}{\pi^2} - \frac{1}{k^2}\right)\bar{\alpha}'^2\left(\frac{c}{\pi}t\right).$$

Hence, if $\bar{\alpha}(t)$ is a first quasi-elementary phase of the carrier $\bar{q}(t)$ in the interval $(\bar{a}, \bar{b})$ then the function

$$\alpha(t) = \frac{\pi}{k}\bar{\alpha}\left(\frac{c}{\pi}t\right)$$

represents in the interval (a, b), $a = (c/\pi)\bar{a}$, $b = (c/\pi)\bar{b}$ a first elementary phase of the carrier

$$q(t) = \frac{c^2}{\pi^2}\bar{q}\left(\frac{c}{\pi}t\right) + c^2\left(\frac{1}{\pi^2} - \frac{1}{k^2}\right)\bar{\alpha}'^2\left(\frac{c}{\pi}t\right).$$

9 Relations between first phases of two differential equations (q), (Q)

9.1 Introduction

Let us consider two differential equations (q), (Q) in the intervals $j = (a, b)$, $J = (A, B)$.

Our investigation will depend in general on the type and kind of the differential equations (q), (Q). In the case when these differential equations admit of left or right 1-fundamental sequences, we shall denote them by

$$(a <)\, a_1 < a_2 < \cdots, \qquad (b >)\, b_{-1} > b_{-2} > \cdots$$

and

$$(A <)\, A_1 < A_2 < \cdots, \qquad (B >)\, B_{-1} > B_{-2} > \cdots$$

9.2 Linked phases

Let α, $\mathbf{A}$ be arbitrary (first) phases of the differential equations (q), (Q) and let their boundary values be denoted by c, d and C, D respectively.

We shall call the phases α, $\mathbf{A}$ *linked* if simultaneously there hold the following relations between them

$$\min(c, d) < \max(C, D); \quad \min(C, D) < \max(c, d). \tag{9.1}$$

We shall show that *the phases* α, $\mathbf{A}$ *have common values if and only if they are linked.* In other words the relation $\alpha(j) \cap \mathbf{A}(J) \neq \varnothing$ implies and is implied by the inequalities (1).

Proof. (a) Let the first inequality (1) be not satisfied; then both the numbers c, d are greater than or equal to each of the numbers C, D. Consequently $\alpha(t) > \mathbf{A}(T)$ for all $t \in j$, $T \in J$.

(b) From (1) it follows that $\min(C, D) \leqslant \min(c, d) < \max(C, D)$ or $\min(c, d) < \min(C, D) < \max(c, d)$. In the first case, we have $\min(C, D) < \alpha(t) < \max(C, D)$ at a certain point $t \in j$, and since in the interval J the function $\mathbf{A}$ takes all values between $\min(C, D)$ and $\max(C, D)$, there is a $T \in J$ for which $\alpha(t) = \mathbf{A}(T)$. In the second case, we have $\min(c, d) < \mathbf{A}(T) < \max(c, d)$ at a point $T \in J$ and moreover, as above, $\mathbf{A}(T) = \alpha(t)$ for some $t \in j$, and the proof is complete.

In what follows we shall assume that the phases α, $\mathbf{A}$ *are linked.* Then L, given by

$$L = \alpha(j) \cap \mathbf{A}(J), \tag{9.2}$$

is an open interval. This is obviously the range of the function α in an open interval $k\ (\subset j)$ and also of $\mathbf{A}$ in an open interval $K\ (\subset J)$. That is to say, $L = \alpha(k) = \mathbf{A}(K)$.

We now wish to determine the intervals

$$k = \alpha^{-1}(L), \quad K = \mathbf{A}^{-1}(L). \tag{9.3}$$

Let $\bar{c}$, $\bar{d}$ and $\bar{C}$, $\bar{D}$ be the normalized boundary values of the phases α, $\mathbf{A}$ and moreover let $\bar{a}$, $\bar{b}$ and $\bar{A}$, $\bar{B}$ be the normalized end points of the intervals j and J with respect to these phases (§ 7.3). Then we have

$$\left.\begin{aligned} \lim_{t\to\bar{a}} \alpha(t) &= \bar{c}, \qquad \lim_{t\to\bar{b}} \alpha(t) = \bar{d}, \\ \lim_{T\to\bar{A}} \mathbf{A}(T) &= \bar{C}, \quad \lim_{T\to\bar{B}} \mathbf{A}(T) = \bar{D} \end{aligned}\right\} \tag{9.4}$$

and moreover

$$\bar{c} < \bar{d}, \qquad \bar{C} < \bar{D}. \tag{9.5}$$

The inequalities (1) can be written as follows

$$\bar{c} < \bar{D}, \qquad \bar{C} < \bar{d}. \tag{9.6}$$

Now, on examining the conditions (5), (6) it is clear that the following five cases, and only these, can occur:

1. $\bar{C} \leqslant \bar{c} < \bar{D} < \bar{d}$, hence $L = (\bar{c}, \bar{D})$;
2. $\bar{C} < \bar{c} < \bar{d} \leqq \bar{D}$, hence $L = (\bar{c}, \bar{d})$;
3. $\bar{c} < \bar{C} < \bar{D} \leqq \bar{d}$, hence $L = (\bar{C}, \bar{D})$;
4. $\bar{c} \leqslant \bar{C} < \bar{d} < \bar{D}$, hence $L = (\bar{C}, \bar{d})$;
5. $\bar{C} = \bar{c} < \bar{D} = \bar{d}$, hence $L = (\bar{c}, \bar{d}) = (\bar{C}, \bar{D})$.

The intervals, k, K are consequently (taking account of (4)) determined in the individual cases as follows

1. $k = (\bar{a}, \alpha^{-1}(\bar{D}))$,
   $K = (\mathbf{A}^{-1}(\bar{c}), \bar{B})$ or $= (\bar{A}, \bar{B})$, according as $\bar{C} < \bar{c}$ or $\bar{C} = \bar{c}$
2. $k = (\bar{a}, \bar{b})$,
   $K = (\mathbf{A}^{-1}(\bar{c}), \mathbf{A}^{-1}(\bar{d}))$ or $= (\mathbf{A}^{-1}(\bar{c}), \bar{B})$, according as $\bar{d} < \bar{D}$ or $\bar{d} = \bar{D}$
3. $k = (\alpha^{-1}(\bar{C}), \alpha^{-1}(\bar{D}))$ or $= (\alpha^{-1}(\bar{C})), \bar{b})$, according as $\bar{D} < \bar{d}$ or $\bar{D} = \bar{d}$
   $K = (\bar{A}, \bar{B})$;
4. $k = (\alpha^{-1}(\bar{C}), \bar{b})$ or $= (\bar{a}, \bar{b})$, according as $\bar{c} < \bar{C}$ or $\bar{c} = \bar{C}$
   $K = (\bar{A}, \mathbf{A}^{-1}(\bar{d}))$;
5. $k = (\bar{a}, \bar{b})$, $K = (\bar{A}, \bar{B})$.

Combining these, we obtain the following result:

Either: at least one of the end points of the interval k coincides with an end point of j, and at the same time at least one of the end points of the interval K coincides with an end point of J (1; 2, $\bar{d} = \bar{D}$; 3, $\bar{D} = \bar{d}$; 4; 5)

or the interval k coincides with j, and the end points of K lie in the interval J (2, $\bar{d} < \bar{D}$)

or the interval K coincides with J and the end points of k lie in the interval j (3, $\bar{D} < \bar{d}$).

We see in particular that *the interval k coincides with j and simultaneously the interval K coincides with J, if and only if* $\bar{c} = \bar{C}$, $\bar{d} = \bar{D}$. In other words: the ranges of the phases α, $\mathbf{A}$ coincide, over their definition intervals j, J, if and only if $C = c$, $D = d$ or $C = d$, $D = c$.

We call the differential equations (q), (Q) *of the same character* if either (i) both are of the same finite type (m), $m \geqslant 1$, and of the same kind (therefore both general or special) or (ii) each is one-sided oscillatory or (iii) both are oscillatory.

We then have (§ 7.2).

The ranges of the phases α, $\mathbf{A}$ *in their definition intervals j, J can coincide only when the differential equations* (q), (Q) *are of the same character.*

In §§ 9.3–9.6 we shall assume this property to hold for the differential equations (q), (Q); that is, (q), (Q) are of the same character.

9.3 Associated numbers

We call two numbers $t_0 \in j$ and $T_0 \in J$ *directly associated* with respect to the differential equations (q), (Q) (or, more shortly, directly associated) if they stand in the same relationship with respect to the numbers a_ν, $b_{-\nu}$ and A_ν, $B_{-\nu}$. Here $\nu = 0, 1, \ldots$; $a_0 = a$, $b_0 = b$; $A_0 = A$, $B_0 = B$.

That is to say:

I. In the case when the differential equations (q), (Q) are of finite type (m) $m \geqslant 1$:

(a) $m = 1$: $t_0 \in j$ arbitrary, $T_0 \in J$ arbitrary;

(b) $m \geqslant 2$:
1. $t_0 = a_{r+1}$, $T_0 = A_{r+1}$;
2. $t_0 = b_{-m+r+1}$, $T_0 = B_{-m+r+1}$;
3. $a_r < t_0 < b_{-m+r+1}$, $A_r < T_0 < B_{-m+r+1}$;
4. $b_{-m+r+1} < t_0 < a_{r+1}$, $B_{-m+r+1} < T_0 < A_{r+1}$;

$(r = 0, 1, \ldots, m - 1)$.

II. In the case when the differential equations (q), (Q) are of infinite type;

(a) Both differential equations (q), (Q) being right oscillatory:

1. $t_0 = a_{r+1}$, $T_0 = A_{r+1}$;
2. $a_r < t_0 < a_{r+1}$, $A_r < T_0 < A_{r+1}$ $(r = 0, 1, \ldots)$.

(b) Both being left oscillatory:

1. $t_0 = b_{-r-1}$, $T_0 = B_{-r-1}$;
2. $b_{-r-1} < t_0 < b_{-r}$, $B_{-r-1} < T_0 < B_{-r}$ $(r = 0, 1, \ldots)$.

(c) Both being oscillatory: $t_0 \in j$ and $T_0 \in J$ are arbitrary.

Further, we call two numbers $t_0 \in j$ and $T_0 \in J$ *indirectly associated* with respect to the differential equations (q), (Q) (more briefly, indirectly associated) if they stand in converse relationship with respect to the numbers a_ν, $b_{-\nu}$ and A_ν, $B_{-\nu}$. Here $\nu = 0, 1, \ldots$; $a_0 = a$, $b_0 = b$: $A_0 = A$, $B_0 = B$. That is to say:

I. In the case when the differential equations (q), (Q) are of finite type (m) $m \geqslant 1$:

(a) $m = 1$: $t_0 \in j$ arbitrary, $T_0 \in J$ arbitrary;

(b) $m \geqslant 2$:
1. $t_0 = a_{r+1}$, $T_0 = B_{-r-1}$;
2. $t_0 = b_{-m+r+1}$; $T_0 = A_{m-r-1}$;
3. $a_r < t_0 < b_{-m+r+1}$, $A_{m-r-1} < T_0 < B_{-r}$;
4. $b_{-m+r+1} < t_0 < a_{r+1}$, $B_{-r-1} < T_0 < A_{m-r-1}$;

$(r = 0, 1, \ldots, m-1)$.

II. In the case when the differential equations (q), (Q) are of infinite type:

(a) The differential equation (q) being right oscillatory and (Q) being left oscillatory:

1. $t_0 = a_{r+1}$, $T_0 = B_{-r-1}$;
2. $a_r < t_0 < a_{r+1}$, $B_{-r-1} < T_0 < B_{-r}$ $(r = 0, 1, \ldots)$.

(b) The differential equation (q) being left oscillatory and (Q) being right oscillatory:

1. $t_0 = b_{-r-1}$, $T_0 = A_{r+1}$;
2. $b_{-r-1} < t_0 < b_{-r}$, $A_r < T_0 < A_{r+1}$ $(r = 0, 1, \ldots)$.

(c) Both differential equations (q), (Q) being oscillatory: $t_0 \in j$ and $T_0 \in J$ are arbitrary.

If therefore the differential equations (q), (Q) are of type (1) or oscillatory, then every two numbers $t_0 \in j$ and $T_0 \in J$ are both directly and indirectly associated. But it is possible in other cases also to have two numbers $t_0 \in j$, $T_0 \in J$ which possess this property. To be precise, this occurs in special differential equations (q), (Q), when m (> 0) is even and $t_0 = a_{\frac{1}{2}m} = b_{-\frac{1}{2}m}$; $T_0 = A_{\frac{1}{2}m} = B_{-\frac{1}{2}m}$; it also occurs in general differential equations (q), (Q) if m is odd and $a_{\frac{1}{2}(m-1)} < t_0 < b_{-\frac{1}{2}(m-1)}$, $A_{\frac{1}{2}(m-1)} < T_0 < B_{-\frac{1}{2}(m-1)}$ or if m (> 0) is even and $b_{-\frac{1}{2}m} < t_0 < a_{\frac{1}{2}m}$, $B_{-\frac{1}{2}m} < T_0 < A_{\frac{1}{2}m}$.

We also observe that if $t_0 \in j$ is a singular number of the differential equation (q) (§ 3.10), then there is precisely one directly associated number or one indirectly associated number $T_0 \in J$ which is a singular number of (Q). Any non-singular number $t_0 \in j$ has always ∞^1 directly or indirectly associated numbers $T_0 \in J$, the set of which represents an open subinterval of J.

9.4 Characteristic triples of two differential equations

Let $t_0 \in j$, $T_0 \in J$ be directly or indirectly associated with respect to the differential equations (q), (Q). Then we have the following:

Theorem. If t_0*;* c*,* d *is a characteristic triple for the differential equation* (q), *then* T_0*;* c*,* d *or* T_0*;* d*,* c *is a characteristic triple for the differential equation* (Q).

Proof. Let t_0; c, d be a characteristic triple for the differential equation (q). The numbers t_0; c, d therefore satisfy one of the relationships (I–IIc) obtained in § 7.13, according to the type and kind of the differential equation (q).

Let $C = c$, $D = d$ or $C = d$, $D = c$ according as t_0, T_0 are directly or indirectly associated.

The theorem will be proved if we can show that the values T_0; A_ν, $B_{-\nu}$; C, D; $\mathbf{E}$ $(= \operatorname{sgn}(D - C))$ satisfy the appropriate conditions I–IIc of § 7.13 corresponding to the type and kind of the differential equation (Q), $(\nu = 0, 1, \ldots)$.

(a) Let the numbers t_0, T_0 be directly associated. Then the number T_0 stands in the same relation to A_ν, $B_{-\nu}$ as does t_0 in relation to a_ν, $b_{-\nu}$. Since moreover $C = c$, $D = d$; $\mathbf{E} = \varepsilon$, the condition which has to be satisfied by T_0; A_ν, $B_{-\nu}$; C, D; $\mathbf{E}$ is a consequence of the corresponding condition satisfied by t_0; a_ν, $b_{-\nu}$; c, d; ε.

(b) Let the numbers t_0, T_0 be indirectly associated. Then the number T_0 stands in the converse relationship to A_ν, $B_{-\nu}$ as does t_0 in relation to a_ν, $b_{-\nu}$. Moreover we have $C = d$, $D = c$; $\mathbf{E} = -\varepsilon$. Let us consider, for definiteness, the case I(a), $m \geqslant 2$ and

$$a_r < t_0 < b_{-m+r+1}; \qquad -(r+1)\pi\varepsilon \lesseqgtr c \lesseqgtr -r\pi\varepsilon;$$
$$(m - r - 1)\pi\varepsilon \lesseqgtr d \lesseqgtr (m - r)\pi\varepsilon. \tag{9.7}$$

Since the numbers t_0, T_0 are indirectly associated, we have

$$A_{m-r-1} < T_0 < B_{-r}. \tag{9.8}$$

From the relations (7) it follows that

$$-(r+1)\pi\varepsilon \lesseqgtr D \lesseqgtr -r\pi\varepsilon; \quad (m - r - 1)\pi\varepsilon \lesseqgtr C \lesseqgtr (m - r)\pi\varepsilon.$$

In these formulae, for $\varepsilon = 1$ and $\varepsilon = -1$ (that is, for $\mathbf{E} = -1$ and $\mathbf{E} = 1$), we take the signs $<$ and $>$ respectively.

We have therefore

$$(r+1)\pi\mathbf{E} \gtreqless D \gtreqless r\pi\mathbf{E}; \quad -(m - r - 1)\pi\mathbf{E} \gtreqless C \gtreqless -(m - r)\pi\mathbf{E},$$

and these formulae can be written as:

$$-(m - r)\pi\mathbf{E} \lesseqgtr C \lesseqgtr -(m - r - 1)\pi\mathbf{E}; \quad r\pi\mathbf{E} \lesseqgtr D \lesseqgtr (r+1)\pi\mathbf{E}. \tag{9.9}$$

If in (8) and (9) we write r in place of $m - r - 1$, we then have

$$A_r < T_0 < B_{-m+r+1}; \qquad -(r+1)\pi\mathbf{E} \lesseqgtr C \lesseqgtr -r\pi\mathbf{E};$$
$$(m - r - 1)\pi\mathbf{E} \lesseqgtr D \lesseqgtr (m - r)\pi\mathbf{E}.$$

This is precisely the relationship (7), written with capital instead of small letters.

The proof in other cases proceeds similarly.

9.5 Similar phases

We call two phases α, $\mathbf{A}$ of the differential equations (q), (Q) *similar* if their normalized boundary values $\bar{c}$, $\bar{d}$ and $\bar{C}$, $\bar{D}$ coincide: $\bar{c} = \bar{C}$, $\bar{d} = \bar{D}$. This obviously occurs if and only if the boundary values c, d of α and C, D of $\mathbf{A}$ are related by either $C = c$, $D = d$ or $C = d$, $D = c$.

If $C = c$, $D = d$ then to be more precise we call the phases α, $\mathbf{A}$ *directly similar*; in this case we have $\operatorname{sgn} \alpha' \operatorname{sgn} \dot{\mathbf{A}} = 1$. If $C = d$, $D = c$, we call the phases α, $\mathbf{A}$ *indirectly similar*; in this case we have $\operatorname{sgn} \alpha' \operatorname{sgn} \dot{\mathbf{A}} = -1$.

If, for instance, the differential equations (q), (Q) are oscillatory, then for all their phases α, $\mathbf{A}$: $\bar{c} = \bar{C} = -\infty$, $\bar{d} = \bar{D} = \infty$. From this it follows that every two phases α, $\mathbf{A}$ of the differential equations (q), (Q) are similar; more precisely, they are directly similar if both phases increase or both decrease, and indirectly similar if one increases and the other decreases.

In particular we have (§ 9.2)

The ranges of two phases α, $\mathbf{A}$ *of the differential equations* (q), (Q) *coincide in their intervals of definition if and only if the phases* α, $\mathbf{A}$ *are similar.*

Now let α, $\mathbf{A}$ be directly or indirectly similar phases. We prove the following results:

1. *The phases* α, $\mathbf{A}$ *take the same value at two directly or indirectly associated singular points of the differential equations* (q), (Q).

Proof. We apply the formulae (7.10) to the left or right null phases $\alpha - c$, $\mathbf{A} - C$ or $\alpha - d$, $\mathbf{A} - D$ of the differential equations (q), (Q) and obtain

$$\left.\begin{aligned} \alpha(a_\nu) &= c + \varepsilon\nu\pi, & \alpha(b_{-\nu}) &= d - \varepsilon\nu\pi; \\ \mathbf{A}(A_\nu) &= C + \mathbf{E}\nu\pi, & \mathbf{A}(B_{-\nu}) &= D - \mathbf{E}\nu\pi \\ (\nu = 1, 2, \ldots;\quad \varepsilon &= \operatorname{sgn} \alpha',\quad \mathbf{E} = \operatorname{sgn} \dot{\mathbf{A}}). \end{aligned}\right\} \tag{9.10}$$

(a) Let the phases α, $\mathbf{A}$ be directly similar:

$$C = c, \qquad D = d; \qquad \mathbf{E} = \varepsilon. \tag{9.11}$$

From § 9.3 any two directly associated singular points t_0, T_0 of the differential equations (q), (Q) must be either $t_0 = a_\nu$, $T_0 = A_\nu$ or $t_0 = b_{-\nu}$, $T_0 = B_{-\nu}$ ($\nu = 1, 2, \ldots$). In both cases there follows from (10) and (11) the relationship $\alpha(t_0) = \mathbf{A}(T_0)$.

(b) Let the phases α, $\mathbf{A}$ be indirectly similar:

$$C = d, \qquad D = c; \qquad \mathbf{E} = -\varepsilon. \tag{9.12}$$

From § 9.3 any two indirectly associated singular points t_0, T_0 of the differential equations (q), (Q) must be either $t_0 = a_\nu$, $T_0 = B_{-\nu}$ or $t_0 = b_{-\nu}$ and $T_0 = A_\nu$ ($\nu = 1, 2, \ldots$). In both cases, from (10) and (11) it follows that $\alpha(t_0) = \mathbf{A}(T_0)$. This completes the proof.

We now assume that α, $\mathbf{A}$ are similar normal phases. We denote their zeros by t_0, T_0.

2. *According as the phases* α, $\mathbf{A}$ *are directly or indirectly similar, their zeros* t_0, T_0 *are directly or indirectly associated.*

Proof. (a) Let the phases α, $\mathbf{A}$ be directly similar. In this case formulae (11) hold and we have relationships similar to those in the theorem of § 7.13 for t_0; c, d; ε and similarly for T_0; C, D; $\mathbf{E}$. Consequently the numbers t_0 and T_0 stand in the same relationship with respect to the numbers a_ν, $b_{-\nu}$ and A_ν, $B_{-\nu}$ respectively. ($\nu = 0, 1, \ldots$).

(b) Let the phases α, $\mathbf{A}$ be indirectly similar; then formulae (12) hold and the reader will easily convince himself that in all possible cases, whether the differential equations (q), (Q) are of finite or infinite type, the numbers t_0 and T_0 stand respectively in converse relationship with respect to the numbers a_ν, $b_{-\nu}$ and A_ν, $B_{-\nu}$ ($\nu = 0, 1, \ldots$).

For instance, consider the case when the differential equations (q), (Q) are of a finite type (m), $m \geqslant 2$, and $t_0 = a_{r+1}$. Then from the theorem of § 7.13, $-(r+1)\pi\varepsilon = c$. From this, and (12), it follows that $(r+1)\pi\mathbf{E} = D$ and moreover (from the same theorem) $T_0 = B_{-r-1}$.

9.6 Existence of similar phases

We now consider the question whether the differential equations (q), (Q), of the same character possess similar phases and, if so, how many such there are. We shall establish the following theorem:

Theorem. Let α *be a normal phase of the differential equation* (q) *and* t_0 *its zero. Let* T_0 *be a number which is directly or indirectly associated with* t_0 *with respect to the differential equations* (q), (Q). *Then there always exist normal phases* $\mathbf{A}$ *of the differential equation* (Q) *with the zero* T_0 *which are directly or indirectly similar to the phase* α. *According to the type and kind of the differential equations* (q), (Q) *and according to whether the numbers* t_0, T_0 *are singular or not, there is either one normal phase* $\mathbf{A}$ *or there is one* 1- *or* 2-*parameter system of normal phases* $\mathbf{A}$.

Proof. Let t_0; c, d be the boundary characteristic of α. Then from § 9.4 T_0; c, d or T_0; d, c is a characteristic triple for the differential equation (Q). From § 7.15 there exist normal phases $\mathbf{A}$ of (Q) with the boundary characteristic T_0; c, d or T_0; d, c. According to the type and kind of the differential equation (Q) and according as the number T_0 is singular or not, there is either one normal phase $\mathbf{A}$ or one 1- or 2-parameter system of normal phases of the differential equation (Q) with the boundary characteristic mentioned. Obviously, each normal phase $\mathbf{A}$ is directly or indirectly similar to the phase α and T_0 is its zero. This completes the proof.

More precisely (from § 7.15) the situation is as follows:

There is one normal phase $\mathbf{A}$ if the differential equations (q), (Q) are general either of type (1) or of type (m), $m \geqslant 2$, the numbers t_0, T_0 not being singular.

There are precisely ∞^1 normal phases $\mathbf{A}$, if the differential equations (q), (Q) are general of type (m), $m \geqslant 2$, and the numbers t_0, T_0 are singular; also, if (q), (Q) are special of type (1) or of type (m), $m \geqslant 2$, the numbers t_0, T_0 not being singular, and finally, if (q), (Q) are 1-sided oscillatory and the numbers t_0, T_0 are not singular.

There are precisely ∞^2 normal phases **A**, if the differential equations (q), (Q) are special of type (m), $m \geqslant 2$, the numbers t_0, T_0 being singular; also if (q), (Q) are 1-sided oscillatory and t_0, T_0 are singular, and finally, if (q), (Q) are oscillatory.

In § 9.2 we saw that the ranges of two phases α, **A** in the definition intervals j, J of the latter can only coincide when the differential equations (q), (Q) are of the same character. This observation, when taken together with the above result, shows that *the differential equations* (q), (Q) *admit of similar phases if and only if they are of the same character.*

10 Algebraic structure of the set of phases of oscillatory differential equations (q) in the interval $(-\infty, \infty)$

In this section we investigate the algebraic structure of the set of first phases of oscillatory differential equations (q), with the definition interval $j = (-\infty, \infty)$. The term *phase function* will here always mean a phase function (§ 5.7) *of class* C_3.

We know from §§ 5.5, 5.7, 5.4, that every first phase of a differential equation (q) represents a phase function which in the oscillatory case is unbounded on both sides. We also know (§§ 5.7, 5.4) that conversely every phase function α which is unbounded on both sides in its definition interval j represents a first phase of the oscillatory differential equation (q) constructed according to the formula (5.16). This section will therefore be concerned with the algebraic structure of the set of all phase functions defined in the interval $j = (-\infty, \infty)$ and unbounded on both sides. We shall call this set the *phase set* of the oscillatory differential equations (q) in the interval ($j =$) $(-\infty, \infty)$ or, more briefly, the phase set.

Instead of phase functions we shall speak more briefly of phases, and we shall call the carrier q of an oscillatory differential equation (q) an *oscillatory carrier*. The oscillatory carriers are therefore formed by making use of the formula (5.16) or (5.18), using a phase function α which is unbounded on both sides.

We remark that the power of the set M formed from all oscillatory carriers is equal to the power of the continuum. For, since M is composed of continuous functions, card $M \leqslant \aleph$ and since it contains all carriers formed from arbitrary constants $-k^2$ $(\neq 0)$, we have also card $M \geqslant \aleph$; we thus have card $M = \aleph$.

10.1 The phase group $\mathfrak{G}$

Let G be the phase set of the oscillatory differential equations (q) in the interval $j = (-\infty, \infty)$.

The phase set G obviously includes the identity phase $\phi_0(t) = t$. Moreover, the function $\alpha[\gamma(t)]$, which is the composition of two arbitrary phases $\alpha, \gamma \in G$ is also an element of G, and so is the function α^{-1} inverse to α. We now introduce into the set G a binary operation, which we call multiplication, by means of composition of functions; for arbitrary phases $\alpha, \gamma \in G$ we define the product $\alpha\gamma$ as being the composite function $\alpha[\gamma(t)]$. The set G with this multiplication thus forms a group $\mathfrak{G}$ with the unit element $\phi_0(t)$. We shall call $\mathfrak{G}$ the *phase group*.

The inverse element α^{-1} corresponding to any element $\alpha \in \mathfrak{G}$ represents an increasing or decreasing phase according as α is an increasing or decreasing function. Moreover, the product $\alpha\gamma$ of two elements $\alpha, \gamma \in \mathfrak{G}$ is an increasing function if both phases α, γ increase or decrease and is a decreasing function if one increases and the other decreases.

The set $\mathfrak{N}$ formed from all increasing phases is a normal sub group of $\mathfrak{G}$: $\alpha^{-1}\mathfrak{N}\alpha = \mathfrak{N}$; $\alpha \in \mathfrak{G}$. The factor group $\mathfrak{G}/\mathfrak{N}$ consists of two elements, namely $\mathfrak{N}$ and the class A of all decreasing phases.

10.2 The equivalence relation Q

Our next step is to introduce into the phase group $\mathfrak{G}$ an equivalence relation, as follows: two phases $\alpha, \gamma \in \mathfrak{G}$ are equivalent if they are linked by means of a relationship of the form

$$\tan \gamma(t) = \frac{c_{11} \tan \alpha(t) + c_{12}}{c_{21} \tan \alpha(t) + c_{22}}, \tag{10.1}$$

where the $c_{11}, c_{12}, c_{21}, c_{22}$ are constants with a non-zero determinant, i.e. $|c_{ij}| \neq 0$, and the relation (1) must hold for all values $t \in j$ except for the singular points of the functions $\tan \alpha(t)$, $\tan \gamma(t)$. It is easy to see that the relation determined by (1) in the phase group $\mathfrak{G}$ is reflexive, symmetric and transitive, and consequently is an equivalence relation. We shall denote this relation by Q.

The phase group $\mathfrak{G}$ is therefore split up into a system of equivalence classes mod Q, which we denote by $\bar{Q}$. $\bar{Q}$ is therefore a partition of the phase group $\mathfrak{G}$; every element $\bar{a} \in \bar{Q}$ consists of those phases which are equivalent to each other, while no phases lying in different elements $\bar{a}, \bar{b} \in \bar{Q}$ are equivalent.

Now, two arbitrary phase functions $\alpha, \gamma \in \mathfrak{G}$ represent first phases of appropriate carriers q, p determined by formulae such as (5.18). If α, γ are equivalent, they belong to the same element $\bar{a} \in \bar{Q}$, so there holds a relationship of the form (1) and from this and the theorem of § 1.8 it follows that

$$p(t) = -\{\tan \gamma, t\} = -\left\{\frac{c_{11} \tan \alpha + c_{12}}{c_{21} \tan \alpha + c_{22}}, t\right\} = -\{\tan \alpha, t\} = q(t),$$

so that $p(t) = q(t)$. Conversely (§ 5.17) the relation (1) holds for any two first phases α, γ of a carrier $q(t)$, $t \in j$; follows that the phase functions α, γ are equivalent and so belong to the same element $\bar{a} \in \bar{Q}$. Thus *every element $\bar{a} \in \bar{Q}$ comprises all first phases of one and the same carrier $q(t)$*. Let us associate with every element $\bar{a} \in \bar{Q}$ the corresponding carrier $q(t)$. We then have a simple mapping $\mathscr{A}$ of the partition $\bar{Q}$ onto the set of all oscillatory carriers. The power of the partition $\bar{Q}$ is that of the continuum, card $\bar{Q} = \aleph$.

In connection with this concept of equivalence of phases we note that: if one of two equivalent phases α, γ is elementary, then since α, γ are first phases of the same carrier q, the other phase is also elementary (§ 8.2), i.e. all phases equivalent to an elementary phase are themselves elementary.

10.3 The fundamental subgroup $\mathfrak{E}$

Next we investigate the algebraic structure of the partition $\bar{Q}$.

With this objective, we consider that element $\mathfrak{E} \in \bar{Q}$ which contains the unit element ϕ_0 of $\mathfrak{G}$. This element obviously consists of all phases $\zeta(t)$ which are equivalent to the unit element ϕ_0, that is to say all phases of the form

$$\tan \zeta(t) = \frac{c_{11} \tan t + c_{12}}{c_{21} \tan t + c_{22}}. \tag{10.2}$$

In view of (2), it is clear that the composite function $\zeta_1\zeta_2$ formed from two phases $\zeta_1, \zeta_2 \in \mathfrak{E}$ also belongs to the class $\mathfrak{E}$, and so does the function ζ_1^{-1} inverse to ζ_1. The elements of $\mathfrak{E}$ thus form a subgroup of $\mathfrak{G}$: $\mathfrak{E} \subset \mathfrak{G}$; we call this subgroup $\mathfrak{E}$ the *fundamental subgroup of* $\mathfrak{G}$.

We now show that *the partition $\bar{Q}$ coincides with the right residue class partition $\mathfrak{G}/_r\mathfrak{E}$ of the phase group $\mathfrak{G}$ with respect to $\mathfrak{E}$.*

Let $\bar{a} \in \bar{Q}$ be an arbitrary element and $\alpha \in \bar{a}$ a phase lying in it. We have to show that $\bar{a} = \mathfrak{E}\alpha$.

Now, for every element $\zeta(t) \in \mathfrak{E}$ there holds a formula such as (2). If, in that, we replace t by $\alpha(t)$ then it is clear that $\zeta\alpha$ is equivalent to α. Consequently $\zeta\alpha \in \bar{a}$ and we have $\mathfrak{E}\alpha \subset \bar{a}$.

Moreover, for every element $\gamma \in \bar{a}$ there holds a formula such as (1). If, in this, we replace t by $\alpha^{-1}(t)$, then it is clear that $\gamma\alpha^{-1}$ is equivalent to t. Hence $\gamma\alpha^{-1} \in \mathfrak{E}$, moreover, $\gamma \in \mathfrak{E}\alpha$, and we have $\bar{a} \subset \mathfrak{E}\alpha$. This establishes the fact that $\bar{a} = \mathfrak{E}\alpha$.

By a result from the theory of groups, the power of all right residue classes in a group with respect to a subgroup is always the same. Consequently, the power of all elements of the partition $\bar{Q}$ is always the same, and consequently equal to the power of $\mathfrak{E}$. This latter is obviously equal to $\aleph$, so card $\mathfrak{E} = \aleph$.

Thus, *the power of the set $\bar{a}$ of all first phases of a carrier $q(t)$ is the power of the continuum*: card $\bar{a} = \aleph$.

We remark that the unit element $\phi_0(t) = t \in \mathfrak{E}$ obviously represents an elementary phase. Since all phases equivalent to an elementary phase are themselves equivalent, it follows that:

The fundamental subgroup $\mathfrak{E}$ is comprised only of elementary phases.

We observe that the mapping $\mathscr{A}$ maps the fundamental subgroup $\mathfrak{E}$ onto the oscillatory carrier $q(t) = -1$.

10.4 The subgroup $\mathfrak{H}$ of elementary phases

Let $\mathfrak{H}$ be the set comprising all elementary phases. We wish to show that *$\mathfrak{H}$ is a subgroup of $\mathfrak{G}$*; $\mathfrak{H} \subset \mathfrak{G}$.

Proof. We have already noted that the unit element ϕ_0 is an elementary phase.

Let $\alpha, \gamma \in \mathfrak{G}$ be arbitrary elementary phases. We note first that the composite phase $\alpha\gamma$ is also elementary, since

$$\alpha\gamma(t+\pi) = \alpha(\gamma(t+\pi)) = \alpha(\gamma(t) + \varepsilon_2\pi) = \alpha(\gamma(t)) + \varepsilon_1\varepsilon_2\pi = \alpha\gamma(t) + \operatorname{sgn}(\alpha\gamma)'\pi$$
$$(\varepsilon_1 = \operatorname{sgn}\alpha', \ \varepsilon_2 = \operatorname{sgn}\gamma').$$

Further, the function inverse to α, namely $\bar{\alpha}$ $(= \alpha^{-1})$ is also an elementary phase. For, when $\bar{t} \in j$ we have

$$\bar{\alpha}(\alpha(\bar{t}) + \pi) = \bar{\alpha}(\alpha(\bar{t} + \varepsilon\pi)) = \bar{t} + \varepsilon\pi \quad (\varepsilon = \operatorname{sgn}\alpha' = \operatorname{sgn}\bar{\alpha}'),$$

and from this it follows for $t = \alpha(\bar{t})$, $\bar{t} = \bar{\alpha}(t)$, that

$$\bar{\alpha}(t+\pi) = \bar{\alpha}(t) + \varepsilon\pi.$$

This completes the proof.

We have shown above that the fundamental subgroup $\mathfrak{E}$ consists only of elementary phases. It follows that $\mathfrak{E}$ is a subgroup of $\mathfrak{H}$, consequently

$$\mathfrak{E} \subset \mathfrak{H}. \tag{10.3}$$

Now let $\mathfrak{G}/_r\,\mathfrak{H}$ be the right residue class partition of the group $\mathfrak{G}$ with respect to $\mathfrak{H}$. The elements of this partition are therefore the right residue classes $\mathfrak{H}\alpha$, $\alpha \in \mathfrak{G}$ with respect to the subgroup $\mathfrak{H}$.

From (3) it follows that the partition $\mathfrak{G}/_r\,\mathfrak{H}$ represents a covering of $\mathfrak{G}/_r\mathfrak{E}$; consequently ([81])

$$\mathfrak{G}/_r\mathfrak{H} \geqq \mathfrak{G}/_r\mathfrak{E}. \tag{10.4}$$

Formula (4) asserts that every element $\mathfrak{H}\alpha \in \mathfrak{G}/_r\,\mathfrak{H}$ is the union of some elements of the partition $\mathfrak{G}/_r\mathfrak{E}$. In particular, the subgroup $\mathfrak{H}$ is also the union of some elements of the partition $\mathfrak{G}/_r\mathfrak{E}$. From a known result in group theory, the power of the set of all elements of $\mathfrak{G}/_r\mathfrak{E}$ whose union gives rise to the element $\mathfrak{H}\alpha$, is independent of the choice of this element. In other words, for every choice of the element $\mathfrak{H}\alpha \in \mathfrak{G}/_r\,\mathfrak{H}$ the power of the set of elements of the partition $\mathfrak{G}/_r\mathfrak{E}$ which give the element $\mathfrak{H}\alpha$ by their union, is always the same.

We now consider the subgroup $\mathfrak{H}$ and a further element $\mathfrak{H}\alpha$ of the partition $\mathfrak{G}/_r\,\mathfrak{H}$ of $\mathfrak{G}$. Let $\bar{A}_0$ and $\bar{A}$ be the sets of all elements of the partition $\mathfrak{G}/_r\mathfrak{E}$ whose unions respectively produce the elements $\mathfrak{H}$ and $\mathfrak{H}\alpha$: $\cup\bar{A}_0 = \mathfrak{H}$, $\cup\bar{A} = \mathfrak{H}\alpha$. From the above, the power of the two sets $\bar{A}_0$, $\bar{A}$ are the same: card $\bar{A}_0$ = card $\bar{A}$. By means of the mapping $\mathscr{A}$ the sets $\bar{A}_0$, $\bar{A}$ are mapped onto certain sets $\mathscr{A}\bar{A}_0$, $\mathscr{A}\bar{A}$ of oscillatory carriers. From the definition of $\mathscr{A}$, the sets $\mathscr{A}\bar{A}_0$ and $\mathscr{A}\bar{A}$ consist of just those carriers whose first phases lie in $\mathfrak{H}$ and $\mathfrak{H}\alpha$ respectively. In particular, the set $\mathscr{A}\bar{A}_0$ comprises those carriers of which the first phases are elementary; that is to say, the elementary carriers. From § 7.6 the power of $\mathscr{A}\bar{A}_0$ is equal to that of the continuum: card $\mathscr{A}\bar{A}_0 = \aleph$. Now, the mapping $\mathscr{A}$ is simple; it follows that card $\mathscr{A}\bar{A}_0$ = card $\bar{A}_0$, card $\mathscr{A}\bar{A}$ = card $\bar{A}$. We have therefore

$$\text{card}\,\mathscr{A}\bar{A} = \text{card}\,\bar{A} = \text{card}\,\bar{A}_0 = \text{card}\,\mathscr{A}\bar{A}_0 = \aleph,$$

which gives the following result:

The power of the set of all oscillatory carriers, whose first phases lie in one and the same element $\mathfrak{H}\alpha \in \mathfrak{G}/_r\,\mathfrak{H}$, is the same for all elements of $\mathfrak{G}_r/\,\mathfrak{H}$ and is equal to the power of the continuum.

II Dispersion theory

Part II of this book is devoted to building up the theory of those functions which we call *dispersions*. These are certain functions of an independent variable which occur in the transformations of *oscillatory* differential equations (q), and can be defined as solutions of certain non-linear differential equations of the third order. They are of two kinds, known as *central dispersions* and *general dispersions*. Chapters A and B of this Part are devoted to building up the theory and applications of central dispersions, while general dispersions are handled in Chapter C.

Theory of central dispersions

In the theory of central dispersions we make contact for the first time in this book with transformations of linear differential equations of the second order (§ 13.5). For this reason we begin by setting out the transformation problem itself, which may appear rather isolated at this point but as our studies continue will come more and more into the foreground.

11 The transformation problem

11.1 Historical background

The transformation problem for ordinary linear differential equations of the second order originated with the German mathematician E. E. Kummer and so can conveniently be referred to as the Kummer Transformation Problem.

In his exposition "*De generali quadam aequatione differentiali tertii ordinis*", which was first given in the year 1834 in the programme of the Evangelical Royal and State Gymnasium in Liegnitz and later, in the year 1887, was re-issued in the J. für die reine und angewandte Math. (Vol. 100), Kummer considered the non-linear third order differential equation

$$2\frac{d^3z}{dz\,dx^2} - 3\left(\frac{d^2z}{dz\,dx}\right)^2 - Z\frac{dz^2}{dx^2} + X = 0. \tag{11.1}$$

It may perhaps be interesting to reproduce (in translation) the starting point of Kummer's study,

"We first notice that our equation, which is of the third order, can be reduced to two linear equations of the second order

$$\frac{d^2y}{dx^2} + p\frac{dy}{dx} + qy = 0, \tag{11.2}$$

$$\frac{d^2v}{dz^2} + P\frac{dv}{dz} + Qv = 0, \tag{11.3}$$

in which p and q are functions of the variable x, P and Q functions of the variable z. However, we shall grasp this inherently difficult problem more clearly, if we derive instead the equation (1) from the equations (2) and (3). To achieve this, let us consider z as a function of the variable x and assume that the variable $y = wv$, where w is a given function of the variable x, satisfies equation (2). Then by differentiation it follows, when we hold the differential dx constant, that

$$y = wv,$$

$$\frac{dy}{dx} = \frac{dw}{dx}v + w\frac{dv}{dz}\frac{dz}{dx},$$

$$\frac{d^2y}{dx^2} = \frac{d^2w}{dx^2}v + 2\frac{dw}{dx}\frac{dz}{dx}\frac{dv}{dz} + w\frac{d^2z}{dx^2}\frac{dv}{dz} + w\frac{dz^2}{dx^2}\frac{d^2v}{dz^2},$$

and when we substitute these values in the equation (2) we obtain

$$w\frac{dz^2}{dx^2}\frac{d^2v}{dz^2}+\left(2\frac{dw}{dx}\frac{dz}{dx}+w\frac{d^2z}{dx^2}+pw\frac{dz}{dx}\right)\frac{dv}{dz}+\left(\frac{d^2w}{dx^2}+p\frac{dw}{dx}+qw\right)v=0. \tag{11.4}$$

This is a second order linear equation in the variable v, and must be identical with equation (3) which has the same form; this will be the case if we set

$$2\frac{dw}{dx}\frac{dz}{dx}+w\frac{d^2z}{dx^2}+pw\frac{dz}{dx}-Pw\frac{dz^2}{dx^2}=0, \tag{11.5}$$

$$\frac{d^2w}{dx^2}+p\frac{dw}{dx}+\left(q-Q\frac{dz^2}{dx^2}\right)w=0. \tag{11.6}$$

From these equations (5) and (6) there follows by elimination of the variable w and its derivatives the third order equation:

$$2\frac{d^3z}{dz\,dx^2}-3\left(\frac{d^2z}{dz\,dx}\right)^2-\left(2\frac{dP}{dz}+P^2-4Q\right)\frac{dz^2}{dx^2}+\left(2\frac{dp}{dx}+p^2-4q\right)=0, \tag{11.7}$$

which for

$$2\frac{dP}{dz}+P^2-4Q=Z; \qquad Q=\frac{1}{4}\left(2\frac{dP}{dz}+P^2-Z\right), \tag{11.8}$$

$$2\frac{dp}{dx}+p^2-4q=X; \qquad q=\frac{1}{4}\left(2\frac{dp}{dx}+p^2-X\right) \tag{11.9}$$

goes over into our equation.

We find, therefore, that equation (1) gives the relationship necessary between the variables z and x in order that $y = wv$ should be an integral of equation (2), in the case when the variables y and v are determined by means of equations (2) and (3) and q and Q by equations (8) and (9).

Moreover, the quantity w, which we shall call the multiplier, is obtained from equation (5); when we divide the latter by $w\,dz/dx$, thus separating the three variables w, z and x, and integrate, it gives the formula

$$w^2=c\cdot e^{\int P dz}\cdot e^{-\int p dz}\cdot\frac{dx}{dz}; \tag{11.10}$$

in this, e denotes the base of natural logarithms and c an arbitrary constant".

11.2 Formulation of the transformation problem

The transformation problem which we wish to consider is as follows:

Let two linear differential equations of the second order be given, namely

$$y''=q(t)y, \tag{q}$$
$$\ddot{Y}=Q(T)Y \tag{Q}$$

in the (open) bounded or unbounded intervals $j = (a, b)$, $J = (A, B)$. We assume that the carriers q, Q of these differential equations are continuous in their intervals of definition j, J.

By a *transformation* of the differential equation (Q) into the differential equation (q) we mean an ordered pair $[w, X]$ of functions $w(t)$, $X(t)$, defined in an open interval i ($i \subset j$), such that for every integral Y of the differential equation (Q) the function

$$y(t) = w(t) \cdot Y[X(t)] \tag{11.11}$$

is a solution of the differential equation (q).

We make the following assumptions regarding the functions w, X:

1. $w \in C_2, \quad X \in C_3$;
2. $wX' \neq 0$ for all $t \in i$;
3. $X(i) \subset J$.

The function X we call the *transformation function* of the differential equations (q), (Q) (note the order), or more shortly the transformation; we also conveniently call it the *kernel of the transformation* $[w, X]$. The function w we shall call the *multiplier* of the transformation $[w, X]$. Naturally, these definitions comprise also the concept of transformation of the differential equation (q) into (Q).

The transformation problem which we have described above in an introductory fashion can now be formulated as follows:

To determine all reciprocal transformations of the differential equations (q), (Q) *and to describe their properties.*

Let $[w, X]$ be a transformation of (Q) into (q) and Y an integral of (Q), then we shall designate the solution of (q) defined in the interval $i \subset j$ by means of formula (11) as the *image* and the integral of the differential equation (q) including this image as the *image integral* of Y under the transformation $[w, X]$. More briefly we call these simply the *image* and *image integral* of Y.

Turning back to the above study by E. E. Kummer, we take formulae (7) and (10) with $p = P = 0$ and write t, T, $-q$, $-Q$ in place of x, z, q, Q; this then yields the following result:

Every transformation function X of the differential equations (q), (Q) *is, in its definition interval i, a solution of the non-linear third order differential equation*

$$-\{X, t\} + Q(X)X'^2 = q(t). \tag{Qq}$$

The multiplier w of each transformation $[w, X]$ of the differential equations (q), (Q), *is determined uniquely by means of its kernel X up to a multiplicative constant $k \neq 0$:*

$$w(t) = \frac{k}{\sqrt{|X'(t)|}}. \tag{11.12}$$

12 Introduction to the theory of central dispersions

We now wish to study certain functions of an independent variable which we shall call central dispersions of the first, second, third and fourth kinds. The central dispersion of the κ-th kind ($\kappa = 1, 2, 3, 4$) occur only in differential equations with κ-conjugate numbers. In order to simplify our study we shall for the rest of this Chapter A always assume that the differential equation (q) under consideration has conjugate numbers of all four kinds; we shall also assume that the carrier q is always negative in its interval of definition: $q \leqslant 0$. This assumption is not necessary when considering conjugate numbers of the first kind.

12.1 Some preliminaries

We consider a differential equation (q), $t \in j = (a, b)$. According to our assumption the differential equation (q) admits of conjugate numbers of all four kinds, and we have $q < 0$ for all $t \in j$. According to § 3.11, for each kind κ ($= 1, 2, 3, 4$), the numbers $t \in j$ which possess a ν-th left or right κ-conjugate number form an open interval $i_{\kappa,\nu}$ or $j_{\kappa,\nu}$; $\nu = 1, 2, \ldots$. These intervals $i_{\kappa,\nu}$, $j_{\kappa,\nu}$ were fully described in that paragraph. We know that each interval $i_{\kappa,\nu}$, $j_{\kappa,\nu}$ is a sub-interval of j, and we recall the following property: if the differential equation (q) is left or right oscillatory, then all the intervals $i_{\kappa,\nu}$ or $j_{\kappa,\nu}$ respectively coincide with j; if the differential equation (q) is oscillatory then all the intervals $i_{\kappa,\nu}$, $j_{\kappa,\nu}$ coincide with j ($\kappa = 1, 2, 3, 4$; $\nu = 1, 2, \ldots$).

12.2 Definition of the central dispersions

Let κ be one of the numbers 1, 2, 3, 4 and, let n_κ ($= n$) be a positive integer; we assume that in the interval j there are numbers for which the n-th right or left κ-conjugate number exists; such numbers consequently make up the interval $j_{\kappa,n}$ or $i_{\kappa,n}$. If, for instance, the differential equation (q) is of finite type (m), $m \geqslant 2$, then we have $n_1 \leqslant m$.

1. Let $\kappa = 1$. In the interval $j_{1,n}(i_{1,n})$ we define the function ϕ_n (ϕ_{-n}) as follows: For $t \in j_{1,n}$ ($t \in i_{1,n}$) let $\phi_n(t)$ ($\phi_{-n}(t)$) be the n-th right (left) number conjugate of the first kind with t. $\phi_n(t)$ ($\phi_{-n}(t)$) is therefore the n-th zero, lying to the right (left) of t, of any integral of the differential equation (q) which vanishes at the point t.

We call the function ϕ_n (ϕ_{-n}) the n-th ($-n$-th) *central dispersion of the first kind* or the 1-*central dispersion with the index n* ($-n$). In the particular case $n = 1$ we speak of the *fundamental dispersion of the first kind.* The fundamental dispersion of the first kind, ϕ_1, is therefore defined in the interval $j_{1,1}$ and its value $\phi_1(t)$ represents the first zero after t of every integral of the differential equation (q) which vanishes at the point t.

2. Let $\kappa = 2$. In the interval $j_{2,n}$ ($i_{2,n}$) we define the function ψ_n (ψ_{-n}) as follows: For $t \in j_{2,n}$ ($t \in i_{2,n}$) let $\psi_n(t)$ ($\psi_{-n}(t)$) be the n-th right (left) number conjugate of the second kind with t. $\psi_n(t)$ ($\psi_{-n}(t)$) is therefore the n-th zero lying to the right (left) of t of the derivative of any integral of the differential equation (q) whose derivative vanishes at the point t.

We call the function ψ_n (ψ_{-n}) the n-th ($-n$-th) *central dispersion of the second kind or the 2-central dispersion with the index n ($-n$)*. For $n = 1$ we speak of *the fundamental dispersion of the second kind.* The fundamental dispersion of the second kind, ψ_1, is therefore defined in the interval $j_{2,1}$, and its value $\psi_1(t)$ represents the first zero following t of the derivative of every integral of the differential equation (q) whose derivative vanishes at the point t.

We see that *if the differential equation* (q) *has the associated differential equation* ($\hat{q}_1$) *then the 2-central dispersions of* (q) *coincide with the 1-central dispersions of* ($\hat{q}_1$).

3. Let $\kappa = 3$. In the interval $j_{3,n}$ ($i_{3,n}$) we define the functions χ_n (χ_{-n}) as follows: For $t \in j_{3,n}$ ($t \in i_{3,n}$) let $\chi_n(t)$ ($\chi_{-n}(t)$) be the n-th right (left) conjugate number of the third kind with t. $\chi_n(t)$ ($\chi_{-n}(t)$) is therefore the n-th zero to the right (left) of t of the derivative of any integral of the differential equation (q) which vanishes at the point t.

We call the function χ_n (χ_{-n}) the n-th ($-n$-th) *central dispersion of the third kind* or the 3-*central dispersion with the index n* ($-n$). For $n = 1$ we speak of the *fundamental dispersion of the third kind.* The fundamental dispersion of the third kind, χ_1 is therefore defined in the interval $j_{3,1}$ and its value $\chi_1(t)$ represents the first zero occurring after t of the derivative of every integral of the differential equation (q) which vanishes at the point t.

4. Finally let $\kappa = 4$. In the interval $j_{4,n}$ ($i_{4,n}$) we define the function ω_n (ω_{-n}) as follows: For $t \in j_{4,n}$ ($t \in i_{4,n}$) let $\omega_n(t)$ ($\omega_{-n}(t)$) be the n-th right (left) conjugate number of the fourth kind with t. $\omega_n(t)$ ($\omega_{-n}(t)$) is therefore the n-th zero lying to the right (left) of t of every integral of the differential equation (q) whose derivative vanishes at the point t.

We call the function ω_n (ω_{-n}) the n-th ($-n$-th) *central dispersion of the fourth kind* or the 4-*central dispersion with the index n* ($-n$). For $n = 1$ we speak of the *fundamental dispersion of the fourth kind.* The fundamental dispersion of the fourth kind, ω_1, is therefore defined in the interval $j_{4,1}$ and its value $\omega_1(t)$ represents the first zero after t of each integral of the differential equation (q) whose derivative vanishes at the point t.

The terminology used for central dispersions is intended as a reminder of the distribution or dispersion of the zeros of integrals of the differential equation (q) and their derivatives. The adjective "central" refers to certain properties of central dispersions of the first and second kinds which are related to the group-theoretical concept of the "centre" (§ 21.6, 4, § 21.7).

12.3 Central dispersions of oscillatory differential equations (q)

The central dispersions which we have just defined exist in various different intervals, according to the kind and index, these intervals generally being proper sub-intervals of j. If the differential equation (q) is oscillatory then the interval of definition of every central dispersion coincides with j. Because of this simplification we shall concern ourselves in what follows with oscillatory differential equations only. We shall therefore assume that the differential equation (q) is oscillatory in its interval of definition $j = (a, b)$, also that $q < 0$ for all $t \in j$.

In this case the integrals of the differential equation (q) have infinitely many zeros which cluster towards a and b. Moreover in the interval j there exist four countable systems of central dispersions, namely the central dispersions of the first, second, third and fourth kinds, $\phi_\nu, \psi_\nu, \chi_\nu, \omega_\nu$: $\nu = \pm 1, \pm 2, \ldots$. It is convenient also to introduce the zero-th central dispersions of the first and second kinds by setting $\phi_0(t) = t$, $\psi_0(t) = t$ for all $t \in j$.

By the above definitions, the values of the central dispersions at an arbitrary point $t \in j$ represent the zeros of integrals of the differential equation (q) or of their derivatives; specifically, of integrals which either vanish or have their derivatives vanishing at the point t. See Fig. 3.

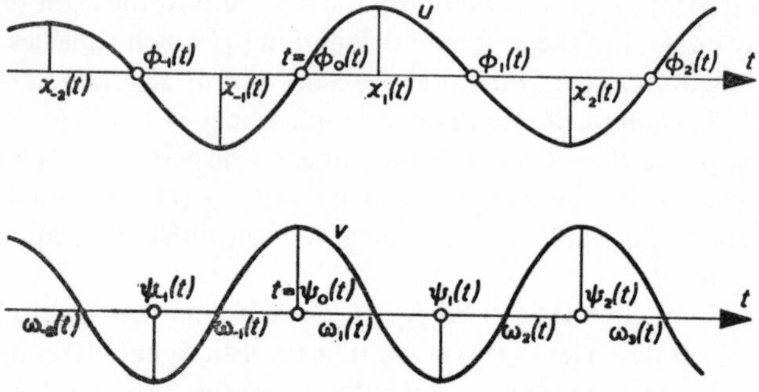

Figure 3

12.4 Relations between central dispersions

Obviously, for every point $t \in j$ we have the following relations

$$\left.\begin{aligned} \cdots < \chi_{-2}(t) < \phi_{-1}(t) < \chi_{-1}(t) < t < \chi_1(t) < \phi_1(t) < \chi_2(t) < \cdots, \\ \cdots < \omega_{-2}(t) < \psi_{-1}(t) < \omega_{-1}(t) < t < \omega_1(t) < \psi_1(t) < \omega_2(t) < \cdots. \end{aligned}\right\} \quad (12.1)$$

From now on we shall often employ the following notation: For two functions f, g defined in the interval j, we shall denote the composite function $f[g(t)]$, by fg. Also, f^{-1} will denote the inverse function to f, when this exists. If ν is an integer, then f^ν denotes the ν-th or $-\nu$-th iterated function f or f^{-1}, according as $\nu > 0$ or $\nu < 0$; that is $\underbrace{ff\ldots f}_{\nu}$ or $\underbrace{f^{-1}f^{-1}\ldots f^{-1}}_{-\nu}$. Finally we set $f^0 = t$.

Moreover let Φ, Ψ, X, Ω denote respectively the set of all central dispersions of the first, second, third, fourth kinds and Γ the union of all these sets: $\Gamma = \Phi \cup \Psi \cup X \cup \Omega$.

Between central dispersions of the same kind and those of different kinds there exist various relationships resulting from the composition of these functions. We set out these relationships schematically in the following "multiplication table":—

	Φ	Ψ	X	Ω
Φ	Φ	—	—	Ω
Ψ	—	Ψ	X	—
X	X	—	—	Ψ
Ω	—	Ω	Φ	—

The significance of this table is as follows: Composition of two central dispersions $a \in A$, $b \in B$ (A, B each denoting one of the sets Φ, Ψ, X, Ω) either gives a function which is not a central dispersion or gives a central dispersion ab from the set C which stands at the intersection of the A row and B column; i.e. $ab \in C$.

Now we give these relationships more precisely.

Let μ, ν and $\rho \neq 0$, $\sigma \neq 0$ be arbitrary integers.

1. $\phi_\mu \phi_\nu = \phi_{\mu+\nu}$. From this relation, it follows that

$$\left.\begin{aligned} \phi_0\phi_\nu &= \phi_\nu\phi_0 = \phi_\nu, & \phi_\nu\phi_{-\nu} &= \phi_0\,(= t), \\ \phi_1\phi_\nu &= \phi_\nu\phi_1 = \phi_{\nu+1}, & \phi_\nu &= \phi_1^\nu. \end{aligned}\right\} \tag{12.2}$$

2. $\psi_\mu \psi_\nu = \psi_{\mu+\nu}$. Hence

$$\left.\begin{aligned} \psi_0\psi_\nu &= \psi_\nu\psi_0 = \psi_\nu, & \psi_\nu\psi_{-\nu} &= \psi_0\,(= t). \\ \psi_1\psi_\nu &= \psi_\nu\psi_1 = \psi_{\nu+1}, & \psi_\nu &= \psi_1^\nu. \end{aligned}\right\} \tag{12.3}$$

3. $$\varphi_\nu \omega_\rho = \begin{cases} \omega_{\nu+\rho} & \text{for } \rho > 0,\ \nu \geqslant -\rho+1 \text{ and for } \rho < 0,\ \nu \leqq -\rho-1; \\ \omega_{\nu+\rho-1} & \text{for } \rho > 0,\ \nu \leqslant -\rho; \\ \omega_{\nu+\rho+1} & \text{for } \rho < 0,\ \nu \geqslant -\rho. \end{cases}$$

4. $$\psi_\nu \chi_\rho = \begin{cases} \chi_{\nu+\rho} & \text{for } \rho > 0,\ \nu \geqslant -\rho+1 \text{ and for } \rho < 0,\ \nu \leqq -\rho-1; \\ \chi_{\nu+\rho-1} & \text{for } \rho > 0,\ \nu \leqslant -\rho; \\ \chi_{\nu+\rho+1} & \text{for } \rho < 0,\ \nu \geqslant -\rho. \end{cases}$$

5. $$\chi_\rho \varphi_\nu = \begin{cases} \chi_{\rho+\nu} & \text{for } \rho > 0,\ \nu \geqslant -\rho+1 \text{ and for } \rho < 0,\ \nu \leqq -\rho-1; \\ \chi_{\rho+\nu-1} & \text{for } \rho > 0,\ \nu \leqslant -\rho; \\ \chi_{\rho+\nu+1} & \text{for } \rho < 0,\ \nu \geqslant -\rho. \end{cases}$$

6. $$\omega_\rho \psi_\nu = \begin{cases} \omega_{\rho+\nu} & \text{for } \rho > 0,\ \nu \geqslant -\rho+1 \text{ and for } \rho < 0,\ \nu \leqq -\rho-1; \\ \omega_{\rho+\nu-1} & \text{for } \rho > 0,\ \nu \leqslant -\rho; \\ \omega_{\rho+\nu+1} & \text{for } \rho < 0,\ \nu \geqslant -\rho. \end{cases}$$

7. $\chi_\rho\omega_\sigma = \begin{cases} \psi_{\rho+\sigma} & \text{for } \rho > 0,\ \sigma < 0 \text{ and for } \rho < 0,\ \sigma > 0; \\ \psi_{\rho+\sigma-1} & \text{for } \rho > 0,\ \sigma > 0; \\ \psi_{\rho+\sigma+1} & \text{for } \rho < 0,\ \sigma < 0. \end{cases}$

In particular, for $\sigma = -\rho$

$$\chi_\rho\omega_{-\rho} = \psi_0\ (= t). \tag{12.4}$$

8. $\omega_\rho\chi_\sigma = \begin{cases} \phi_{\rho+\sigma} & \text{for } \rho > 0,\ \sigma < 0 \text{ and for } \rho < 0,\ \sigma > 0; \\ \phi_{\rho+\sigma-1} & \text{for } \rho > 0,\ \sigma > 0; \\ \phi_{\rho+\sigma+1} & \text{for } \rho < 0,\ \sigma < 0. \end{cases}$

In particular, for $\sigma = -\rho$

$$\omega_\rho\chi_{-\rho} = \phi_0\ (= t). \tag{12.5}$$

From the above relations we have the following corollaries:

$$\phi_\nu\phi_{-\nu} = t, \quad \psi_\nu\psi_{-\nu} = t, \quad \chi_\rho\omega_{-\rho} = t, \quad \omega_\rho\chi_{-\rho} = t, \tag{12.6}$$

and moreover

$$\left.\begin{aligned} \phi_n &= \phi_1\phi_{n-1}, & \psi_n &= \psi_1\psi_{n-1}, & \chi_n &= \chi_1\phi_{n-1}, & \omega_n &= \omega_1\psi_{n-1}, \\ \phi_{-n} &= \phi_{-1}\phi_{-n+1}, & \psi_{-n} &= \psi_{-1}\psi_{-n+1}, & \chi_{-n} &= \chi_1\phi_{-n}, & \omega_{-n} &= \omega_1\psi_{-n} \end{aligned}\right\} \tag{12.7}$$

$$\left.\begin{aligned} \phi_n &= \phi_1^n, & \psi_n &= \psi_1^n, \quad \chi_n = \chi_1\phi_1^{n-1} = \psi_1^{n-1}\chi_1, \\ & & \omega_n &= \omega_1\psi_1^{n-1} = \phi_1^{n-1}\omega_1, \\ \phi_{-n} &= \phi_{-1}^n, & \psi_{-n} &= \psi_{-1}^n, \quad \chi_{-n} = \chi_{-1}\phi_{-1}^{n-1} = \psi_{-1}^{n-1}\chi_{-1}, \\ & & \omega_{-n} &= \omega_{-1}\psi_{-1}^{n-1} = \phi_{-1}^{n-1}\omega_{-1} \\ & & & (n = 1, 2, \ldots). \end{aligned}\right\} \tag{12.8}$$

The formulae 6 show that to every central dispersion there corresponds another central dispersion which is its inverse, and more precisely the following central dispersions are pairs of inverses: ϕ_ν, $\phi_{-\nu}$; ψ_ν, $\psi_{-\nu}$; χ_ρ, $\omega_{-\rho}$; ω_ρ, $\chi_{-\rho}$. From (8) it follows that every central dispersion can be obtained by composition of the fundamental dispersions and their inverses.

12.5 Algebraic structure of the set of central dispersions

In the set $\Gamma = \Phi \cup \Psi \cup X \cup \Omega$ we introduce a binary operation (multiplication) by defining the product of two elements as their composition. It is clear from the above table that certain ordered pairs of central dispersions $a \in A$, $b \in B$ (A, B each representing one of the sets Φ, Ψ, X, Ω) have a product $ab = c \in \Gamma$, while other ordered pairs of central dispersions do not possess any product in the set Γ. The set Γ under this operation forms an algebraic structure, a so-called *semi-groupoid*.

From the formulae (6), (8) we see that the set Φ, under the multiplication considered, forms an infinite cyclic group generated by the element ϕ_1. The unit element $\underline{1}$ of the group Φ is the element ϕ_0 $(= t)$. For every integer ν, the central dispersions ϕ_ν and $\phi_{-\nu}$ are inverse elements of the group Φ.

The structure of the set Ψ is similar; it forms an infinite cyclic group generated by the element ψ_1. The unit element $\underline{1}$ of the group Ψ is ϕ_0 $(= t)$ and so coincides with that of the group Φ. For every integer ν, ψ_ν and $\psi_{-\nu}$ are inverse elements of the group Ψ. The groups Φ, Ψ consequently have the unit element 1 in common.

Further, formula (6) shows that each of the two sets X, Ω consists of elements which are the inverses of elements of the other. Any two elements $\chi_\rho \in X$ and $\omega_{-\rho} \in \Omega$ are inverse to each other; that is, their product gives the unit element: $\chi_\rho \omega_{-\rho} = \omega_{-\rho}\chi_\rho = \underline{1}$. Obviously the sets X, Ω have no element in common with the groups Φ, Ψ.

To sum up: the semi-groupoid Γ is formed from two infinite cyclic groups Φ, Ψ, which have the unit element $\underline{1} = t$ in common, and also from two countable sets X, Ω, disjoint from the former two, whose elements are inverses of each other in pairs. Moreover, multiplication in the semi-groupoid is given by the formulae of § 12.4.

13 Properties of central dispersions

In this section we shall investigate some elementary properties of central dispersions, particularly their behaviour, their continuity and other properties associated with the existence of derivatives of central dispersions.

13.1 Monotonicity and continuity

1. *The range of each central dispersion of any kind is the interval j.*

For, if Δ is a central dispersion of any kind and $t \in j$ an arbitrary number, then the function Δ takes the value t at the point $\Delta^{-1}(t)$.

2. *Every central dispersion of any kind is an increasing function.*

Proof. As every central dispersion of any kind can be constructed by composition of the fundamental dispersion and its inverse (§ 12.4) it is sufficient to show the truth of this statement for the fundamental dispersion. Let, therefore, δ be a fundamental dispersion of arbitrary kind. In this paragraph we shall denote the fundamental dispersions briefly by ϕ, ψ, χ, ω.

Let $t < x$ be an arbitrary number in the interval j. We have to show that $\delta(t) < \delta(x)$. From (12.1) we have $t < \delta(t)$, $x < \delta(x)$. If $\delta(t) < x$, then we already have $\delta(t) < \delta(x)$. It will therefore be sufficient to show that the inequality $t < x < \delta(t)$ implies the inequality $\delta(t) < \delta(x)$. This we achieve by showing, on the basis of the ordering theorems, the impossibility of the inequality $t < x < \delta(x) < \delta(t)$. The equality relation $\delta(t) = \delta(x)$ is obviously impossible, because the inverse function δ^{-1} exists.

We therefore assume that $t < x < \delta(x) < \delta(t)$, and consider the four kinds of dispersion separately.

(a) $\delta = \phi$. From the definition of the function ϕ, the numbers t, $\delta(t)$ and x, $\delta(x)$ are respectively neighbouring zeros of integrals u, v of the differential equation (q). Obviously these integrals u, v, are independent. From the first ordering theorem (§ 2.3) precisely one zero of the integral v lies between t and $\delta(t)$, which is obviously inconsistent with the above inequalities.

(b) $\delta = \psi$. The proof proceeds on the same lines as (a), using the second ordering theorem.

(c) $\delta = \chi$. From the definition of the function χ, the numbers t, $\delta(t)$ are two neighbouring zeros of an integral v of the differential equation (q) and its derivative v'. The function v' is consequently non-zero between t and $\delta(t)$. Similarly, x, $\delta(x)$ are two neighbouring zeros of an integral u of the differential equation (q) and its derivative u'. Obviously the integrals u, v are independent. From the fourth ordering theorem the function v' has a zero between t and $\delta(x)$, and so (by the above inequalities) a zero between t and $\delta(t)$, which yields a contradiction.

(d) $\delta = \omega$. The proof is similar to that of case (c), using the third ordering theorem, and the proof is complete.

3. *Every central dispersion of arbitrary kind is everywhere continuous.*

This is an immediate consequence of the above theorems 1 and 2.

13.2 The functional equation of the central dispersions

In this paragraph we denote by ϕ, ψ, χ, ω an arbitrary central dispersion of the first, second, third and fourth kinds respectively. Let u, v be arbitrary independent integrals of the differential equation (q).

From the theorem in § 3.12 we conclude that: *In the interval j there hold the following identical relationships*

$$\left.\begin{aligned} u(t)v[\phi(t)] - u[\phi(t)]v(t) &= 0,\\ u'(t)v'[\psi(t)] - u'[\psi(t)]v'(t) &= 0,\\ u(t)v'[\chi(t)] - u'[\chi(t)]v(t) &= 0,\\ u'(t)v[\omega(t)] - u[\omega(t)]v'(t) &= 0. \end{aligned}\right\} \tag{13.1}$$

These relationships are called the *functional equations of the central dispersions.*

We see that the ratios u/v, u'/v' of the integrals u, v and their derivatives u', v' are invariant in the interval j with respect to composition of central dispersions, in the sense of the following formulae:

$$\left.\begin{aligned} \frac{u(t)}{v(t)} &= \frac{u[\phi(t)]}{v[\phi(t)]}, &\quad \frac{u'(t)}{v'(t)} &= \frac{u'[\psi(t)]}{v'[\psi(t)]},\\ \frac{u(t)}{v(t)} &= \frac{u'[\chi(t)]}{v'[\chi(t)]}, &\quad \frac{u'(t)}{v'(t)} &= \frac{u[\omega(t)]}{v[\omega(t)]}. \end{aligned}\right\} \tag{13.2}$$

13.3 Derivatives of central dispersions

Again let ϕ, ψ, χ, ω denote arbitrary central dispersions of the first, second, third and fourth kinds.

We now prove the following theorem:

(1) *All central dispersions of any kind have at each point $t \in j$ continuous (first) derivatives. These can be represented in terms of arbitrary independent integrals*

u, v of the differential equation (q) *and their derivatives u', v' by the following formulae:*

$$\left.\begin{aligned} \phi'(t) &= -\frac{u'(t)\,v[\phi(t)] - u[\phi(t)]\,v'(t)}{u(t)\,v'[\phi(t)] - u'[\phi(t)]\,v(t)}; \\ \psi'(t) &= -\frac{q(t)}{q[\psi(t)]}\cdot\frac{u(t)\,v'[\psi(t)] - u'[\psi(t)]\,v(t)}{u'(t)\,v[\psi(t)] - u[\psi(t)]\,v'(t)}; \\ \chi'(t) &= -\frac{1}{q[\chi(t)]}\cdot\frac{u'(t)\,v'[\chi(t)] - u'[\chi(t)]\,v'(t)}{u(t)\,v[\chi(t)] - u[\chi(t)]\,v(t)}; \\ \omega'(t) &= -q(t)\,.\,\frac{u(t)\,v[\omega(t)] - u[\omega(t)]\,v(t)}{u'(t)\,v'[\omega(t)] - u'[\omega(t)]\,v'(t)}. \end{aligned}\right\} \tag{13.3}$$

Proof. We shall confine ourselves to the case of the central dispersion ϕ. Let u, v be two linearly independent integrals of the differential equation (q). We consider the basis function

$$F(t, x) = u(t)v(x) - u(x)v(t).$$

(see § 2.6; $Q = q$).

Let $t_0 = j$ be arbitrary and $x_0 = \phi(t_0)$ the corresponding value of ϕ. Then we have $F(t_0, x_0) = 0$. (§ 3.12). From § 2.6 there is precisely one continuous function $x(t)$ defined in a neighbourhood i ($\subset j$) of t_0, which takes the value x_0 at the point t_0 and satisfies the equation $F[t, x(t)] = 0$ in the interval i. This function x possesses, in the interval i, the continuous derivative

$$x'(t) = -\frac{u'(t)\,v[x(t)] - u[x(t)]\,v'(t)}{u(t)\,v'[x(t)] - u'[x(t)]\,v(t)}. \tag{13.4}$$

But the function ϕ is defined and continuous in the interval j, and consequently also in the interval i, and satisfies the above equation $F[t, \phi(t)] = 0$ in i; consequently $x(t) = \phi(t)$ for $t \in i$. The existence of $\phi'(t_0)$ follows, and hence, taking account of (4), the first formula (3).

2. *The derivatives of central dispersions may be represented as follows in terms of an integral u of the differential equation* (q) *and its derivative u'*:—

$$\phi'(t) = \begin{cases} \dfrac{u^2[\phi(t)]}{u^2(t)} & \text{for} \quad u(t) \neq 0, \\[2ex] \dfrac{u'^2(t)}{u'^2[\phi(t)]} & \text{for} \quad u(t) = 0; \end{cases} \tag{13.5}$$

$$\psi'(t) = \begin{cases} \dfrac{q(t)}{q[\psi(t)]}\cdot\dfrac{u'^2[\psi(t)]}{u'^2(t)} & \text{for} \quad u'(t) \neq 0, \\[2ex] \dfrac{q(t)}{q[\psi(t)]}\cdot\dfrac{u^2(t)}{u^2[\psi(t)]} & \text{for} \quad u'(t) = 0; \end{cases} \tag{13.6}$$

$$\chi'(t) = \begin{cases} -\dfrac{1}{q[\chi(t)]} \cdot \dfrac{u'^2[\chi(t)]}{u^2(t)} & \text{for} \quad u(t) \neq 0, \\[2ex] -\dfrac{1}{q[\chi(t)]} \cdot \dfrac{u'^2(t)}{u^2[\chi(t)]} & \text{for} \quad u(t) = 0; \end{cases} \tag{13.7}$$

$$\omega'(t) = \begin{cases} -q(t) \,.\, \dfrac{u^2[\omega(t)]}{u'^2(t)} & \text{for} \quad u'(t) \neq 0, \\[2ex] -q(t) \,.\, \dfrac{u^2(t)}{u'^2[\omega(t)]} & \text{for} \quad u'(t) = 0. \end{cases} \tag{13.8}$$

Proof. We shall restrict ourselves to the proof of (5); essentially it is obtained by a transformation of the first formula (3).

In the case $u(t) \neq 0$ we multiply the numerator and denominator of the first formula of (3) by $u(t)$ and in the numerator make use of the relationship $u(t)v[\phi(t)] = u[\phi(t)]v(t)$; then we proceed analogously in the denominator, multiplying by $u[\phi(t)]$ and making use of the same relationship. After division by the Wronskian $w(t) = w[\phi(t)]$ of u, v we obtain the first formula (5).

In the case $u(t) = 0$ we have $u'(t)\, u'[\phi(t)] \neq 0$. We multiply, as above, the numerator and denominator by $u'(t)$, then by $u'[\phi(t)]$ and so obtain the second formula (5).

3. *The derivatives of the central dispersions can be represented as follows by means of the first or second amplitudes r, s of an arbitrary basis (u, v) of the differential equation* (q):—

$$\left.\begin{aligned} \phi'(t) &= \frac{r^2[\phi(t)]}{r^2(t)}, & \psi'(t) &= \frac{q(t)}{q[\psi(t)]} \cdot \frac{s^2[\psi(t)]}{s^2(t)}, \\ \chi'(t) &= -\frac{1}{q[\chi(t)]} \cdot \frac{s^2[\chi(t)]}{r^2(t)}, & \omega'(t) &= -q(t) \,.\, \frac{r^2[\omega(t)]}{s^2(t)}. \end{aligned}\right\} \tag{13.9}$$

Proof. We give the proof of the first formula as an example.

Let (u, v) be a basis of the differential equation (q) and r, s the corresponding first and second amplitudes. Let $t \in j$ be arbitrary. At least one of the two numbers $u(t)$, $v(t)$ is non-zero; let us assume, for definiteness, that it is $u(t)$. Considering the function $\lambda = v/u$ we have (from (2)) $\lambda(t) = \lambda[\phi(t)]$, and consequently

$$\phi'(t) = \frac{u^2[\phi(t)]}{u^2(t)} = \frac{1 + \lambda^2(t)}{1 + \lambda^2[\phi(t)]} \cdot \frac{r^2[\phi(t)]}{r^2(t)} = \frac{r^2[\phi(t)]}{r^2(t)},$$

which is the first formula of (9).

13.4 Higher derivatives

We see from the above results that

If the carrier q of the differential equation (q) *is continuous in the interval j, all central dispersions of the first kind have continuous derivatives of the third order in that interval and all other central dispersions have continuous derivatives of the first order.*

More precisely we have

If the carrier q of the differential equation (q) *belongs to the class C_k ($k = 0, 1, \ldots$) then all central dispersions of the first kind $\in C_{k+3}$ and all other central dispersions $\in C_{k+1}$.*

13.5 The connection between central dispersions and the transformation problem

The formulae (5)–(8) have an important bearing on the transformation problem (§ 11).

Let $\phi_\nu, \psi_\nu, \chi_\rho, \omega_\rho$ ($\nu = 0, \pm 1, \ldots; \rho = \pm 1, \ldots$) be arbitrary central dispersions of the first, second, third and fourth kinds, and let u be an arbitrary integral of (q). Then the above formulae (5)–(8) hold for these central dispersions.

Now, on taking account of the ordering theorems, (§ 2.3), these formulae give the following relationships for all $t \in j$:

$$u(t) = \frac{(-1)^\nu}{\sqrt{\phi_\nu'(t)}}\, u[\phi_\nu(t)], \tag{13.10}$$

$$\frac{u'(t)}{\sqrt{-q(t)}} = \frac{(-1)^\nu}{\sqrt{\psi_\nu'(t)}}\,\frac{u'[\psi_\nu(t)]}{\sqrt{-q[\psi_\nu(t)]}}, \tag{13.11}$$

$$u(t) = \frac{(-1)^\rho}{\sqrt{\chi_\rho'(t)}}\,\frac{u'[\chi_\rho(t)]}{\sqrt{-q[\chi_\rho(t)]}}, \tag{13.12}$$

$$\frac{u'(t)}{\sqrt{-q(t)}} = \frac{(-1)^\rho}{\sqrt{\omega_\rho'(t)}}\, u[\omega_\rho(t)]. \tag{13.13}$$

By formula (10), therefore, the ordered pair $[(-1)^\nu/\sqrt{\phi_\nu'(t)}, \phi_\nu(t)]$ represents a transformation (§ 11.2) of the differential equation (q) into itself in which every integral u of (q) is transformed into itself.

Similarly the formulae (11)–(13) show that if we assume that the function q $(< 0) \in C_2$ then the differential equation (q) admits of the associated differential equation ($\hat{q}_1$). We know (§ 1.9) that the function $u_1(t) = u'(t)/\sqrt{(-q(t))}$ represents an integral of the differential equation ($\hat{q}_1$), namely the integral associated with u.

The ordered pair $[(-1)^\nu/\sqrt{\psi_\nu'(t)}, \psi_\nu(t)]$ obviously represents a transformation of the associated differential equation ($\hat{q}_1$) into itself, in which every integral u_1 of ($\hat{q}_1$) is transformed into itself.

The ordered pair $[(-1)^\rho/\sqrt{\chi_\rho'(t)}, \chi_\rho(t)]$ represents a transformation of ($\hat{q}_1$) into (q). In this transformation every integral u_1 of the differential equation ($\hat{q}_1$) is transformed into the associated integral u of (q).

The ordered pair $[(-1)^\rho/\sqrt{\omega_\rho'(t)}, \omega_\rho(t)]$ represents a transformation of (q) into ($\hat{q}_1$). In this transformation every integral u of the differential equation (q) is transformed into the associated integral u_1 of ($\hat{q}_1$).

To sum up; *the central dispersions of oscillatory differential equations* (q) *are particular solutions of the Kummer transformation problem for the differential equation* (q) *and its associated differential equation* ($\hat{q}_1$).

13.6 Relations between derivatives of the central dispersions and the values of the carrier q

1. Let now ϕ, ψ, χ, ω be the *fundamental* dispersions of the first, second, third and fourth kinds. We have the following result:

Theorem. The first derivatives of the fundamental dispersions at every point $t \in j$ may be expressed as ratios of appropriate values taken by the carrier q, as follows

$$\phi'(t) = \frac{q(t_1)}{q(t_3)}, \qquad \psi'(t) = \frac{q(t)\ q(t_4)}{q[\psi(t)]q(t_2)},$$
$$\chi'(t) = \frac{q(t_1)}{q[\chi(t)]}, \quad \omega'(t) = \frac{q(t)}{q(t_2)};$$

in which t_1, t_2, t_3, t_4 denote appropriate numbers ordered as follows

$$t < t_1 < \chi(t) < t_3 < \phi(t); \quad t < t_2 < \omega(t) < t_4 < \psi(t).$$

Proof. For every integral u of the differential equation (q) and arbitrary numbers t, $x \in j$ we have obviously the formula

$$u'^2(x) - u'^2(t) = \int_t^x q(\sigma)[u^2(\sigma)]'\,d\sigma, \tag{13.14}$$

which provides the basis for our proof.

(a) Let u be any integral of the differential equation (q), which vanishes at the point t, so that $u(t) = u'[\chi(t)] = 0$. In formula (14) we set $x = \chi(t)$, so obtaining

$$-u'^2(t) = \int_t^{\chi(t)} q(\sigma)[u^2(\sigma)]'\,d\sigma.$$

Now, in the interval $(t, \chi(t))$ the function uu' and consequently also the function $(u^2)'$ is positive. By the mean value theorem we have

$$\int_t^{\chi(t)} q(\sigma)\cdot[u^2(\sigma)]'\,d\sigma = q(t_1)u^2[\chi(t)],$$

where $t < t_1 < \chi(t)$. The last two formulae give

$$-u'^2(t) = q(t_1)u^2[\chi(t)],$$

and consequently, using (7),

$$\chi'(t) = \frac{q(t_1)}{q[\chi(t)]} \qquad (t < t_1 < \chi(t)).$$

(b) Now let u be any integral of the differential equation (q), whose derivative u' vanishes at the point t: $u'(t) = u[\omega(t)] = 0$.

In formula (14) we set $x = \omega(t)$ giving

$$u'^2[\omega(t)] = \int_t^{\omega(t)} q(\sigma)[u^2(\sigma)]'\,d\sigma.$$

In the interval $(t, \omega(t))$ the function uu', and consequently also $(u^2)'$ is negative. By the mean value theorem we have

$$\int_t^{\omega(t)} q(\sigma)[u^2(\sigma)]'\, d\sigma = -q(t_2)u^2(t),$$

in which $t < t_2 < \omega(t)$. So

$$u'^2[\omega(t)] = -q(t_2)u^2(t),$$

from which, using (8) we obtain

$$\omega'(t) = \frac{q(t)}{q(t_2)} \qquad (t < t_2 < \omega(t)).$$

(c) From the identical relationship

$$\phi(t) = \omega[\chi(t)]$$

we obtain, on applying the above results,

$$\phi'(t) = \omega'[\chi(t)]\chi'(t) = \frac{q[\chi(t)]}{q(t_3)} \cdot \frac{q(t_1)}{q[\chi(t)]} = \frac{q(t_1)}{q(t_3)},$$

in which $\chi(t) < t_3 < \phi(t)$. Consequently we have

$$\phi'(t) = \frac{q(t_1)}{q(t_3)} \qquad (\chi(t) < t_3 < \phi(t)).$$

(d) Similarly, we obtain from the identity

$$\psi(t) = \chi[\omega(t)]$$

the formula

$$\psi'(t) = \frac{q(t)}{q(t_2)} \cdot \frac{q(t_4)}{q[\psi(t)]} \qquad (\omega(t) < t_4 < \psi(t)).$$

2. The above formulae apply, as we have said, to the fundamental dispersions of appropriate kinds. More generally, for $n = 1, 2, \ldots$ we have the following relations.

$$\phi_n'(t) = \frac{q(t_1)}{q(t_3)} \cdot \frac{q(t_5)}{q(t_7)} \cdots \frac{q(t_{4n-3})}{q(t_{4n-1})};$$

$$\chi_n'(t) = \frac{q(t_1)}{q(t_3)} \cdot \frac{q(t_5)}{q(t_7)} \cdots \frac{q(t_{4n-7})}{q(t_{4n-5})} \cdot \frac{q(t_{4n-3})}{q[\chi_n(t)]};$$

$$\phi_{\mu-1}(t) < t_{4\mu-3} < \chi_\mu(t) < t_{4\mu-1} < \phi_\mu(t); \quad \mu = 1, 2, \ldots, n.$$

$$\psi_n'(t) = \frac{q(t)}{q(t_2)} \cdot \frac{q(t_4)}{q(t_6)} \cdots \frac{q(t_{4n-4})}{q(t_{4n-2})} \cdot \frac{q(t_{4n})}{q[\psi_n(t)]};$$

$$\omega_n'(t) = \frac{q(t)}{q(t_2)} \cdot \frac{q(t_4)}{q(t_6)} \cdots \frac{q(t_{4n-4})}{q(t_{4n-2})};$$

$$\psi_{\mu-1}(t) < t_{4\mu-2} < \omega_\mu(t) < t_{4\mu} < \psi_\mu(t); \quad \mu = 1, 2, \ldots, n.$$

$$\phi'_{-n}(t) = \frac{q(t_{-1})}{q(t_{-3})} \cdot \frac{q(t_{-5})}{q(t_{-7})} \cdots \frac{q(t_{-(4n-3)})}{q(t_{-(4n-1)})};$$

$$\chi'_{-n}(t) = \frac{q(t_{-1})}{q(t_{-3})} \cdot \frac{q(t_{-5})}{q(t_{-7})} \cdots \frac{q(t_{-(4n-7)})}{q(t_{-(4n-5)})} \cdot \frac{q(t_{-(4n-3)})}{q[\chi_{-n}(t)]};$$

$$\phi_{-\mu}(t) < t_{-(4n-1)} < \chi_{-\mu}(t) < t_{-(4\mu-3)} < \phi_{-\mu+1}(t); \quad \mu = 1, 2, \ldots, n.$$

$$\psi'_{-n}(t) = \frac{q(t)}{q(t_{-2})} \cdot \frac{q(t_{-4})}{q(t_{-6})} \cdots \frac{q(t_{-(4n-4)})}{q(t_{-(4n-2)})} \cdot \frac{q(t_{-4n})}{q[\psi_{-n}(t)]};$$

$$\omega'_{-n}(t) = \frac{q(t)}{q(t_{-2})} \cdot \frac{q(t_{-4})}{q(t_{-6})} \cdots \frac{q(t_{-(4n-4)})}{q(t_{-(4n-2)})};$$

$$\psi_{-\mu}(t) < t_{-4\mu} < \omega_{-\mu}(t) < t_{-(4\mu-2)} < \psi_{-\mu+1}(t); \quad \mu = 1, 2, \ldots, n.$$

These formulae for ϕ'_n, χ'_n, ψ'_n, ω'_n are easily proved by induction, using the relations (12.7) and the above formulae for ϕ', χ', ψ', ω'.

Moreover, from the relations $\phi\phi_{-1}(t) = t$, $\psi\psi_{-1}(t) = t$, and the formulae for ϕ', ψ' there follow the further relations

$$\phi'_{-1}(t) = \frac{q(t_{-1})}{q(t_{-3})}; \qquad \psi'_{-1}(t) = \frac{q(t)}{q(t_{-2})} \cdot \frac{q(t_{-4})}{q[\psi_{-1}(t)]}; \tag{13.15}$$

in which $\phi_{-1}(t) < t_{-3} < \chi_{-1}(t) < t_{-1} < t$; $\psi_{-1}(t) < t_{-4} < \omega_{-1}(t) < t_{-2} < t$. The formulae for ϕ'_{-n}, χ'_{-n}, ψ'_{-n}, ω'_{-n} are easily proved by induction, making use of the relations (12.7), (13.15), and the formulae for χ', ω'.

13.7 Relations between central dispersions and phases

Let α, β be respectively a first and second phase of the basis (u, v) of the differential equation (q). We assume, for definiteness, that these phases are adjacent in the mixed phase system of the basis (u, v) (§ 5.14); that is to say, one of the relations $0 < \beta - \alpha < \pi$, $-\pi < \beta - \alpha < 0$ is satisfied in the interval j. By virtue of our assumption that $q < 0$ we have $\operatorname{sgn} \alpha' = \operatorname{sgn} \beta'$ $(= \varepsilon)$. In what follows ϕ, ψ, χ, ω denote *fundamental* dispersions of the four kinds.

Let $x \in j$ be an arbitrary number.

First we consider an integral y of the differential equation (q), which vanishes at the point x. From (5.27) we have

$$\left.\begin{aligned} y(t) &= k \cdot r(t) \cdot \sin[\alpha(t) - \alpha(x)], \\ y'(t) &= \pm k \cdot s(t) \cdot \sin[\beta(t) - \alpha(x)], \end{aligned}\right\} \tag{13.16}$$

where naturally k $(\neq 0)$ is an appropriate constant.

For simplicity, let us set $A(t) = \alpha(t) - \alpha(x)$. Obviously, we have $A(x) = 0$. In the interval j the function A tends monotonically to $+\infty$ or $-\infty$ according as $\varepsilon = 1$ or $\varepsilon = -1$. Consequently, there is a number x_1 $(> x)$, for which the function A takes the value $\varepsilon\pi$. From (16), however, x_1 is the first zero of the integral y to the right of x; we have therefore $x_1 = \phi(x)$ and moreover

$$\alpha\phi(x) = \alpha(x) + \varepsilon\pi. \tag{13.17}$$

Now we consider the function $B(t) = \beta(t) - \alpha(x)$. This function also tends monotonically in the interval j to $+\infty$ or $-\infty$, according as $\varepsilon = 1$ or $\varepsilon = -1$. Consequently there is a number x_3 ($> x$), for which the function B takes the value $\frac{1}{2}(\varepsilon + 1)\pi$ or $\frac{1}{2}(\varepsilon - 1)\pi$, according as $0 < \beta(x) - \alpha(x) < \pi$ or $-\pi < \beta(x) - \alpha(x) < 0$. From (16), however, x_3 is the first zero of the function y' to the right of x. We have therefore $x_3 = \chi(x)$ and moreover

$$\left.\begin{aligned} \beta\chi(x) &= \alpha(x) + \frac{1}{2}(\varepsilon + 1)\pi \quad \text{if} \quad 0 < \beta(x) - \alpha(x) < \pi; \\ \beta\chi(x) &= \alpha(x) + \frac{1}{2}(\varepsilon - 1)\pi \quad \text{if} \quad -\pi < \beta(x) - \alpha(x) < 0. \end{aligned}\right\} \tag{13.18}$$

In the second place we consider an integral y of the differential equation (q) whose derivative y' vanishes at the point x. From (5.27) we have

$$\begin{aligned} y(t) &= k \cdot r(t) \cdot \sin[\alpha(t) - \beta(x)], \\ y'(t) &= \pm k \cdot s(t) \cdot \sin[\beta(t) - \beta(x)]. \end{aligned}$$

By an analogous method to that used above, we deduce from formula (17) the relationship

$$\beta\psi(x) = \beta(x) + \varepsilon\pi, \tag{13.19}$$

and then

$$\left.\begin{aligned} \alpha\omega(x) &= \beta(x) + \frac{1}{2}(\varepsilon - 1)\pi \quad \text{if} \quad 0 < \beta(x) - \alpha(x) < \pi, \\ \alpha\omega(x) &= \beta(x) + \frac{1}{2}(\varepsilon + 1)\pi \quad \text{if} \quad -\pi < \beta(x) - \alpha(x) < 0. \end{aligned}\right\} \tag{13.20}$$

The formulae (17), (18), (19), (20) are known as the *Abel functional equations* for the fundamental dispersions.

If one combines these relations with the relations (12.8), one obtains the *general Abel functional equations for the central dispersions*, or more briefly the *Abel functional equations*

$$\left.\begin{aligned} \alpha\phi_\nu(x) &= \alpha(x) + \varepsilon\nu\pi, \\ \beta\psi_\nu(x) &= \beta(x) + \varepsilon\nu\pi. \end{aligned}\right\} \tag{13.21}$$

Further, in the case $0 < \beta(x) - \alpha(x) < \pi$, we have

$$\left.\begin{aligned} \beta\chi_\mu(x) &= \alpha(x) + \frac{1}{2}((2\mu - \operatorname{sgn}\mu)\varepsilon + 1)\pi, \\ \alpha\omega_\mu(x) &= \beta(x) + \frac{1}{2}((2\mu - \operatorname{sgn}\mu)\varepsilon - 1)\pi \end{aligned}\right\} \tag{13.22}$$

and, in the case $-\pi < \beta(x) - \alpha(x) < 0$,

$$\left.\begin{aligned} \beta\chi_\mu(x) &= \alpha(x) + \frac{1}{2}((2\mu - \operatorname{sgn}\mu)\varepsilon - 1)\pi, \\ \alpha\omega_\mu(x) &= \beta(x) + \frac{1}{2}((2\mu - \operatorname{sgn}\mu)\varepsilon + 1)\pi, \\ (x \in j;\quad \varepsilon = \operatorname{sgn}\alpha' &= \operatorname{sgn}\beta';\quad \nu = 0, \pm 1, \pm 2, \ldots;\quad \mu = \pm 1, \pm 2, \ldots). \end{aligned}\right\} \tag{13.23}$$

13.8 Representation of central dispersions and their derivatives by means of phases

The Abel functional equations for the central dispersions obviously give a representation, in the interval j, of the central dispersions in terms of the phases α, β.

The representation of the central dispersions of the first and second kinds is obtained from the formulae (21), and is

$$\phi_\nu(t) = \alpha^{-1}[\alpha(t) + \nu\pi \cdot \operatorname{sgn}\alpha'], \qquad \psi_\nu(t) = \beta^{-1}[\beta(t) + \nu\pi \cdot \operatorname{sgn}\beta'] \qquad (\nu = 0, \pm 1, \pm 2, \ldots). \tag{13.24}$$

A similar representation is obviously possible for the central dispersions of the third and fourth kinds, χ_μ, ω_μ ($\mu = \pm 1, \pm 2, \ldots$), and this is obtained by application of the formulae (22) and (23); an explicit statement of the corresponding formulae is, however, not needed here. These representations have, as immediate consequences, the properties 1–3 of central dispersions given in § 13.1 as well as the continuous differentiability of all central dispersions. For, every first or second phase takes all real values, is continuously increasing or decreasing, and belongs to the class C_3 or C_1.

By differentiation of the Abel functional equations we obtain the following representations, in the interval j, of the derivatives of central dispersions

$$\left.\begin{aligned} \phi_\nu'(t) &= \frac{\alpha'(t)}{\alpha'[\phi_\nu(t)]}, & \psi_\nu'(t) &= \frac{\beta'(t)}{\beta'[\psi_\nu(t)]}, \\ \chi_\mu'(t) &= \frac{\alpha'(t)}{\beta'[\chi_\mu(t)]}, & \omega_\mu'(t) &= \frac{\beta'(t)}{\alpha'[\omega_\mu(t)]} \\ &(\nu = 0, \pm 1, \pm 2, \ldots;\quad \mu = \pm 1, \pm 2, \ldots). \end{aligned}\right\} \tag{13.25}$$

13.9 Structure of the Abel functional equations

Every Abel functional equation for the central dispersions obviously has the form

$$\gamma(X) = \bar{\gamma}(t); \tag{13.26}$$

in which each of the symbols γ, $\bar{\gamma}$ represents a first or second phase of the arbitrary basis (u, v) of the differential equation (q) and X is a central dispersion.

If, conversely, we choose a first or second phase γ of the basis (u, v) and allow $\bar{\gamma}$ to run through all the phases of the first or second phase system of the basis (u, v) then the function

$$X(t) = \gamma^{-1}[\bar{\gamma}(t)] \tag{13.27}$$

runs through all central dispersions of a particular kind κ $(= 1, 2, 3, 4)$. If γ is a first (second) phase and $\bar{\gamma}$ runs through the first or second phase system, then X runs through the central dispersions of the first or fourth (third or second) kinds.

13.10 Representation of the central dispersions by normalized polar functions

For simplicity, we shall restrict ourselves to the representation of the fundamental dispersions, which we again here denote by ϕ, ψ, χ, ω. The extension to central dispersions with arbitrary index presents no difficulty.

1 *Representation of the fundamental dispersions* ϕ, ω *by* 1-*normalized polar functions*

Let $h(\alpha)$ be a 1-normalized polar function of the differential equation (q), and $\alpha(t)$, $\beta(t)$ the corresponding phases. We have therefore $\beta = \alpha + h(\alpha)$ at every point $t \in j$. Because of the oscillatory character of the differential equation (q) the definition interval of h coincides with the interval $(-\infty, \infty)$ and, by (6.30), we have in this interval

1. $h \in C_1$;
2. $n\pi < h < (n + 1)\pi$ (n integral)
3. $h^{\backslash} > -1$.

We now choose a number $t_0 \in j$ and set $\alpha_0 = \alpha(t_0)$, $\alpha' = \alpha'(t_0)$. Then (6.28) shows that at two homologous points $\alpha(t) = \alpha$ and $\alpha^{-1}(\alpha) = t \in j$ we have the formula

$$t = t_0 + \frac{1}{\alpha_0'} \int_{\alpha_0}^{\alpha} \left(\exp 2 \int_{\alpha_0}^{\sigma} \cot h(\rho)\, d\rho \right) d\sigma. \tag{13.28}$$

We apply this formula at the point $\phi(t)$ and make use of the Abel functional equation $\alpha[\phi(t)] = \alpha(t) + \varepsilon\pi$ ($\varepsilon = \operatorname{sgn} \alpha'$). Then, for two homologous numbers t, α, we have

$$\phi(t) = t_0 + \frac{1}{\alpha_0'} \int_{\alpha_0}^{\alpha + \varepsilon\pi} \left(\exp 2 \int_{\alpha_0}^{\sigma} \cot h(\rho)\, d\rho \right) d\sigma, \tag{13.29}$$

which can obviously be written

$$\phi(t) = t + \frac{1}{\alpha_0'} \int_{\alpha}^{\alpha + \varepsilon\pi} \left(\exp 2 \int_{\alpha_0}^{\sigma} \cot h(\rho)\, d\rho \right) d\sigma. \tag{13.30}$$

From this we have, by differentiation

$$\phi'(t) = \exp 2 \int_{\alpha}^{\alpha + \varepsilon\pi} \cot h(\rho)\, d\rho, \tag{13.31}$$

and further

$$\frac{\phi''(t)}{\phi'(t)} = 2\alpha_0'[\cot h(\alpha + \varepsilon\pi) - \cot h(\alpha)] \exp\left(-2\int_{\alpha_0}^{\alpha} \cot h(\rho)\, d\rho\right). \quad (13.32)$$

Similarly, formula (28) gives, for two homologous numbers t, α

$$\omega(t) = t + \frac{1}{\alpha_0'}\int_{\alpha}^{\alpha + h(\alpha) - n\pi - \frac{1}{2}(1-\varepsilon)\pi} \left(\exp 2\int_{\alpha_0}^{\sigma} \cot h(\rho)\, d\rho\right) d\sigma. \quad (13.33)$$

2 *Representation of the fundamental dispersions* ψ, χ *by* 2*-normalized polar functions*

Now let $-k(\beta)$ be a 2-normalized polar function of the differential equation (q), and $\alpha(t)$, $\beta(t)$ the corresponding phases; we have therefore $\alpha = \beta + k(\beta)$ at every point $t \in j$. Because of the oscillatory character of the differential equation (q) the interval of definition of $-k$ coincides with the interval $(-\infty, \infty)$ and we have in this interval (from (6.37))

1. $-k \in C_1$;
2. $n\pi < -k < (n+1)\pi$ (n integral);
3. $-k^{\backslash} < 1$.

We choose an arbitrary number $t_0 \in j$ and put $\beta_0 = \beta(t_0)$, $\alpha_0' = \alpha'(t_0)$. Then (by (6.35)) at any two homologous points $\beta(t) = \beta$ and $\beta^{-1}(\beta) = t \in j$ we have

$$t = t_0 + \frac{1}{\alpha_0'}\int_{\beta_0}^{\beta} [1 + k^{\backslash}(\sigma)] \cdot \exp\left(-2\int_{\beta_0}^{\sigma} [1 + k^{\backslash}(\rho)] \cot k(\rho)\, d\rho\right) d\sigma. \quad (13.34)$$

We apply this formula at the point $\psi(t)$ and make use of the Abel functional equation $\beta[\psi(t)] = \beta(t) + \varepsilon\pi$ ($\varepsilon = \operatorname{sgn} \alpha' = \operatorname{sgn} \beta'$) and so obtain, for any two homologous points t, β

$$\psi(t) = t_0 + \frac{1}{\alpha_0'}\int_{\beta_0}^{\beta + \varepsilon\pi} [1 + k^{\backslash}(\sigma)] \exp\left(-2\int_{\beta_0}^{\sigma} [1 + k^{\backslash}(\rho)] \cot k(\rho)\, d\rho\right) d\sigma. \quad (13.35)$$

This may obviously also be written

$$\psi(t) = t + \frac{1}{\alpha_0'}\int_{\beta}^{\beta + \varepsilon\pi} [1 + k^{\backslash}(\sigma)] \exp\left(-2\int_{\beta_0}^{\sigma} [1 + k^{\backslash}(\rho)] \cot k(\rho)\, d\rho\right) d\sigma, \quad (13.36)$$

and there follows, by differentiation, the result

$$\psi'(t) = \frac{1 + k^{\backslash}(\beta + \varepsilon\pi)}{1 + k^{\backslash}(\beta)} \exp\left(-2\int_{\beta}^{\beta + \varepsilon\pi} [1 + k^{\backslash}(\rho)] \cot k(\rho)\, d\rho\right). \quad (13.37)$$

Similarly, formula (34) gives, for any two homologous numbers t, β, the following formula

$$\chi(t) = t + \frac{1}{\alpha_0'} \int_{\beta}^{\beta + k(\beta) + n\pi + \frac{1}{2}(1+\varepsilon)\pi} [1 + k^{\backprime}(\sigma)] \exp\left(-2 \int_{\beta_0}^{\sigma} [1 + k^{\backprime}(\rho)] \cot k(\rho)\, d\rho\right) d\sigma. \tag{13.38}$$

13.11 Differential equations of the central dispersions

The central dispersions of the first kind of the differential equation (q) satisfy a non-linear differential equation of the third order; the same is true for central dispersions of all higher kinds when the carrier $q \in C_2$. These non-linear differential equations of the third order are, as we shall see, special cases of the Kummer differential equation (11.1), whose significance is fundamental for the theory of transformations. We now wish to derive these third order differential equations from the Abel functional equation.

Let X be a central dispersion of the kind κ $(= 1, 2, 3, 4)$. We choose an arbitrary basis (u, v) of the differential equation (q). Then, as we have seen in (26), we have the Abel functional equation

$$\gamma(X) = \bar{\gamma}(t), \tag{13.39}$$

holding in the interval j for appropriate first or second phases γ, $\bar{\gamma}$ of the basis (u, v).

When $\kappa = 1$ or 2 respectively this equation holds if both γ, $\bar{\gamma}$ are first phases or if both are second phases of the basis (u, v); in the cases $\kappa = 3$ or 4 it holds if γ is a second and $\bar{\gamma}$ a first phase, or γ is a first and $\bar{\gamma}$ a second phase respectively.

From formula (39) it follows that for all $t \in j$, apart from the singular points where the functions $\gamma(X)$, $\bar{\gamma}(t)$ are odd multiples of $\frac{\pi}{2}$,

$$\tan \gamma(X) = \tan \bar{\gamma}(t). \tag{13.40}$$

If the central dispersion X is of class C_3, then at every non-singular point $t \in j$ we can take the Schwarzian derivative of this relation. When we take account of (1.17), this gives

$$-\{X, t\} - \{\tan \gamma, X\} \cdot X'^2(t) = -\{\tan \bar{\gamma}, t\}.$$

Now, from (5.18), (5.24), we have

$$-\{\tan \gamma, t\} = q(t) \qquad \text{or} \qquad -\{\tan \gamma, t\} = \hat{q}_1(t),$$

according as γ is a first or second phase of the basis (u, v) and an analogous formula for $\tan \bar{\gamma}$, the right-hand side of (40). Here, $\hat{q}_1$ denotes of course the carrier of the first associated differential equation $(\hat{q}_1)$ of (q):

$$\hat{q}_1(t) = q(t) + \sqrt{|q(t)|} \left(\frac{1}{\sqrt{|q(t)|}}\right)''. \tag{13.41}$$

We come thus to the following theorem:

Theorem. All the central dispersions ϕ of the first kind, with arbitrary indices, satisfy in the interval j the non-linear third order differential equation

$$-\{\phi, t\} + q(\phi) \cdot \phi'^2(t) = q(t). \qquad \text{(qq)}$$

Moreover, if the carrier $q \in C_2$ all the central dispersions ψ, χ, ω of the second, third and fourth kinds, with arbitrary indices, satisfy in the interval j the equations:

$$-\{\psi, t\} + \hat{q}_1(\psi) \cdot \psi'^2(t) = \hat{q}_1(t). \qquad (\hat{q}_1\hat{q}_1)$$

$$-\{\chi, t\} + \hat{q}_1(\chi) \cdot \chi'^2(t) = q(t), \qquad (\hat{q}_1 q)$$

$$-\{\omega, t\} + q(\omega) \cdot \omega'^2(t) = \hat{q}_1(t). \qquad (q\hat{q}_1)$$

These are the so-called *non-linear third order differential equations for central dispersions*; more precisely for central dispersions of the first, second, third and fourth kinds.

13.12 Solutions of the Abel functional equations with unknown phase functions α, β

The problem of determining the differential equation (q) from a knowledge of its central dispersions leads us to consider the Abel functional equations for the central dispersions with unknown phase functions α, β.

Let ϕ_ν or ψ_ν be a central dispersion of the first or second kind of the differential equation (q). A phase function α, β (§ 5.7) of class C_3 or C_1 respectively, which satisfies the Abel functional equation (21) in the interval j, represents a first or second phase of a differential equation ($\bar{q}$), whose ν-th central dispersion of the first or second kind coincides with the function ϕ_ν or ψ_ν. The carrier $\bar{q}$ is thus determined, respectively, in these two cases by (5.16) or by means of a certain solution of a non-linear second order differential equation (§ 5.12).

Now let χ_μ or ω_μ be a central dispersion of the third or fourth kind of a differential equation (q). If we have two phase functions $\alpha, \beta \in C_3, C_1$ respectively, which are related in the interval j by a formula such as (5.34), and satisfy the Abel functional equation (22), say, then these represent a first and second phase of the same basis of a differential equation ($\bar{q}$), whose μ-th central dispersion of the third or fourth kind coincides with the function χ_μ or ω_μ. The carrier $\bar{q}$ is determined uniquely from the phase α, by the formula (5.16).

We shall restrict ourselves from now on to the solution of the Abel functional equation by phase functions, of class C_3, of a given fundamental dispersion of the first kind.

First we observe that by (12.1) and §§ 13.1, 13.3, 13.4 the fundamental dispersion of the first kind ϕ of every differential equation (q) has the following properties in the interval $j = (a, b)$.

$$\left.\begin{array}{l} 1.\ \phi(t) > t, \\ 2.\ \lim\limits_{t \to a+} \phi(t) = a, \quad \lim\limits_{t \to b-} \phi(t) = b, \\ 3.\ \phi \in C_3, \\ 4.\ \phi'(t) > 0. \end{array}\right\} \qquad (13.42)$$

Now we have the following result due to E. Barvínek [2].

Theorem. Corresponding to every function ϕ defined in the interval j and with properties 1 *to* 4 *above, there exist infinitely many phase functions $\alpha \in C_3$ which are solutions of the Abel functional equation*

$$\alpha(\phi) = \alpha(t) + \pi \cdot \operatorname{sgn} \alpha' \tag{13.43}$$

and which may, moreover, be obtained constructively.

Proof. Let ϕ be a function defined in the interval j with the above properties 1–4. We shall, as an illustration, construct an increasing solution $\alpha \in C_3$ of (43).

We select a number $t_0 \in j$ and put $t_\nu = \phi^\nu(t_0)$ for $\nu = 0, \pm 1, \pm 2, \ldots$. Then the interval j separates into sub-intervals $j_\nu = [t_\nu, t_{\nu+1})$.

Now in the interval j_0 we choose any function $f \in C_3$ with a continuous positive derivative f', which has the following behaviour in a left neighbourhood of t_1:

$$\lim_{t \to t_1 -} f(t) = f(t_0) + \pi,$$

$$\lim_{t \to t_1 -} f'(t) = \frac{f'^+(t_0)}{\phi'(t_0)}.$$

$$\lim_{t \to t_1 -} f''(t) = \frac{1}{\phi'(t_0)} \left(\frac{f'^+(t)}{\phi'(t)}\right)'^+_{t_0},$$

$$\lim_{t \to t_1 -} f'''(t) = \frac{1}{\phi'(t_0)} \left[\frac{1}{\phi'(t)} \left(\frac{f'^+(t)}{\phi'(t)}\right)'^+\right]'^+_{t_0}.$$

Here, the $+$ symbol indicates a right derivative so that, for instance, $f'^+(t_0)$ is the right derivative of f at the point t_0.

By means of this function f, we can now define in the interval j the function α as follows

$$\alpha(t) = \begin{cases} f(t) & \text{for} \quad t \in j_0, \\ \alpha[\phi^{-1}(t)] + \pi & \text{for} \quad t \in j_\nu, \quad \nu > 0, \\ \alpha[\phi(t)] - \pi & \text{for} \quad t \in j_\nu, \quad \nu < 0. \end{cases}$$

This function α is obviously an increasing phase function $\in C_3$ and satisfies the Abel functional equation (43), so the proof is complete.

We know that when a first phase α of the differential equation (q) is given, the carrier q is uniquely determined by the formula (5.16). From this fact and the above result we expect that to every function ϕ defined in the interval j with the properties (42) there will correspond in general infinitely many oscillatory differential equations (q) having the function ϕ as fundamental dispersion of the first kind. Later, (§ 15.10), we shall show that the power of the set of all differential equations (q) in the interval $(-\infty, \infty)$ with the same fundamental dispersion of the first kind ϕ is independent of the choice of the latter and is equal to the power $\aleph$ of the continuum.

For the solution of the Abel functional equations, by means of phase functions $\in C_1$, with given fundamental dispersions of the second, third or fourth kinds, we refer to the papers by J. Chrasitna [38] and F. Neuman [54]. In general, the situation is as follows:

Let $\lambda(t)$ be a function defined in the interval $j = (a, b)$ with the following properties

$$\left.\begin{array}{l} 1.\ \lambda(t) > t, \\ 2.\ \lim_{t\to a+} \lambda(t) = a, \quad \lim_{t\to b-} \lambda(t) = b, \\ 3.\ \lambda \in C_1, \\ 4.\ \lambda'(t) > 0. \end{array}\right\} \tag{13.44}$$

(a) There are in the interval j infinitely many oscillatory differential equations (q) with $q < 0$, whose fundamental dispersion of the second kind ψ coincides with λ; $\psi(t) = \lambda(t)$ for $t \in j$.

(b) Let $t_0 \in j$ be an arbitrary number. There are in the interval $[t_0, b)$ infinitely many right oscillatory differential equations (q) with $q < 0$, whose fundamental dispersion χ of the third kind coincides with λ: $\chi(t) = \lambda(t)$ for $t \in [t_0, b)$.

(c) Let $t_0 \in j$ be an arbitrary number. There are in the interval $(a, t_0]$ infinitely many left oscillatory differential equations (q) with $q < 0$, whose fundamental dispersion ω of the fourth kind coincides with λ in the interval $(a, \lambda^{-1}(t_0)]$: $\omega(t) = \lambda(t)$ for $t \in (a, \lambda^{-1}(t_0)]$.

It should be noted that problems of this kind are associated with the solutions of appropriate non-linear differential equations of the second order with delayed argument.

13.13 Consequences of the above results

Now we derive some consequences of the above theory.

2. *Monotonic character of the differences.* $\phi_\nu(t) - t$, $\psi_\nu(t) - t$, $\chi_\nu(t) - t$, $\omega_\nu(t) - t$. Consider the differences

$$\phi_n(t) - t, \quad \psi_n(t) - t, \quad \chi_n(t) - t, \quad \omega_n(t) - t, \tag{13.45}$$

$$\phi_{-n}(t) - t, \quad \psi_{-n}(t) - t, \quad \chi_{-n}(t) - t, \quad \omega_{-n}(t) - t \tag{13.46}$$

$$(n = 1, 2, \ldots)$$

in the interval j.

From the formulae in § 13.6 we deduce that if the carrier q is a non-increasing or a decreasing function in the interval j, then the quantities

$$\phi_n'(t) - 1, \quad \psi_n'(t) - 1, \quad \chi_n'(t) - 1, \quad \omega_n'(t) - 1$$

are respectively $\leqslant 0$ or < 0, and the quantities

$$\phi_{-n}'(t) - 1, \quad \psi_{-n}'(t) - 1, \quad \chi_{-n}'(t) - 1, \quad \omega_{-n}'(t) - 1$$

are $\geqslant 0$ or > 0 respectively.

Hence

If the carrier q is a non-increasing or a decreasing function in the interval j, then the same is true of all the differences (45), *while the differences* (46) *are non-decreasing or increasing functions.*

Similarly we show that

If the carrier q is a non-decreasing or an increasing function in the interval j, then the same is true of the differences (45), *while the differences* (46) *are non-increasing or decreasing functions.*

Corollary. Let

$$\cdots < t_{-2} < t_{-1} < t_0 < t_1 < t_2 < \cdots$$

be the sequence of zeros of any integral of the differential equation (q) and let

$$\cdots < t_{-2m} < t_{-m} < t_0 < t_m < t_{2m} < \cdots \quad (m \geqq 1) \qquad (13.47)$$

be a subsequence of it.

If the carrier q in the interval j is non-increasing or if it is decreasing, then (47) denotes respectively a concave or a strictly concave sequence. If the carrier q in the interval j is non-decreasing or increasing then (47) is a convex or a strictly convex sequence. For the case $m = 1$ this gives a result of Sturm-Szegö.

An analogous result holds for the sequence of zeros of the derivative of an arbitrary integral of the differential equation (q).

2. *Derivatives of composite functions.* According to a classical result, the composition of two or more functions of a class C_k ($k = 0, 1, \ldots$) is a function of the same class. It can happen, however, that the function arising from such composition belongs to a higher class than the original functions. Our results on derivatives of central dispersions lead to many situations of this kind. We shall content ourselves with a few remarks on this topic, since a deeper investigation would be beyond the scope of this book.

We shall show that:

Let q be a negative continuous function in the interval $j = (a, b)$, such that the differential equation (q) *is oscillatory. Then there are in the interval j two functions X, Y satisfying the inequality $t < X(t) < Y(t)$, such that the function $q[X(t)]/q[Y(t)]$ belongs to the class C_2. If the function q is strictly monotonic, then there are continuous functions X, Y with the property quoted.*

For we obtain functions X, Y of the kind described if we evaluate $X(t)$, $Y(t)$ at every point $t \in j$ according to the theorem of § 13.6, taking $X(t) = t_1$, $Y(t) = t_3$; $t < t_1 < t_3$. The function $q[X(t)]/q[Y(t)]$ $(= \phi'(t))$ has at the point t a continuous derivative of the second order, as we know from § 13.4. If the function q is strictly monotonic, then it follows from the result in § 13.6 and the relationship $\omega\chi = \phi$ that

$$X(t) = q^{-1}[q[\chi(t)] \cdot \chi'(t)]; \qquad Y(t) = q^{-1}[q[\chi(t)] : \omega'[\chi(t)]],$$

where naturally q^{-1} denotes the inverse function of q. It follows from these formulae that the functions X, Y are continuous in the interval j.

Special problems of central dispersions

This chapter is devoted to a study of special problems arising in the theory of linear oscillatory differential equations of the second order. We shall be concerned with problems which are related to the concept of central dispersions and which can be solved by application of the theory developed in Chapter A.

14 Extension of solutions of a differential equation (q) and their derivatives

In this paragraph we shall continue to make the previous assumptions, namely that (q) is oscillatory, $j = (a, b)$ and $q < 0$ for all $t \in j$. The last assumption will however not be needed in § 14.1 (on extension of solutions) but is first required in § 14.2 (on extension of derivatives of solutions).

14.1 Extension of solutions of the differential equation (q)

The elementary theory of linear differential equations of the second order shows that for every integral v of (q) the function defined by $v(t) \int_x^t d\sigma/v^2(\sigma)$ is a solution of (q) independent of v, in a neighbourhood of every point $x \in j$ which is not a zero of the integral v. (§ 1.2). We now wish to extend this solution over the entire interval j, in terms of values of the integral v.

Let, therefore, v be an arbitrary integral of the differential equation (q), and let t_0 denote a zero of v. We denote by

$$\cdots < t_{-2} < t_{-1} < t_0 < t_1 < t_2 < \cdots \tag{14.1}$$

the set of zeros of v. Then in the above notation we have $v(t_\nu) = 0$, $t_\nu = \phi_\nu(t_0)$; $\nu = 0, \pm 1, \pm 2, \ldots$.

Let $j_\nu = (t_\nu, t_{\nu+1})$. Moreover let $x_0 \in j_0$ be an arbitrary number and $x_\nu = \phi_\nu(x_0)$; we have therefore $x_\nu \in j_\nu$.

Now we define, in the interval j, a function u, which we shall conveniently denote by

$$v(t) \int_{(x)}^t \frac{d\sigma}{v^2(\sigma)} \tag{14.2}$$

as follows:

$$u(t) = \begin{cases} v(t) \displaystyle\int_{x_\nu}^t \frac{d\sigma}{v^2(\sigma)} & \text{for} \quad t \in j_\nu, \\ -\dfrac{1}{v'(t_\nu)} & \text{for} \quad t = t_\nu. \end{cases} \tag{14.3}$$

We note first that the function u represents a solution of the differential equation (q) in every interval j_ν, and in fact the solution with the initial values

$$u(x_\nu) = 0, \qquad u'(x_\nu) = \frac{1}{v(x_\nu)}. \tag{14.4}$$

Further, u clearly satisfies the limiting conditions

$$\lim_{t \to t_\nu -} u(t) = -\frac{1}{v'(t_\nu)} = \lim_{t \to t_\nu +} u(t).$$

Hence the function u is everywhere continuous and in every interval j_ν clearly represents the solution of the differential equation (q) determined by the initial values (4).

Now let $U(t)$, $t \in j$, be the integral of the differential equation (q) determined by the initial values

$$U(x_0) = 0, \qquad U'(x_0) = \frac{1}{v(x_0)}.$$

Then, at every point x_ν,

$$U(x_\nu) = 0,$$

and further, from (13.5),

$$U'(x_\nu) = U'[\phi_\nu(x_0)] = (-1)^\nu \frac{U'(x_0)}{\sqrt{\phi_\nu(x_0)}} =$$

$$\frac{1}{(-1)^\nu v(x_0)\sqrt{\phi_\nu(x_0)}} = \frac{1}{v[\phi_\nu(x_0)]} = \frac{1}{v(x_\nu)}.$$

Consequently the integral U and its derivative U' take the same values at the point x_ν as the functions u, u', hence the functions u, U coincide in every interval j_ν. We have, therefore, $u(t) = U(t)$ in the entire interval j, with the possible exceptions of the points t_ν. But by the continuity of the functions u, U in the interval j it follows that $u(t) = U(t)$ at each point t_ν; hence the function u is an integral of the differential equation (q) in the interval j.

To sum up:

The function

$$u(t) = v(t) \int_{(x)}^{t} \frac{d\sigma}{v^2(\sigma)}$$

represents in the interval j the integral of the differential equation (q) *determined by the initial values* $u(x_0) = 0$, $u'(x_0) = \dfrac{1}{v(x_0)}$. *The integrals u, v are independent; the Wronskian of the basis* (u, v) *has the value* -1.

A consequence of this result is worth noting. Every (first) phase α of the basis (u, v) satisfies the relationship

$$\tan \alpha(t) = \int_{(x)}^{t} \frac{d\sigma}{v^2(\sigma)}$$

in the interval j, with the exception of the points t_ν. Consider, in particular, the phase α_0 with the zero x_0. This is obviously given by

$$\alpha_0(t) = \text{Arctan} \int_{(x)}^{t} \frac{d\sigma}{v^2(\sigma)}, \qquad \alpha_0(t_\nu) = (2\nu - 1)\frac{\pi}{2}$$

where the symbol Arctan denotes that branch of the function, in the interval j_ν, which takes the value $\nu\pi$ at the point x_ν. The initial values of the phase α_0 at the point x_0 are

$$\alpha_0(x_0) = 0, \qquad \alpha_0'(x_0) = \frac{1}{v^2(x_0)}, \qquad \alpha_0''(x_0) = -2\frac{v'(x_0)}{v^2(x_0)}.$$

The formula (5.18) gives

$$q(t) = -\left\{\int_{(x)} \frac{d\sigma}{v^2(\sigma)}, t\right\}.$$

14.2 Extension of derivatives of solutions of the differential equation (q)

Again, let v be an arbitrary integral of the differential equation (q) and let t_0' be a zero of its derivative v'. Analogously to the above study, we define $v'(t_\nu') = 0$, $t_\nu' = \varphi_\nu(t_0')$; $j_\nu' = (t_\nu', t_{\nu+1}')$; $\nu = 0, \pm 1, \pm 2, \ldots$. We choose an arbitrary number $x_0' \in j_0'$ and set $x_\nu' = \varphi_\nu(x_0')$. Our supposition that $q < 0$ for all $t \in j$ implies that $x_\nu' \in j_\nu'$.

In the interval j we define the function u', which we conveniently denote by

$$v'(t) \int_{(x')}^{t} \frac{q(\sigma)}{v'^2(\sigma)}\, d\sigma$$

as follows:

$$u'(t) = \begin{cases} v'(t) \displaystyle\int_{x_\nu'}^{t} \frac{q(\sigma)}{v'^2(\sigma)}\, d\sigma & \text{for} \quad t \in j_\nu', \\ -\dfrac{1}{v(t_\nu')} & \text{for} \quad t = t_\nu'. \end{cases}$$

Then we show similarly that:

The function

$$u'(t) = v'(t) \int_{(x')}^{t} \frac{q(\sigma)}{v'^2(\sigma)}\, d\sigma$$

represents in the interval j the derivative of the integral u of (q) *determined by the initial values $u(x_0') = \dfrac{1}{v'(x_0')}$, $u'(x_\nu') = 0$. The integrals u, v are independent, the Wronskian of the basis (u, v) being equal to* 1.

15 Differential equations with the same central dispersions of the first kind

In § 13.12, we stated that to every function ϕ defined in the interval $(j =)$ $(-\infty, \infty)$, with the properties (13.42), there correspond infinitely many oscillatory differential equations (q), (with power $\aleph$), whose fundamental dispersion of the first kind is precisely the function ϕ.

The set of all oscillatory differential equations (q) defined in an interval $j = (a, b)$ is partitioned into classes, each being formed of all differential equations (q) with the same fundamental dispersion of the first kind. Now, for a given oscillatory differential equation (q), every 1-central dispersion ϕ_ν is obtained by iteration of the corresponding fundamental dispersion of the first kind or its inverse, i.e. $\phi_\nu = \phi_1^\nu$ for $\nu = 0, \pm 1, \pm 2, \ldots$ (12.2). It follows that for all differential equations (q) contained in a given class the 1-central dispersions ϕ_ν are the same.

In what follows we shall be concerned with differential equations (q) with the same fundamental dispersion of the first kind. For brevity, instead of speaking of differential equations (q), ($\bar{q}$), with the same fundamental dispersion of the first kind we shall speak of carriers q, $\bar{q}$ with the same fundamental dispersion; also, instead of ϕ_1 we shall write simply ϕ.

We let (q) be, therefore, an oscillatory differential equation in the interval $j = (a, b)$.

15.1 Integral strips

Let ϕ be a function defined in the interval j with the properties (13.42).

By $Q\phi$, or briefly Q, we denote the set of all carriers with the same fundamental dispersion ϕ, and by $(Q\phi)$, more briefly (Q), the set of all differential equations (q), $q \in Q\phi$. All differential equations (q) $\in$ (Q) have therefore the same 1-central dispersions ϕ_ν.

Further, let $J\phi$, more briefly J, denote the set of all integrals of all differential equations (q) contained in the set (Q). Since these differential equations (q) all have the same 1-central dispersions ϕ_ν, J consists of functions with the same zeros; that is to say, two arbitrary integrals y, $\bar{y} \in J$, having one zero in common, have all their zeros in common.

Let $c \in j$ be an arbitrary number. By the *integral strip* of the set (Q) with the *node* c, or more briefly the integral strip (c), we mean the set of all elements of J which vanish at the point c, and we denote this set by Bc. The integral strip Bc thus consists of all integrals vanishing at the point c, of all differential equations (q) which have the common fundamental dispersion of the first kind ϕ. The elements of Bc have therefore the same zeros; these zeros are called the *nodes* of the integral strip Bc. Clearly, the integral strip Bc is uniquely determined by any one of its nodes c': i.e. $Bc = Bc'$.

15.2 Statement of the problem

We are now in a position to describe more precisely the contents of this paragraph; it consists essentially of a study of the following topics:

1. Properties of the integral strips of the set $(Q\phi)$.
2. Relations between carriers with the same fundamental dispersion ϕ; that is, between functions $q, \bar{q} \in Q\phi$.
3. Explicit formulae for carriers with the same fundamental dispersion ϕ.
4. The power of the set $Q\phi$ in the case $j = (-\infty, \infty)$.

15.3 Properties of the integral strips of the set $(Q\phi)$

Let Bc be an integral strip of the set $(Q\phi)$. First we show that:

For every integral $y \in Bc$ the following formulae hold:

$$\int_x^{\phi(x)} \left[\frac{y'^2(c)}{y^2(\sigma)} - \frac{1}{(\sigma - c)^2}\right] d\sigma = \frac{1}{c - x} + \frac{1}{\phi(x) - c}; \tag{15.1}$$

$$\int_c^{\phi(c)} \left[\frac{y'^2(c)}{y^2(\sigma)} - \frac{1}{(\sigma - c)^2} - \frac{1}{(\sigma - \phi(c))^2}\right] d\sigma = \frac{1 + \phi'(c)}{\phi(c) - c} - \frac{1}{2}\frac{\phi''(c)}{\phi'(c)}; \tag{15.2}$$

where x is an arbitrary number satisfying the inequalities $x < c < \phi(x)$.

For, let $(q) \in (Q\phi)$ be an arbitrary differential equation. We have already derived the formula (1) (5.45), and so have only to prove (2).

Let $y \in Bc$ be an integral of the differential equation (q). We choose an arbitrary number t, such that $c < t < \phi(c)$, and apply the formulae (5.43), (5.42) on the intervals $[c, t]$ and $[t, \phi(c)]$

$$\int_c^t \left[\frac{y'^2(c)}{y^2(\sigma)} - \frac{1}{(\sigma - c)^2}\right] d\sigma = -\cot \alpha_0(t) + \frac{1}{t - c},$$

$$\int_t^{\phi(c)} \left[\frac{y'^2\phi(c)}{y^2(\sigma)} - \frac{1}{(\sigma - \phi(c))^2}\right] d\sigma = \cot \alpha_1(t) - \frac{1}{t - \phi(c)}.$$

Here α_0 and α_1 are the first phases of the differential equation (q) determined by the initial values:

$$\alpha_0(c) = 0, \quad \alpha_0'(c) = 1, \quad \alpha_0''(c) = 0; \quad \alpha_1\phi(c) = 0, \quad \alpha_1'\phi(c) = 1, \quad \alpha_1''\phi(c) = 0.$$

From the Abel functional equation $\alpha_1\phi = \alpha_1 + \pi$ and from the derivative of this relation, we see that at the point c the functions $\alpha_1, \alpha_1', \alpha_1''$ take the following values: $\alpha_1(c) = -\pi$, $\alpha_1'(c) = \phi'(c)$, $\alpha_1''(c) = \phi''(c)$.

Then, from (5.39), we see that the values taken by the phases $\alpha_0(t)$, $\alpha_1(t)$ at the point t are related by:

$$-\cot \alpha_0(t) + \phi'(c) \cdot \cot \alpha_1(t) = -\frac{1}{2}\frac{\phi''(c)}{\phi'(c)}. \tag{15.3}$$

The above formulae, which can also be written in the form,

$$\int_c^t \left[\frac{y'^2(c)}{y^2(\sigma)} - \frac{1}{(\sigma - c)^2} - \phi'(c)\frac{1}{(\sigma - \phi(c))^2}\right] d\sigma$$

$$= -\cot \alpha_0(t) + \frac{1}{t - c} + \phi'(c)\left[\frac{1}{t - \phi(c)} - \frac{1}{c - \phi(c)}\right],$$

$$\int_t^{\phi(c)} \left[\frac{y'^2(c)}{y^2(\sigma)} - \phi'(c)\frac{1}{(\sigma - \phi(c))^2} - \frac{1}{(\sigma - c)^2}\right] d\sigma$$

$$= \phi'(c)\left[\cot \alpha_1(t) - \frac{1}{t - \phi(c)}\right] + \frac{1}{\phi(c) - c} - \frac{1}{t - c}$$

give, by addition,

$$\int_c^{\phi(c)} \left[\frac{y'^2(c)}{y^2(\sigma)} - \frac{1}{(\sigma - c)^2} - \phi'(c)\frac{1}{(\sigma - \phi(c))^2}\right] d\sigma$$

$$= -\cot \alpha_0(t) + \phi'(c)\cdot \cot \alpha_1(t) + \frac{1}{\phi(c) - c}[1 + \phi'(c)],$$

and when we apply (3) to this, it gives (2).

As a supplement to this result we remark that by using the relations (5.46), or (5.49) we can derive formulae which generalize (1), (2) above to the case of 1-central dispersions ϕ_v with arbitrary index v $(= 0, \pm 1, \pm 2, \ldots)$ or derive similar formulae for the 2-central dispersions ψ_v. We shall not concern ourselves further with this, since the relations (1), (2) are sufficient for our purposes.

We now have some further results:

The (Riemann) integrals which occur on the left of the formulae (1), (2) *are independent of the elements y of the integral strip Bc.*

In other words: these integrals are invariant with respect to the elements y of the integral strip Bc.

A further property of the integral strip Bc is that *the ratio of the derivatives $y', \bar{y}'$ of two arbitrary elements y, $\bar{y}$ in the integral strip Bc takes the same value* $(= k)$ *at all nodes of this strip.*

For, let c' be an arbitrary node of Bc. Then we have $c' = \phi_v(c)$, v being some integer, and (13.5) gives

$$(-1)^v \frac{y'(c)}{y'(c')} = \sqrt{\phi_v'(c)} = (-1)^v \frac{\bar{y}'(c)}{\bar{y}'(c')},$$

which proves our statement.

We show further that:

Every two elements y, $\bar{y}$ of the integral strip Bc are related in the interval j by:

$$\int_t^{\phi(t)} \left[\frac{y'^2(c)}{y^2(\sigma)} - \frac{\bar{y}'^2(c)}{\bar{y}^2(\sigma)}\right] d\sigma = 0. \tag{15.4}$$

To establish this, we first observe that formulae (1) and (2) show that (4) is valid if the number t $(= x)$ satisfies the inequalities $t \leqslant c < \phi(t)$. Now let t be an arbitrary number $\in j$; obviously, there is one node c' of Bc which satisfies the inequalities $t \leqslant c' < \phi(t)$, so that formula (4) holds with c' in place of c. Now, from the above results we have $y'^2(c') = \lambda y'^2(c)$, $\bar{y}'^2(c') = \lambda \bar{y}'^2(c)$, in which λ (> 0) is some constant, and this fact completes the proof.

15.4 A sufficient condition for two differential equations to have the same fundamental dispersion

We now wish to examine how far the above properties characterize the integral strips of the set $(Q\phi)$; to that end, let us consider two oscillatory differential equations (q), ($\bar{\text{q}}$) in the interval $j = (a, b)$ with the fundamental dispersions ϕ, $\bar{\phi}$ of the first kind.

We assume that the differential equations (q), ($\bar{\text{q}}$) possess integrals y, $\bar{y}$ which have the same zeros and also the property that the ratio of their derivatives $y'/\bar{y}'$ ($= k$) is the same at each of these zeros x. Finally, let at least one (always the same one) of the two relations

$$\int_t^{\phi(t)} \left[\frac{k}{y^2(\sigma)} - \frac{1}{k}\frac{1}{\bar{y}^2(\sigma)}\right] d\sigma = 0, \qquad \int_t^{(\bar{\phi}t)} \left[\frac{k}{y^2(\sigma)} - \frac{1}{k}\frac{1}{\bar{y}^2(\sigma)}\right] d\sigma = 0. \quad (15.5)$$

be satisfied at every $t \in j$ except for the zeros x.

Then the fundamental dispersions ϕ, $\bar{\phi}$ coincide, $\phi = \bar{\phi}$, and consequently y, $\bar{y}$ are elements of the integral strip Bx of the set $(Q\phi)$.

To show this, let us assume that the first relation (5) holds and let $t \in j$ be arbitrary. If t is a (common) zero of the integrals y, $\bar{y}$, then from the definition of ϕ, $\bar{\phi}$ it follows immediately that $\phi(t) = \bar{\phi}(t)$. We therefore assume that $y(t) \neq 0$, $\bar{y}(t) \neq 0$.

Let c_{-1}, c, c_1 be successive zeros of y and $\bar{y}$ determined by the inequalities $c_{-1} < t < c < c_1$; then $\phi(t)$, $\bar{\phi}(t)$ lie between c and c_1.

From (5) we have

$$\int_t^{\phi(t)} \left[\frac{y'^2(c)}{y^2(\sigma)} - \frac{1}{(\sigma - c)^2}\right] d\sigma = \int_t^{\phi(t)} \left[\frac{\bar{y}'^2(c)}{\bar{y}^2(\sigma)} - \frac{1}{(\sigma - c)^2}\right] d\sigma.$$

From this and (5.44) it follows that $\cot \bar{\alpha}\phi(t) = \cot \bar{\alpha}(t)$; $\bar{\alpha}$ being the phase of the differential equation ($\bar{\text{q}}$) determined by the initial values $\bar{\alpha}(c) = 0$, $\bar{\alpha}'(c) = 1$, $\bar{\alpha}''(c) = 0$. We have therefore $\bar{\alpha}\phi(t) = \bar{\alpha}(t) + m\pi$, m being an integer. But since $\phi(t)$ lies between c and c_1, $m = 1$, and consequently $\bar{\alpha}\phi(t) = \bar{\alpha}(t) + \pi$. When we compare this relationship with the Abel functional equation $\bar{\alpha}\bar{\phi}(t) = \bar{\alpha}(t) + \pi$, then we obtain $\bar{\phi}(t) = \phi(t)$, and the proof is complete.

15.5 Ratios of elements of an integral strip

We consider again an integral strip Bc of the set $(Q\phi)$. Let $y, \bar{y} \in Bc$ be arbitrary elements of Bc and w the Wronskian of y, $\bar{y}$:

$$w = y\bar{y}' - y'\bar{y}. \quad (15.6)$$

At every point $t \in j$ the function w has the derivative

$$w' = (\bar{q} - q)y\bar{y}. \tag{15.7}$$

Clearly, the functions w, w' vanish at every node c_ν $(= \phi_\nu(c))$ of Bc, $(\nu = 0, \pm 1, \pm 2, \ldots; c_0 = c)$, i.e.

$$w(c_\nu) = 0, \qquad w'(c_\nu) = 0. \tag{15.8}$$

Further, from (13.5) and (7) we have

$$w\phi_\nu = w, \tag{15.9}$$

$$(\bar{q}\phi_\nu - q\phi_\nu)\phi_\nu'^2 = \bar{q} - q. \tag{15.10}$$

Now we define the function p in the interval j by:

$$p(t) = \begin{cases} \dfrac{\bar{y}(t)}{y(t)} & \text{for} \quad t \neq c_\nu; \quad \nu = 0, \pm 1, \pm 2, \ldots; \quad c_0 = c, \\[2mm] \dfrac{\bar{y}'(c)}{y'(c)} & \text{for} \quad t = c_\nu, \end{cases} \tag{15.11}$$

and the remainder of this paragraph is devoted to studying the properties of the function p.

First, from (13.5), we have the following relationship

$$p\phi_\nu(t) = p(t), \tag{15.12}$$

holding in the interval j; further, the function p is either always positive or always negative, according as $\bar{y}'(c)/y'(c) > 0$ or < 0. For let us assume, for definiteness, that $\bar{y}'(c)/y'(c) > 0$. Then both the functions y, $\bar{y}$ are positive or both are negative in the interval $(c, \phi(c))$, and consequently p is positive. Thus $p(t) > 0$ for $t \in [c, \phi(c)]$ and consequently, from (12), for all $t \in j$.

Also, the function p is continuous in the interval j. This follows from the fact that by definition it is continuous at every point $t \neq c_\nu$ and for $t \to c_\nu$ it tends to the limit $p(c_\nu)$ $(= p(c))$.

The function p belongs, indeed, to the class C_2. Obviously it is twice continuously differentiable at every point $t \neq c_\nu$, since

$$p' = \frac{w}{y^2} \tag{15.13}$$

$$p'' = (\bar{q} - q)p - 2\frac{y'}{y}p'. \tag{15.14}$$

Further, applying L'Hôpital's rule,

$$\lim_{t \to c_\nu} p'(t) = 0, \qquad \lim_{t \to c_\nu} p''(t) = \frac{1}{3}[\bar{q}(c_\nu) - q(c_\nu)]p(c_\nu), \tag{15.15}$$

whence the functions p, p' possess derivatives $p'(c_\nu)$, $p''(c_\nu)$ at the point c_ν which are equal to the limiting values (15) and our statement is proved.

Moreover, from (4) and (11), we have

$$\int_t^{\phi(t)} \left[\frac{1}{p^2(\sigma)} - \frac{1}{p^2(c)}\right] \frac{d\sigma}{y^2(\sigma)} = 0 \tag{15.16}$$

in the interval j. Finally we note that the integrand in (16) is continuous in the interval j if we define its value at each point c_ν by the following limit:—

$$\lim_{\sigma \to c_\nu} \left[\frac{1}{p^2(\sigma)} - \frac{1}{p^2(c)}\right] \frac{1}{y^2(\sigma)} = -\frac{p''(c_\nu)}{p^3(c_\nu) y'^2(c_\nu)}. \tag{15.17}$$

Coordinating the above facts, the function p has the following properties:—

$$\left.\begin{array}{l} 1.\ p \neq 0 \quad \text{for} \quad t \in j; \\ 2.\ p\phi(t) = p(t) \quad \text{for} \quad t \in j; \\ 3.\ p \in C_2; \\ 4.\ p'(c) = 0; \\ 5.\ \displaystyle\int_c^{\phi(c)} \left[\frac{1}{p^2(\sigma)} - \frac{1}{p^2(c)}\right] \frac{d\sigma}{y^2(\sigma)} = 0. \end{array}\right\} \tag{15.18}$$

15.6 Relations between carriers with the same fundamental dispersion ϕ

We consider arbitrary elements $q, \bar{q}$ of the set $Q\phi$; thus the differential equations (q) ($\bar{\text{q}}$) have the same fundamental dispersion ϕ. For brevity we write $\Delta = \bar{q} - q$.

Let $c \in j$ be arbitrary. We shall be concerned with the number of zeros of the function Δ lying between two neighbouring nodes of the integral strip Bc; the possibility of an infinite number of such zeros is not excluded.

Formula (10) shows that the relation $\Delta(c) = 0 \Rightarrow \Delta(c_\nu) = 0$ for every node c_ν of Bc. In other words, if the carriers $q, \bar{q}$ take equal values at the point c, then they take equal values at every node of Bc.

We next show that the number of zeros of Δ lying between two neighbouring nodes of Bc is always the same. For, let c, $\phi(c)$ and c', $\phi(c')$ be two neighbouring nodes of Bc and suppose, for definiteness, that $c < c'$. Then we have $c' = \phi_\nu(c)$, $\phi(c') = \phi_\nu\phi(c)$ for some positive index ν. Since the function ϕ_ν is increasing, it maps the interval $(c, \phi(c))$ simply onto $(c', \phi(c'))$. But formula (10) shows that in this mapping zeros of Δ are mapped onto zeros, and the result follows.

Clearly, the number of points lying between two neighbouring nodes of the integral strip Bc at which the carriers $q, \bar{q}$ take the same value is always the same, and consequently does not depend upon the choice of these nodes.

We now wish to prove the following important theorem:

Theorem. The number of the zeros of the function Δ lying in an arbitrary interval $[c, \phi(c))$, $c \in j$, is always at least four.

Proof. Let $y, \bar{y} \in Bc$ be arbitrary elements of Bc and p the function defined in the interval j by means of the formula (11).

First, the relations (18), 1° and 5° show that the function p takes the value $p(c)$ at a point $x \in (c, \phi(c))$. Consequently, the function p assumes the same value $p(c)$ at the points c, x and $\phi(c)$ which are such that $c < x < \phi(c)$. Hence its derivative p' vanishes at least at two points $x_1' \in (c, x)$, $x_2' \in (x, \phi(c))$. Now, from (13), x_1' and x_2' are zeros of the function w. Consequently, w vanishes at points x_1', x_2' such that $c < x_1' < x_2' < \phi(c)$; its derivative w' therefore vanishes at least at three points $x_1 \in (c, x_1')$, $x_2 \in (x_1', x_2')$, $x_3 \in (x_2', \phi(c))$.

From (7) the numbers x_1, x_2, x_3 are zeros of the function Δ. This completes the first part of our proof.

In the second part we show that: if the function $\Delta \neq 0$ at the point c, and consequently also at $\phi(c)$, then it has not less than four zeros in the interval $(c, \phi(c))$.

We assume that $\Delta(c) \neq 0$, and that there are precisely three zeros $x_1 < x_2 < x_3$ of Δ lying in the interval $(c, \phi(c))$. Then we have

$$c < x_1 < x_2 < x_3 < \phi(c) < \phi(x_1) < \phi(x_2) < \phi(x_3)$$

and between any two neighbouring elements in this sequence the function Δ is not zero. Consider now the integral strip Bx_1. Clearly, between two neighbouring nodes x_1, $\phi(x_1)$ of Bx_1 there lie precisely two zeros x_2, x_3 of Δ. This completes the proof.

This result may also be formulated as follows:

Two arbitrary carriers q, $\bar{q}$ with the same fundamental dispersion of the first kind ϕ take the same value at not less than four points in each interval $[c, \phi(c))$, $c \in j$.

We shall soon show (§ 15.8) that this theorem is best-possible in the sense that *there are carriers q, $\bar{q}$ with the same fundamental dispersion ϕ, for which the lower bound 4 in the above theorem is obtained.*

15.7 Explicit formula for carriers with the same fundamental dispersion ϕ

Let q be the carrier of an oscillatory differential equation (q) in the interval $j = (a, b)$ and ϕ its fundamental dispersion of the first kind. Let $c \in j$ be arbitrary and $y \in Bc$ an arbitrary element of the integral strip Bc of the set $(Q\phi)$. Then y is an integral of the differential equation, belonging to the set $(Q\phi)$, vanishing at the point c; its zeros, which are therefore the nodes of Bc, are $c_\nu = \phi_\nu(c)$; $\nu = 0, \pm 1, \pm 2, \ldots$, $c_0 = c$.

We now have the following theorem:—

All carriers $\bar{q}$ with the fundamental dispersion ϕ are given by the formula

$$\bar{q} = q + \frac{p''}{p} + 2\frac{y'}{p} \cdot \frac{p'}{y}. \tag{15.19}$$

In this, p is an arbitrary function defined in the interval j with the properties (18), *and the value of the last term is defined at each point c_ν to be $2p''(c_\nu)/p(c)$.*

Proof. (a) Let $\bar{q}$ be an arbitrary carrier with the fundamental dispersion ϕ, so that $q, \bar{q} \in Q\phi$.

Moreover, let $\bar{y}$ be an integral of the differential equation $(\bar{\text{q}})$ contained in the integral strip Bc, and p the function defined in the interval j by the expression (11).

The function p has therefore the properties (18), and at every point $t \in j$ the relation (14) holds; the formula (19) then follows immediately.

(b) Now let $\bar{q}$ be a function defined in the interval j by (19) where of course p is a function with the properties (18) and the value of the last term at each point c_ν is as explained above.

By elementary calculation, we show that the function

$$\bar{y}(t) = p(t)y(t)$$

is a solution of the differential equation ($\bar{\text{q}}$), and indeed is the (unique) solution determined by the intial values $\bar{y}(c) = 0$, $\bar{y}'(c) = p(c)y'(c)$. The functions y, $\bar{y}$ obviously have the same zeros $c_\nu = \phi_\nu(c)$ ($\nu = 0, \pm 1, \pm 2, \ldots$; $c_0 = c$) and from the relationship $\bar{y}' = p'y + py'$ it is clear that the ratio $y'/\bar{y}'$ of their derivatives is the same at all these zeros, taking the value $1/p(c)$.

Now let $F(\sigma)$ denote, in the interval j, the integrand of (18), 5°; its value at each point c_ν is specified as the limiting value (17). This function F is continuous in j. Further, from (18) property 2° and (13.5), we see that the following relationship holds for all $t \in j$:

$$F[\phi(t)]\phi'(t) = F(t).$$

It follows that

$$\left[\int_t^{\phi(t)} F(\sigma)\, d\sigma\right]' = 0,$$

and further from (18), property 5°

$$\int_t^{\phi(t)} F(\sigma)\, d\sigma = \int_c^{\phi(c)} F(\sigma)\, d\sigma = 0.$$

Hence for all $t \in j$ we have the relationship

$$\int_t^{\phi(t)} \left[\frac{k}{y^2(\sigma)} - \frac{1}{k}\frac{1}{\bar{y}^2(\sigma)}\right] d\sigma = 0 \quad (k = 1/p(c)),$$

and application of the result of § 15.4 shows that the fundamental dispersion of the first kind of the differential equation ($\bar{\text{q}}$) coincides with ϕ. This completes the proof.

15.8 Explicit formulae for elementary carriers

In § 8.4 we determined all the elementary carriers in an interval j by means of the formula (6). The theorem of § 15.7 leads to another and perhaps simpler explicit expression for elementary carriers, still of course in the interval $j = (-\infty, \infty)$.

An elementary carrier q in the interval j $(= (-\infty, \infty))$ is characterized by the fact that all its first phases α are elementary—that is, $\alpha(t + \pi) = \alpha(t) + \pi \operatorname{sgn} \alpha'$ $\forall\, t \in j$. We recall, in this context, that differential equations (q) with elementary carriers, and only such equations, have the zeros of all their integrals separated by a distance π.

Clearly the carrier q defined in j is elementary if and only if its fundamental dispersion of the first kind, ϕ, is linear, with $\phi(t) = t + \pi$.

The elementary carriers in the interval j are therefore precisely those carriers whose fundamental dispersion is $\phi(t) = t + \pi$. Among these, naturally, is included the carrier -1; the integral y of this with initial values $y(c) = 0$, $y'(c) = 1$ (which may be assigned at an arbitrary point $c \in j$) is given by the function $y(t) = \sin(t - c)$.

If we apply the result of § 15.7 we obtain the following theorem:

Theorem. The set of all elementary carriers in the interval $j = (-\infty, \infty)$ is given by the following formula

$$\bar{q}(t) = -1 + \frac{p''(t)}{p(t)} + 2\frac{p'(t)}{p(t)} \cot(t - c); \tag{15.20}$$

in which c is an arbitrary number and p a function with the following properties in the interval j

$$\left.\begin{array}{l} 1.\ p \neq 0 \quad \text{for} \quad t \in j; \\ 2.\ p(t + \pi) = p(t) \quad \text{for} \quad t \in j; \\ 3.\ p \in C_2; \\ 4.\ p'(c) = 0; \\ 5.\ \displaystyle\int_0^\pi \left[\frac{1}{p^2(\sigma)} - \frac{1}{p^2(c)}\right] \frac{d\sigma}{\sin^2(\sigma - c)} = 0. \end{array}\right\} \tag{15.21}$$

If we set $p(t) = p(c) \exp f(t)$, then formula (20) gives

$$\bar{q}(t) = -1 + f''(t) + f'^2(t) + 2f'(t) \cdot \cot(t - c), \tag{15.22}$$

in which f denotes a function defined on $(-\infty, \infty)$ with the following properties:—

$$\left.\begin{array}{l} f(t + \pi) = f(t) \quad \text{for} \quad t \in j; \quad f \in C_2; \quad f(c) = f'(c) = 0; \\ \displaystyle\int_0^\pi \frac{\exp(-2f(\sigma)) - 1}{\sin^2(\sigma - c)}\, d\sigma = 0. \end{array}\right\} \tag{15.23}$$

The above formula (22) is due to F. Neuman ([53]). If, in particular, we choose

$$f(t) = -\frac{1}{2}\log\left[1 - \frac{1}{3}\sin 2(t - c)\sin^2(t - c)\right], \tag{15.24}$$

then we obtain the one-parameter system of elementary carriers $q(t|c)$ which we introduced in (8.7).

This result provides also a simple example of carriers $q, \bar{q}$ with the same fundamental dispersion ϕ $(= t + \pi)$, for which the lower limit 4, referred to in the theorem of § 15.6, is in fact attained. This holds, for instance, for the carriers $q(t) = -1$ and $\bar{q}(t) = q(t|0)$. In this case, $\Delta(t) = \bar{q}(t) - q(t) = \sin 4t + \frac{1}{3}\sin^4 t$, and it is easy to verify that for every number $c \in j$ the number of zeros of the function Δ lying in the interval $[c, c + \pi)$ is precisely 4.

15.9 Relations between first phases of differential equations with the same fundamental dispersion ϕ

Let (q), ($\bar{\text{q}}$) be oscillatory differential equations in the interval $j = (-\infty, \infty)$ and ϕ, $\bar{\phi}$ their fundamental dispersions of the first kind. Also, let α, $\bar{\alpha}$ be arbitrary (first) phases of (q), ($\bar{\text{q}}$) respectively. We then have:

Theorem. The fundamental dispersions ϕ, $\bar{\phi}$ coincide if and only if the phase functions $\gamma = \alpha\bar{\alpha}^{-1}$, $\gamma^{-1} = \bar{\alpha}\alpha^{-1}$, (which are inverse to each other), are elementary; that is, if $\forall t \in j$ we have

$$\gamma(t + \pi) = \gamma(t) + \eta\pi, \quad \gamma^{-1}(t + \pi) = \gamma^{-1}(t) + \eta\pi$$
$$(\eta = \operatorname{sgn} \gamma' = \operatorname{sgn} (\gamma^{-1})').$$

Proof. (a) Let $\phi = \bar{\phi}$. Then in the interval j there hold the Abel functional equations

$$\alpha\phi = \alpha + \varepsilon\pi, \quad \bar{\alpha}\bar{\phi} = \bar{\alpha} + \bar{\varepsilon}\pi \quad (\varepsilon = \operatorname{sgn} \alpha', \bar{\varepsilon} = \operatorname{sgn} \bar{\alpha}'),$$

and consequently

$$\bar{\alpha}^{-1}(\bar{\alpha} + \bar{\varepsilon}\pi) = \alpha^{-1}(\alpha + \varepsilon\pi),$$
$$\gamma(\bar{\alpha} + \bar{\varepsilon}\pi) = \alpha + \varepsilon\pi,$$
$$\gamma(t + \bar{\varepsilon}\pi) = \gamma(t) + \varepsilon\pi \quad (t \in j).$$

From the last formula, we have

$$\gamma(t + \pi) = \gamma(t) + \varepsilon\bar{\varepsilon}\pi,$$

and from the fact that $\operatorname{sgn} \gamma' = \varepsilon\bar{\varepsilon}$, it is clear that the phase function γ is elementary. The same, naturally, is true of γ^{-1}.

(b) Let the phase functions γ, γ^{-1} be elementary. Then we have in the interval j

$$\alpha = \gamma\bar{\alpha},$$
$$\alpha\bar{\phi} = \gamma\bar{\alpha}\bar{\phi} = \gamma(\bar{\alpha} + \bar{\varepsilon}\pi) = \gamma\bar{\alpha} + \varepsilon\pi = \alpha + \varepsilon\pi = \alpha\phi;$$

it follows that $\phi = \bar{\phi}$, and the theorem is proved.

We now apply the above theorem to obtain a further explicit formula for carriers with the same fundamental dispersion of the first kind.

Let q, $\bar{q} \in Q\phi$ be arbitrary carriers with the same fundamental dispersion ϕ in $j = (-\infty, \infty)$. We choose arbitrary first phases α, $\bar{\alpha}$ of (q), ($\bar{\text{q}}$) respectively; then by the above result there is a relationship

$$\bar{\alpha} = \bar{\gamma}\alpha, \tag{15.25}$$

$\bar{\gamma}$ being an appropriate elementary phase function. The latter is obviously of class C_3 and represents a first phase of a differential equation (g). Since $\bar{\gamma}$ is elementary, the carrier $\bar{g}$ is also elementary.

Taking the Schwarz derivative of each side of (25) at an arbitrary point $t \in j$ it follows that

$$\{\tan \bar{\alpha}, t\} = \{\tan \bar{\gamma}, \alpha\}\alpha'^2 + \{\alpha, t\}$$

and hence, using (5.16), (5.18)

$$\bar{q} = q + [1 + \bar{g}\alpha]\alpha'^2.$$

We now apply formula (22), writing t for $t-c$ and $f(t)$ for $f(t+c)$, obtaining

$$\bar{q} = q + [f''\alpha + f'^2\alpha + 2f'\alpha \cdot \cot \alpha]\alpha'^2. \tag{15.26}$$

We have thus the following result:

All carriers $\bar{q}$ with the fundamental dispersion ϕ are given precisely by the formula (26) *in which q is a fixed carrier with the fundamental dispersion ϕ, α is a fixed first phase of the differential equation* (q) *and f is an arbitrary function in the interval j with period π, belonging to the class C_2 and having the properties*

$$f(0) = f'(0) = 0, \qquad \int_0^\pi [\exp(-2f(\sigma)) - 1]\, d\sigma/\sin^2 \sigma = 0$$

15.10 Power of the set $Q\phi$

We now determine the power of the set $Q\phi$. To do this, we go back to the information obtained in § 10, where we studied the algebraic structure of the phase group $\mathfrak{G}$.

As we know, all elementary phases form a sub-group $\mathfrak{H}$ of $\mathfrak{G}$. The power of the set of all carriers, whose first phases lie in one and the same element of the right residue class partition $\mathfrak{G}/_r\mathfrak{H}$ is the same for all elements of $\mathfrak{G}/_r\mathfrak{H}$ and is equal to the power of the continuum $\aleph$.

Now let q be a carrier and ϕ its fundamental dispersion of the first kind, and let α be a first phase of the differential equation (q). From the theorem of § 15.9 it follows that the first phases of the differential equations ($\bar{\text{q}}$) with the fundamental dispersion ϕ are precisely those phases of the form $\bar{\alpha} = \bar{\gamma}\alpha$, in which $\bar{\gamma}$ ranges over all elementary phases; that is, all elements of the sub-group $\mathfrak{H}$. In other words, the first phases with fundamental dispersion ϕ are precisely those elements of $\mathfrak{G}$ lying in the right residue class $\mathfrak{H}\alpha \in \mathfrak{G}/_r\mathfrak{H}$. Consequently, the carriers $\bar{q}$ with the fundamental dispersion ϕ are precisely those carriers whose first phases α lie in the right residue class $\mathfrak{H}\alpha$. From the above, the power of the set of all these carriers $\bar{q}$ is equal to the power of the continuum.

Thus we have the following theorem.

Theorem. The power of the set of all oscillatory differential equations (q) *in the interval $j = (-\infty, \infty)$ with the same fundamental dispersion of the first kind ϕ is independent of the choice of the latter and is equal to the power of the continuum, $\aleph$.*

This result calls for some further comment. In the numerical treatment of differential equations we often have to calculate the zeros of an integral of a particular differential equation (q). Such a calculation naturally depends on the carrier q and in certain circumstances can be very difficult. We now have the possibility of replacing the carrier q by a "representative" $\bar{q}$, that is to say a carrier $\bar{q}$ with the same fundamental dispersion of the first kind as q; then the integrals of the differential equation ($\bar{\text{q}}$) have the same zeros as those of (q). Naturally, we choose the representative $\bar{q}$ in such a way that the calculation of the zeros of its integrals is as simple as possible. The above result ensures that there are always infinitely many representatives $\bar{q}$, (indeed, the set of these has the power $\aleph$) which can replace the given carrier q. This raises the following problem: to develop methods for discovering representatives of a given carrier with advantageous properties for numerical work.

16 Differential equations with coincident central dispersions of the $\varkappa$-th and $(\varkappa + 1)$-th kinds ($\varkappa = 1,3$)

In this paragraph we shall be concerned with differential equations (q) whose central dispersions of the first and second kinds ϕ_ν and ψ_ν or of the third and fourth kinds χ_ρ and ω_ρ coincide ($\nu = 0, \pm 1, \pm 2, \ldots; \rho = \pm 1, \pm 2, \ldots$). Obvious examples of such differential equations are those equations (q) whose carrier q is a negative constant in the interval $(-\infty, \infty)$. A carrier q with the property $\phi_\nu = \psi_\nu$ or $\chi_\rho = \omega_\rho$ we shall call, for brevity, an *F-carrier* or an *R-carrier* respectively.

Consider an oscillatory differential equation (q) in the interval $j = (a, b)$ and assume that $q < 0$ for all $t \in j$. We denote by ϕ, ψ, χ, ω the fundamental dispersions of the corresponding kinds; these are thus defined in the entire interval j.

A convenient starting point for the theory of F- and R-carriers is provided by the properties of normalized polar functions (§ 6). Let $\theta(t) = \beta(t) - \alpha(t)$ be a polar function of the carrier q, and $h(\alpha), -k(\beta), p(\zeta)$ be the corresponding 1-, 2-, 3-normalized polar functions. The functions $h, -k, p$ are therefore defined in the interval $J = (-\infty, \infty)$, and the following relations hold at every point $t \in j$

$$\left.\begin{array}{ll} \beta(t) = \alpha(t) + h\alpha(t), & \alpha(t) = \beta(t) + k\beta(t), \\ \beta(t) - \alpha(t) = p\zeta(t), & \zeta(t) = \alpha(t) + \beta(t), \\ n\pi < h\alpha(t) = -k\beta(t) = p\zeta(t) < (n+1)\pi; & n \text{ integral} \end{array}\right\} \tag{16.1}$$

I. Theory of F-Carriers

16.1 Characteristic properties

First we note that from the formulae (12.2), (12.3) it follows that q is an F-carrier if and only if its fundamental dispersions of the 1st and 2nd kinds coincide; $\phi = \psi$ for all $t \in j$.

In the development of the theory which follows we shall confine ourselves generally to the properties of the 1-normalized polar function h. We can reach the same objective by making use of suitable properties of the 2- or 3-normalized polar functions $-k, p$, but we shall content ourselves in this respect with a few comments as opportunity offers.

Theorem. The carrier q is an F-carrier, if and only if the 1*-normalized polar function h has period π.*

Proof. (a) Let q be an F-carrier. Then in the interval j we have $\phi(t) = \psi(t)$. Then, taking account of (1),

$$h[\alpha(t) + \varepsilon\pi] = h\alpha\phi(t) = \beta\phi(t) - \alpha\phi(t) = \beta\psi(t) - \alpha\phi(t)$$
$$= (\beta(t) + \varepsilon\pi) - (\alpha(t) + \varepsilon\pi) = h\alpha(t)$$
$$(\varepsilon = \operatorname{sgn} \alpha' = \operatorname{sgn} \beta'),$$

and consequently $h(\alpha + \pi) = h(\alpha)$ for $\alpha \in (-\infty, \infty)$.

(b) Let the polar function h have period π, so $h(\alpha + \pi) = h(\alpha)$ for $\alpha \in (-\infty, \infty)$. Then at every point $t \in j$,

$$\beta\phi(t) = \alpha\phi(t) + h\alpha\phi(t) = \alpha(t) + \varepsilon\pi + h[\alpha(t) + \varepsilon\pi] = \alpha(t) + \varepsilon\pi + h\alpha(t) = \beta(t) + \varepsilon\pi,$$

and it follows that $\psi(t) = \phi(t)$.

We have thus determined all F-carriers:

The F-carriers are precisely those which are derived by the formula (6.29) *from normalized polar functions h with period π in the interval* $(-\infty, \infty)$ $(h' > -1)$.

Similarly, the F-carriers can be characterized by periodicity with period π or 2π of the 2- or 3-normalized polar functions $-k$ or p.

We have also the following result (due to M. Laitoch [41].

Theorem. The carrier q is an F-carrier if and only if its fundamental dispersion of the first kind, ϕ, is linear:

$$\phi(t) = ct + k \qquad (c > 0,\ k = \text{const}). \tag{16.2}$$

This follows immediately from the above results and (13.32).

16.2 Domain of definition of *F*-carriers

We now wish to determine the intervals of definition of the F-carriers.

Let q be an F-carrier. The 1-normalized polar function h is therefore periodic with period π, and formula (2) holds. From (13.31) we obtain

$$c = \exp 2 \int_0^{\pi} \cot h(\rho)\, d\rho. \tag{16.3}$$

Now the formula (2) gives, for the ν-th central dispersion $\phi_\nu(t)$, $\nu = 0, \pm1, \pm2, \ldots$.

$$\phi_\nu(t) = c^\nu t + k\,\frac{c^\nu - 1}{c - 1} \qquad \text{or} \qquad \varphi_\nu(t) + \nu k, \tag{16.4}$$

according as $c \neq 1$ or $c = 1$.

From the facts that $\phi_\nu(t) \to b$ as $\nu \to \infty$, and $\phi_\nu(t) \to a$ as $\nu \to -\infty$, we have (from (4)): in the case $c > 1$

$$b = \infty, \quad a = -k/(c - 1), \qquad \text{hence} \qquad j = (a, \infty), \quad a \text{ finite;}$$

in the case $c < 1$

$$b = k/(1-c), \quad a = -\infty, \qquad \text{hence} \qquad j = (-\infty, b), \quad b \text{ finite};$$

in the case $c = 1$

$$k > 0, \quad a = -\infty, \quad b = \infty.$$

We have thus determined the intervals of definition of all F-carriers:

The interval of definition j of the F-carrier q is unbounded on one or both sides according as

$$\int_0^\pi \cot h(\rho)\, d\rho \neq 0 \qquad \text{or} \qquad = 0.$$

16.3 Elementary carriers

We remind the reader that this term is applied to carriers whose first phases are elementary (§ 8.4). Now we show that:

The carrier q is elementary if and only if the 1*-normalized polar function h has period* π *and satisfies the following conditions*

$$\left.\begin{aligned} &\int_0^\pi \cot h(\rho)\, d\rho = 0, \qquad \int_0^{\varepsilon\pi} \left(\exp 2 \int_0^\sigma \cot h(\rho)\, d\rho\right) d\sigma = \pi\alpha_0'. \\ &\qquad (\alpha_0' = \alpha'[\alpha^{-1}(0)]; \qquad \varepsilon = \operatorname{sgn} \alpha_0'). \end{aligned}\right\} \tag{16.5}$$

For, if the carrier q is elementary, then its fundamental dispersion of the 1st kind $\phi(t)$ has the form (2) with $c = 1$, $k = \pi$. The 1-normalized polar function h has therefore period π, and from (13.31), (13.30) the relations (5) follow. The second part of the theorem is proved similarly.

We have thus determined all elementary carriers:

The elementary carriers q are precisely those derived by the formula (6.29) *from* 1*-normalized polar functions h defined in the interval* $(-\infty, \infty)$, *having period* π, *and satisfying the conditions* (5) $(h^\diagdown > -1)$.

Similarly, the elementary carriers may be expressed in terms of 2- or 3-normalized polar functions $-k$ or p, being given explicitly by the formulae (6.36) or (6.41).

16.4 Kinematic properties of *F*-carriers

We now make use of the kinematic significance of integrals of the differential equation (q), described in § 1.5, as applied to an F-carrier q.

Let q be an F-carrier. Consider two points P, P' lying on the oriented straight line G, whose motion is given by integrals, u, v of the differential equation (q).

Since the differential equation (q) is oscillatory, the motion of each of these points consists of an oscillation about the fixed point (the origin) O of the straight line G.

We assume that at any instant t_0 at which the point P passes through O, the point P' does not coincide with O and its velocity is zero. At the instant t_0, therefore, the point P' is at a relative maximum distance from O. The times at which the point P passes through the origin O are obviously $\phi_\nu(t_0)$, and those at which the point P' is at a maximum distance from O are $\psi_\nu(t_0)$; $\nu = \ldots, -1, 0, 1, \ldots$. Since q is an F-carrier, we have $\phi_\nu(t_0) = \psi_\nu(t_0)$.

We see therefore that:

The oscillations of the points P, P' about the origin O are such that the point P passes through the origin O when the point P' is at a relative maximum distance from O.

II. Theory of *R*-Carriers

16.5 Characteristic properties of *R*-carriers

From the formulae in § 12.4 we have, for all $t \in j$,

$$\left.\begin{aligned} \chi\omega &= \psi, & \omega\chi &= \phi, \\ \omega_n &= \phi^{n-1}\omega, & \omega_{-n} &= \phi_{-1}^{n-1}\chi^{-1}, \\ \chi_n &= \psi^{n-1}\chi, & \chi_{-n} &= \psi_{-1}^{n-1}\omega^{-1} \\ (n = 1, 2, \ldots;\ & \phi_{-1} = \phi^{-1}, \psi_{-1} = \psi^{-1}). \end{aligned}\right\} \tag{16.6}$$

Hence, from $\chi = \omega$ it follows that $\phi = \psi$ and $\chi_\rho = \omega_\rho$ for $\rho = \pm 1, \pm 2, \ldots$.

This gives the result:

q is an R-carrier if and only if its fundamental dispersions of the third and fourth kinds coincide: $\chi = \omega$ for $t \in j$. An R-carrier is always an F-carrier.

Theorem. The carrier q is an R-carrier if and only if the 1*-normalized polar function h satisfies the following relation in the interval $J = (-\infty, \infty)$:*

$$h\alpha + h[\alpha + h\alpha - n\pi] = (2n + 1)\pi. \tag{16.7}$$

Proof. If (7) is satisfied, then on applying it at the point $\alpha + h\alpha - n\pi$, there follows the π-periodicity of h:

$$h(\alpha + \pi) = h\alpha. \tag{16.8}$$

We shall now give the proof first for the case $n = 0$. We then have $0 < \beta - \alpha < \pi$, so the corresponding Abel functional equations (13.18), (13.20) hold.

(a) Let q be an R-carrier, so that $\chi = \omega$. Then, in the interval j, we have

$$\beta\chi(t) = \alpha\chi(t) + h\alpha\chi(t)$$

and further, from (13.18), (13.20),

$$\alpha(t) + \pi = \beta(t) + h\left[\beta(t) + \frac{1}{2}(\varepsilon - 1)\pi\right].$$

Since however q is an F-carrier, the function h has period π, and on taking account of (1) the last relationship gives the formula (7) for the case $n = 0$.

(b) Now let the relation (7) be satisfied when $n = 0$; then, from (8), the function h is π-periodic. From (1) and (13.20), we have

$$\beta\omega(t) = \alpha\omega(t) + h\alpha\omega(t) = \beta(t) + \frac{1}{2}(\varepsilon - 1)\pi + h\left[\beta(t) + \frac{1}{2}(\varepsilon - 1)\pi\right]$$
$$= \alpha(t) + \frac{1}{2}(\varepsilon + 1)\pi - \pi + h\alpha(t) + h\left[\alpha(t) + h\alpha(t) + \frac{1}{2}(\varepsilon - 1)\pi\right].$$

Since the function h satisfies (7) and has period π, the last expression, in view of (13.18), equal to $\beta\chi(t)$. We have, therefore $\chi = \omega$ for $t \in j$.

The extension of the proof to the general case, in which n is any integer, is simple. We set

$$h\alpha = h_0\alpha + n\pi. \tag{16.9}$$

Then h_0 is a 1-normalized polar function of the carrier q with the property $0 < h_0 < \pi$.

If q is an R-carrier, then from (a) the function h_0 satisfies the condition

$$h_0\alpha + h_0[\alpha + h_0\alpha] = \pi; \tag{16.10}$$

and from this and (9) the relation (7) follows.

If, conversely, the condition (7) is satisfied, then (10) holds; from that we deduce (using (b)) that q is an F-carrier. This completes the proof.

We have thus determined all the R-carriers;

The R-carriers are precisely those derived by the formula (6.29) *from the* 1-*normalized polar functions h defined in the interval* $(-\infty, \infty)$ *and satisfying* (7) $(h^{\backslash} > -1)$.

Similarly, the R-carriers can be determined by means of 2- or 3-normalized polar functions satisfying the conditions

$$k\beta + k[\beta + k\beta + n\pi] = -(2n + 1)\pi \tag{16.11}$$

and

$$p\zeta + p(\zeta + \pi) = (2n + 1)\pi, \tag{16.12}$$

being given by the formulae (6.36) and (6.41).

16.6 Further properties of *R*-carriers

The following study takes us further into the properties of R-carriers.

Let q be an R-carrier in the interval j $(= (a, b))$.

We consider an integral curve $\mathfrak{K}$ of the differential equation (q) with the parametric co-ordinates $u(t)$, $v(t)$ in which, for precision, we take the Wronskian $w = uv' - u'v < 0$. We denote the origin of the coordinate system by O.

Let $P, \tilde{P} \in \mathfrak{K}$ be points determined by the parameters t, $\chi(t)$ where $t \in j$ is arbitrary. *Our interest will centre upon the area* Δ *of the triangle* $PO\tilde{P}$.

Obviously

$$2\Delta = r(t) \cdot r\chi(t) \cdot \sin\theta(t); \tag{16.13}$$

where $r(t)$, $r\chi(t)$ are the lengths of the vectors $\overrightarrow{OP}$, $\overrightarrow{O\tilde{P}}$ and $\theta(t)$ is the angle formed by the latter.

Let α be a proper first phase of the basis (u, v). Since $-w > 0$ we have $\alpha' > 0$; we also have $\chi(t) > t$, consequently $\alpha\chi(t) > \alpha(t)$ and since $0 \leqslant \theta(t) < 2\pi$,

$$\theta(t) = \alpha\chi(t) - \alpha(t) + 2n\pi, \qquad 0 \geqslant n \text{ integral}. \tag{16.14}$$

Moreover, let β be the proper second phase of (u, v) neighbouring to α, so that $0 < \beta - \alpha < \pi$.

We write the relation (14) as follows

$$\theta(t) = [\alpha\chi(t) - \beta(t)] + [\beta(t) - \alpha(t)] + 2n\pi$$

and apply the formulae (13.20). Since $\varepsilon = 1$ and $0 < \beta - \alpha < \pi$, we have

$$\theta(t) = \beta(t) - \alpha(t), \tag{16.15}$$

so θ is that polar function of the basis (u, v) generated by α and lying between 0 and π.

To help on the development of this study, it is convenient to quote here the following formulae:

$$\theta\chi = -\theta + \pi, \qquad \alpha' = \beta'\chi \cdot \chi', \qquad \beta' = \alpha'\chi \cdot \chi' \qquad [(13.18), (13.20)] \tag{16.16}$$

$$r\chi \cdot r'\chi = -rr' \quad [(16) \text{ and } (6.8)] \tag{16.17}$$

$$\alpha' = \frac{w \cdot q\chi}{s^2\chi}\,\chi', \qquad \beta' = \frac{-w}{r^2\chi}\,\chi' \quad [(16) \text{ and } (5.14), (5.23)] \tag{16.18}$$

Logarithmic differentiation of (13) shows that

$$\frac{\Delta'}{\Delta} = \frac{r'}{r} + \frac{r'\chi}{r\chi}\,\chi' + \cot\theta \cdot \theta',$$

and the formulae (6.8), (5.14) and then (17), (18) give

$$\cot\theta \cdot \theta' = -\frac{1}{w}\,rr'(\beta' - \alpha') = -\frac{1}{w}\,rr' \cdot \beta' - \frac{r'}{r} = -\frac{r'\chi}{r\chi}\,\chi' - \frac{r'}{r}\,.$$

Consequently $\Delta' = 0$, and we have the result:

Theorem. The area Δ of the triangle $PO\tilde{P}$ is constant throughout the curve $\mathfrak{R}$.

16.7 Connection between *R*-carriers and Radon curves

From the relationships [(16) and (5.28)]

$$rs \cdot \sin\theta = -w, \qquad r\chi \cdot s\chi \cdot \sin\theta = -w \tag{16.19}$$

there follows, when we take account of (13),

$$r\chi = \frac{2\Delta}{-w}\,s, \qquad s\chi = \frac{-w}{2\Delta}\,r. \tag{16.20}$$

Moreover we have from (13.20) and (18)

$$\mathrm{W}\alpha\chi = \mathrm{W}\beta, \qquad \mathrm{W}\beta\chi = \mathrm{W}\alpha \pm \pi, \tag{16.21}$$

in which the sign + or − must be taken according as $0 \leqslant \mathrm{W}\alpha < \pi$ or $\pi \leqslant \mathrm{W}\alpha < 2\pi$.

We now apply to the integral curve $\mathfrak{K}$ the transformation $\boldsymbol{R}$ (§ 6.1) which consists of the inversion $\boldsymbol{K}_{\sqrt{2\Delta}}$, followed by a quarter rotation about O in the positive sense.

The curve $\mathfrak{K}$ is then transformed into a curve $\bar{\mathfrak{K}}$: the point $P \in \mathfrak{K}$ goes over into the point $\bar{P} \in \bar{\mathfrak{K}}$, while the corresponding amplitudes $\bar{r}$, s and angles $\mathrm{W}\alpha$, $\mathrm{W}\beta$; $\bar{\alpha}$, $\bar{\beta}$ are transformed as follows [(6.5)]

$$\bar{r} = \frac{2\Delta}{-w} s, \qquad \bar{\alpha} = \mathrm{W}\beta, \quad \bar{\beta} = \mathrm{W}\alpha \pm \pi, \tag{16.22}$$

in which we take the sign + or − according as $0 \leqslant \mathrm{W}\alpha < \pi$ or $\pi \leqslant \mathrm{W}\alpha < 2\pi$.

Comparing this with (20), (21), gives

$$\bar{r} = r\chi, \qquad \bar{\alpha} = \mathrm{W}\alpha\chi, \qquad \bar{\beta} = \mathrm{W}\beta\chi. \tag{16.23}$$

Clearly, the transformation R takes the curve $\mathfrak{K}$ into itself, so

The integral curves of an R-carrier are Radon curves.

16.8 Connection between *R*- and *F*-carriers

The second formula (18), taken together with (5.23) gives the following formula holding in the interval j

$$\frac{\chi'}{r^2\chi} = -\frac{q}{s^2} \tag{16.24}$$

and moreover, using (20),

$$\chi' = -d^2 q \quad \left(d = \frac{2\Delta}{-w}\right). \tag{16.25}$$

This formula is due to E. Barvínek ([2]). It follows that for $t_0, t \in j$,

$$\chi(t) = \chi(t_0) - d^2 \int_{t_0}^{t} q(\sigma)\, d\sigma. \tag{16.26}$$

Similarly the first formula (18) and (5.14) show that

$$\chi' = -\frac{1}{d^2}\frac{1}{q\chi}; \tag{16.27}$$

thus for $t_0, t \in j$,

$$\chi(t) = \chi(t_0) - \frac{1}{d^2}\int_{t_0}^{t} \frac{d\sigma}{q\chi(\sigma)}. \tag{16.28}$$

From (25) and (27) we see that the product of the values of the R-carrier q at any two points t, $\chi(t) \in j$ is constant:

$$q(t)q\chi(t) = \frac{1}{d^4}. \tag{16.29}$$

From the formula (26) we have

$$\chi\chi(t) = \chi(t_0) - d^2 \int_{t_0}^{\chi(t_0)} q(\sigma)\, d\sigma - d^2 \int_{\chi(t_0)}^{\chi(t)} q(\sigma)\, d\sigma.$$

If the last integral is transformed by means of the substitution $\sigma = \chi(\tau)$ and we apply formulae (25) and (29) then we obtain

$$(\phi(t) =)\ \chi\chi(t) = t + k \tag{16.30}$$

with a determinate constant k (> 0). Since $\chi\chi = \phi$, this formula shows that *every R-carrier belongs to the set of F-carriers defined in the interval $j = (-\infty, \infty)$*, (§ 16.2, $c = 1$).

16.9 Kinematic properties of *R*-carriers

Let q be an R-carrier.

We consider two points P, P' lying on the oriented straight line G, whose motions follow the integrals u, v of the differential equation (q). Let the positions of the points P, P' at an instant t_0 be such that the point P passes through the origin O when P' is at a relative maximum distance from O. Since q is an R-carrier, and consequently also an F-carrier, we have the situation described in § 16.4. Now the instants at which the point P is at its greatest distance from O are $\chi_\rho(t_0)$ and those at which the point P' passes through the origin O are $\omega_\rho(t_0)$: $\rho = \ldots, -1, 1, \ldots$. But since q is an R-carrier, we have $\chi_\rho(t_0) = \omega_\rho(t_0)$. Thus:

The oscillations of the points P, P' about the origin O are such that each of these passes through the origin at the instant when the other is at a relative maximum distance from the origin.

17 Bunch curves and Radon curves

In our study of polar functions we have met plane curves with special centro-affine properties, among them the Radon curves (§ 6.1). These are curves which are always intersected by the straight lines of a bunch in at least two points, and such that the tangents at these points of intersection are parallel. This section is devoted to an investigation of these curves from the geometrical standpoint. It will appear that these curves stand in a close relationship to the F-carriers and R-carriers, whose integral curves they in fact are. Our object is to determine the geometrical properties of these curves and to derive finite equations for them. This, as may be expected, involves an application of the analytical apparatus which we have derived above.

17.1 Fundamental concepts

We begin with the definition of the curves to be studied, which we shall call bunch curves.

Let (F) be a straight line bunch with centre O. By a *bunch curve with respect to* (F) or more shortly a *bunch curve* we mean a plane curve $\mathfrak{F}$ with the following property:

The curve $\mathfrak{F}$ is met by each straight line of the bunch (F) in at least two distinct points, and is such that the tangents to the curve at each of the points of intersection are parallel.

By the *pole* of the curve $\mathfrak{F}$ we mean the centre O of (F). In view of the methods to be applied we wish to restrict ourselves to curves $\mathfrak{F}$ which are regular (i.e. locally convex and without turning points). Moreover we shall assume that every curve $\mathfrak{F}$ can be specified by means of parametric coordinates U, V with respect to a coordinate system with origin O; the functions U, V are defined in an open interval and belong to the class C_2.

Obvious examples of such bunch curves are ellipses and logarithmic spirals.

Let $\mathfrak{F}$ be a bunch curve with parametric coordinates U, V in the interval J. We notice first that from the definition of bunch curves, the functions U, V must be linearly independent. It follows that the functions U, V are integrals of a linear differential equation (a) with continuous coefficients $\bar{a}$, $\bar{b}$:

$$Y'' + \bar{a}Y' + \bar{b}Y = 0. \tag{a}$$

We see further that the properties of the curve $\mathfrak{F}$ which have been described are centro-affine invariant and consequently remain valid under every transformation of parameter (§ 4.1). If we choose, in particular, the curve parameter given by the formula (1.1), then the differential equation (a) takes the Jacobian form (q).

Hence, for appropriate choice of the curve parameter, the bunch curve $\mathfrak{F}$ is an integral curve of a differential equation (q). We call q the *carrier of the bunch curve* $\mathfrak{F}$.

17.2 Determination of the carriers of the bunch curves

We now determine the carriers q of the bunch curves.

Since we are only considering regular bunch curves, we can assume from the start that $q(t) \neq 0$ for $t \in j$ (§ 4.6). Let (q) be a differential equation in the interval $j, q(t) \neq 0$ for $t \in j$ and $\mathfrak{K}$ an integral curve of (q) with the parametric coordinates $u(t), v(t)$ with respect to a coordinate system with origin O. Let $P(t) \in \mathfrak{K}$, $t \in j$ be an arbitrary point and $\tau(t)$ the tangent to $\mathfrak{K}$ at the point $P(t)$. Further, let $p(t)$ be the straight line $OP(t)$.

The following theorem is the foundation of further developments:

Theorem. Two points $P(t_1), P(t_2) \in \mathfrak{K}$, $t_1 \neq t_2$, lie on a straight line passing through the point O if and only if the numbers t_1, t_2 are 1-conjugate.

Two tangents $\tau(t_1), \tau(t_2)$ to $\mathfrak{K}$, $t_1 \neq t_2$, are parallel if and only if the numbers t_1, t_2 are 2-conjugate.

The tangent $\tau(t_2)$ to $\mathfrak{K}$ and the straight line $p(t_1)$ are parallel if and only if t_2 is 3-conjugate with t_1 and consequently t_1 is 4-conjugate with t_2.

Proof. Two points $P(t_1), P(t_2) \in \mathfrak{K}$, $t_1 \neq t_2$, lie on a straight line passing through the point O if and only if $u(t_1)v(t_2) - u(t_2)v(t_1) = 0$.

Two tangents $\tau(t_1), \tau(t_2)$ to $\mathfrak{K}$, $t_1 \neq t_2$ are parallel if and only if $u'(t_1)v'(t_2) - u'(t_2)v'(t_1) = 0$.

The tangent $\tau(t_2)$ to $\mathfrak{K}$ and the straight line $p(t_1)$ are parallel if and only if $u'(t_2)v(t_1) - v'(t_2)u(t_1) = 0$.

The result now follows immediately from the above facts and the theorem of § 3.12.

We observe also that:

The tangents $\tau(t_1), \tau(t_2)$, $(t_1 \neq t_2)$, to the curve $\mathfrak{K}$, at two points of intersection of the latter with a straight line passing through the point O are parallel if and only if the numbers t_1, t_2 are both 1-conjugate and 2-conjugate.

The tangents $\tau(t_1), \tau(t_2)$ to the curve $\mathfrak{K}$ and the straight lines $p(t_1), p(t_2)$ passing through the point O and their points of contact $P(t_1), P(t_2)$ with the curve $\mathfrak{K}$ are parallel if and only if each number t_1, t_2 is both 3- and 4-conjugate with the other.

Now let q be a carrier of the bunch curve $\mathfrak{F}$ and $q \neq 0$ for $t \in j$.

Let $t_1 \in j$ be arbitrary and $P(t_1) \in \mathfrak{F}$ the corresponding point of $\mathfrak{F}$. From the definition of a bunch curve we know that the straight line $OP(t_1)$ passing through the point O must have at least one point $P(t_2)$ $(\neq P(t_1))$ where it meets the curve $\mathfrak{F}$. Obviously, $t_2 \neq t_1$, and by the above theorem t_2 is 1-conjugate to t_1. Consequently, every number $t_1 \in j$ has 1-conjugate numbers, so we conclude that every integral of (q) vanishes at least twice in the interval j.

We note that the differential equation (q) is either of finite type (m), $m \geqslant 3$, or of infinite type. It follows, in the first place, that $q < 0$ for $t \in j$ and further that the differential equation (q) admits of fundamental dispersions of the first and second kinds ϕ, ψ. These are defined in appropriate intervals $i \subset j$, $i' \subset j'$.

Now let $t \in i$ be arbitrary. Since $\phi(t)$ is 1-conjugate with t, the points $P(t), P\phi(t)$ lie on a straight line through the point O. Since, moreover, $\mathfrak{F}$ is a bunch curve, the tangents $\tau(t), \tau\phi(t)$ are parallel. Taking account of the above theorem, we conclude that $\phi(t)$ is 2-conjugate with t, so $\phi(t) = \psi_\nu(t)$ for some integer ν, $\nu \geqslant 1$. There is therefore an integral y of (q), whose derivative y' vanishes at the points t, $\phi(t)$ and possibly also at further points $x_1, \ldots, x_{\nu-1}$ lying between t and $\phi(t)$. Now since the function $q < 0$ in j, every interval $(t, x_1), \ldots, (x_{\nu-1}, \phi(t))$ contains precisely one zero of y, so that between t and $\phi(t)$ there are precisely ν zeros. Hence $\nu = 1$, and $\phi(t) = \psi(t)$; that is to say $i \subset i'$ and $\phi(t) = \psi(t)$ for $t \in i$.

Now we show that the opposite relation $i' \subset i$ also holds, and $\psi(t') = \phi(t')$ for $t' \in i'$. For let us suppose, if possible, that $t' \in i'$ but $t' \notin i$. Then the number $\bar{t}$ such that $(t' <)\ \bar{t} = \psi(t')\ (\in j)$ is defined, but there exists no number which is 1-conjugate with t' on the right. We know that the straight line $OP(\bar{t})$ cuts the curve $\mathfrak{F}$ at least twice, whence follows the existence of numbers $\phi_\nu(\bar{t})\ (\in j)$, ν integral, which are 1-conjugate with $\bar{t}$. Since there is no number which is 1-conjugate with t' on the right, and consequently none 1-conjugate with $\bar{t}\ (> t')$, we must have $\nu < 0$, so the number $\phi_{-1}(\bar{t})\ (= t)$ exists. But the tangents to the curve at the points $P(\bar{t})$, $P(t) \in \mathfrak{F}$ are parallel, which shows that t is 2-conjugate with $\bar{t}$. We have therefore $\phi_{-1}(\bar{t}) = \psi_{-n}(\bar{t})$, for some $n \geqslant 1$, and from this we find, as above, that $\phi_{-1}(\bar{t}) = \psi_{-1}(\bar{t})\ (= t')$. We have therefore $t' = \phi_{-1}(\bar{t})$, hence $\bar{t} = \phi(t')$; this contradicts the above assumption that $t' \notin i$, and so establishes the desired result.

To sum up: the carrier q of the bunch curve $\mathfrak{F}$ is always < 0; moreover the differential equation (q) is of finite type (m), $m \geqslant 3$, or of infinite type, and its fundamental dispersions of the first and second kinds, ϕ, ψ, coincide.

We leave it to the reader to convince himself of the validity of the converse statement: every continuous function q with these properties is the carrier of a bunch curve.

From this we have:

Theorem. The carriers of bunch curves are partial functions of F-carriers.

The carrier q of a bunch curve can for instance be represented in the form (6.29), in which h is an arbitrary function, defined in an interval J of length $> 2\pi$, with the following properties:

$$\left.\begin{array}{l} 1.\ h \in C_1; \\ 2.\ n\pi < h < (n+1)\pi, \quad n \text{ integral}; \\ 3.\ h(\alpha + \pi) = h(\alpha) \quad \text{for} \quad \alpha, \alpha + \pi \in J; \\ 4.\ h^{\backprime}(\alpha) > -1 \quad \text{for} \quad \alpha \in J. \end{array}\right\} \tag{17.1}$$

17.3 Centro-affine length of arcs of bunch curves

Let $\mathfrak{F}$ be a bunch curve with the parametric coordinates $u(t)$, $v(t)$, $t \in j$. We shall continue to employ the notation used above.

First we recall that the centro-affine oriented arc of the curve $\mathfrak{F}$ determined by the points $P(t_1)$, $P(t_2) \in \mathfrak{F}$ is given by the formula (4.14). Its length $s(t_1|t_2)$ is therefore

$$s(t_1|t_2) = \left| \int_{t_1}^{t_2} \sqrt{-q(\sigma)}\, d\sigma \right|. \tag{17.2}$$

We now make use of the formulae (6.29) and (13.31) to obtain, for $t \in i$,

$$-q(\phi(t))\phi'^2(t) = -q(t)$$

and thus, taking account of (2),

$$s(\phi(t_1)|\phi(t_2)) = s(t_1|t_2). \tag{17.3}$$

We know that the points $P(t_1)$, $P\phi(t_1)$ lie on a straight line passing through the point O (§ 17.2); the same holds for $P(t_2)$, $P\phi(t_2)$.

The arcs $P(t_1)P(t_2)$ and $P\phi(t_1)P\phi(t_2)$ of the bunch curve $\mathfrak{F}$, cut off by two arbitrary lines $OP(t_1)$, $OP(t_2)$ of the bunch F, have the same centro-affine length.

17.4 Finite equations of bunch curves

In this paragraph we shall derive finite equations for bunch curves. Since the bunch curve carriers are partial functions of the F-carriers, we shall confine ourselves to bunch curves of F-carriers. We shall also assume that the differential equations (q) of the bunch curves under consideration are oscillatory and consequently defined in the intervals given in § 16.2.

Let $\mathfrak{F}$ be a bunch curve and q its carrier in the interval $j = (a, \infty)$ or $(-\infty, b)$ or $(-\infty, \infty)$ where a, b are finite. Moreover, let u, v be parametric coordinates of the curve $\mathfrak{F}$ with respect to a coordinate system with origin O; we assume for definiteness that $(w =)\ uv' - u'v < 0$.

Let $r(t)$ be the amplitude, $\alpha(t)$ be a first phase $(t \in j)$ and $h(\alpha)$ $(\alpha \in (-\infty, \infty))$ one of the 1-normalized polar functions of the basis (u, v) generated by the phase α. The function h has the above properties 1°–4° and without loss of generality we assume that $n = 0$.

We now start from the formula (6.25), taking $\alpha_0 = 0$:

$$r(t) = r_0 \cdot \exp \int_0^{\alpha} \cot h(\rho)\, d\rho. \tag{17.4}$$

From (4) and the fact that $\operatorname{sgn}(-w) = \operatorname{sgn} \alpha' = 1$, we have in the interval j

$$r\phi(t) = r_0 \cdot \exp \int_0^{\alpha+\pi} \cot h(\rho)\, d\rho$$

and moreover,

$$r\phi(t) = \sqrt{c} \cdot r(t);$$

in this c (> 0) is the coefficient of t in the fundamental dispersion $\phi(t) = ct + k$ of the differential equation (q), and the formula (16.3) holds.

We now see that the following function, which is defined for $t \in j$,

$$g(t) = c^{\frac{1}{2\pi}\alpha(t)} \tag{17.5}$$

is transformed by the substitution $t \to \phi(t)$ in the same manner as the function $r(t)$. Consequently, the function

$$f(t) = r(t)/g(t) \tag{17.6}$$

is invariant under this substitution, i.e.

$$f\phi(t) = f(t). \tag{17.7}$$

From (5), (6) it follows that

$$r(t) = C^{\alpha(t)} \cdot f(t) \quad \left(C = c^{\frac{1}{2\pi}}\right). \tag{17.8}$$

Let $F(\alpha)$ (> 0) be the function defined in the interval $(-\infty, \infty)$ by means of the relationship

$$f(t) = F(\alpha);$$

in this, $t = \alpha^{-1}(\alpha) \in j$ and $\alpha = \alpha(t) \in (-\infty, \infty)$ are two homologous values.

If we compare the formulae (4) and (8), we obtain

$$F(\alpha) = r_0 \cdot C^{-\alpha} \cdot \exp \int_0^{\alpha} \cot h(\rho)\, d\rho.$$

Clearly, the function F belongs to the class C_2 and has period π. We have moreover

$$h = \operatorname{arcot}\left(\frac{F^{\backslash}}{F} + \log C\right), \tag{17.9}$$

$$h^{\backslash} = -\frac{FF^{\backslash\backslash} - F^{\backslash 2}}{F^2 + (F^{\backslash} + F \cdot \log C)^2}, \tag{17.10}$$

where the symbol arccot denotes the branch of the above function lying between 0 and π.

The function F has therefore the following properties in its interval of definition $(-\infty, \infty)$

$$\left.\begin{array}{l} 1.\ F > 0, \\ 2.\ F \in C_2, \\ 3.\ F(\alpha + \pi) = F(\alpha), \\ 4.\ \dfrac{F^{\backslash\backslash}}{F} < \dfrac{F^{\backslash 2}}{F^2} + \left(\dfrac{F^{\backslash}}{F} + \log C\right)^2 + 1. \end{array}\right\} \tag{17.11}$$

Clearly, the equation of the bunch curve $\mathfrak{F}$ in polar coordinates is

$$r = C^{\alpha} \cdot F(\alpha), \tag{17.12}$$

where C (> 0) is a constant and F is a function defined in the interval $(-\infty, \infty)$ with the properties (11).

Conversely, it can be shown without difficulty that: if we construct a function h according to the formula (9) using an arbitrary constant C (> 0) and a function F defined in the interval $(-\infty, \infty)$ with the properties (11), then this function h represents a 1-normalized polar function of an F-carrier q, satisfying the relations (1). Moreover, this carrier is specified by means of a formula such as (6.29).

We have thus proved the following theorem:

Theorem. All bunch curves with F-carriers are given in polar coordinates by the equation (12), *in which* C (> 0) *is a constant and* F *is a function defined in the interval* $(-\infty, \infty)$ *with the properties* (11).

17.5 Radon curves

Let q be the carrier of a bunch curve $\mathfrak{F}$ in an interval j; moreover let O be the pole of $\mathfrak{F}$ and (F) the bunch of straight lines with the centre O. We shall use the symbols $P(t)$, $p(t)$, $\tau(t)$ $(t \in j)$ in the same sense as in § 17.2.

The curve $\mathfrak{F}$ determines a simple mapping $\boldsymbol{F}$ of the bunch (F) into itself which is defined as follows: corresponding to every straight line $p \in (F)$, $\boldsymbol{F}p \in (F)$ is the line parallel to the tangents to the curve at their points of intersection with p.

We call $\mathfrak{F}$ a *Radon curve* when this mapping $\boldsymbol{F}$ is involutory, that is when the combined mapping $\boldsymbol{FF}$ is the identity which maps (F) onto itself.

Every Radon curve therefore has the ellipse property (§ 6.1): from $p' = \boldsymbol{F}p$ it follows that $\boldsymbol{F}p' = p$.

Now let $\mathfrak{R}$ be a Radon curve. Then the fundamental dispersions of the first and second kinds, ϕ, ψ of the differential equation (q) are defined in some interval $i \subset j$ and from this we deduce that the fundamental dispersions of the third and fourth kinds of (q), χ and ω, also exist in certain intervals $k, k' \subset i$. Clearly, $t < \chi(t) < \phi(t)$ for $t \in k$ and $t < \omega(t) < \psi(t)$ $(= \phi(t))$ for $t \in k'$.

We show first that $k = k'$ and $\chi(t) = \omega(t)$ for $t \in k$.

Let $t \in k$ be arbitrary. Since $\chi(t)$ is 3-conjugate with t, the tangent to the curve $\tau\chi(t)$ is parallel to the line $p(t)$. Since $\mathfrak{R}$ is a Radon curve, $\tau(t)$ is parallel to $p\chi(t)$; consequently (§ 17.2) t is 3-conjugate with $\chi(t)$ and also $\chi(t)$ is 4-conjugate with t. We have therefore $\chi(t) = \omega_\rho(t)$, ρ being an integer. From $\chi(t) > t$ it follows that $\rho \geqslant 1$; if $\rho \geqslant 2$, then $t < \chi(t) < \phi(t) < \omega_\rho(t)$, but this is impossible. Consequently $\rho = 1$, so $\chi(t) = \omega(t)$, thus $k \subset k'$ and $\chi(t) = \omega(t)$ for $t \in k$. In a similar manner, we find that $k' \subset k$ and $\omega(t) = \chi(t)$ for $t \in k'$. This completes the proof.

We leave it to the reader to convince himself of the truth of the converse theorem: every bunch curve carrier q with the property $\chi = \omega$ is the carrier of a Radon curve.

We thus have the following result:

Theorem. The carriers of Radon curves are partial functions of R-carriers.

In what follows, we shall consider only Radon curves with R-carriers.

Let $\mathfrak{R}$ be a Radon curve with the R-carrier q. From § 16.8 the function q is defined in the interval $j = (-\infty, \infty)$ and the fundamental dispersion ϕ of the first kind of (q) is given by the formula $\phi(t) = t + k$ $(c = 1;$ (16.30)).

From (12) there follows the equation of the curve $\mathfrak{R}$ in polar coordinates:

$$r = F(\alpha); \tag{17.13}$$

where F is a function defined in the interval $(-\infty, \infty)$ with the properties (11).

Since the function F is, from (11), 3°, periodic with period π and every two points $P(\alpha)$, $P(\alpha + \pi) \in \mathfrak{R}$ lie on the same straight line passing through O, we have:

The curve $\mathfrak{R}$ *is closed and has central symmetry.*

We know (§ 16.7) that the function h defined by means of the formula (9), (with $C = 1$), which we may assume to lie in the interval $(0, \pi)$, satisfies the relation $h\alpha + h[\alpha + h\alpha] = \pi$.

It follows that

$$\frac{F^{\diagdown}\alpha}{F\alpha} + \frac{F^{\diagdown}\left[\alpha + \operatorname{arccot}\dfrac{F^{\diagdown}\alpha}{F\alpha}\right]}{F\left[\alpha + \operatorname{arccot}\dfrac{F^{\diagdown}\alpha}{F\alpha}\right]} = 0. \tag{17.14}$$

The equation of the Radon curve $\mathfrak{R}$ in polar coordinates has therefore the form (13); F is a function defined in the interval $(-\infty, \infty)$ with the properties (11) and (14), taking $C = 1$.

Conversely, it can easily be shown that if any function h is constructed by means of (9) using a function F defined in the interval $(-\infty, \infty)$ with the above properties, then this function h represents a 1-normalized polar function of an R-carrier, satisfying the conditions (1) and (16.7) with $n = 0$, and this carrier is specified by means of a formula such as (6.29).

To sum up:

Theorem. All Radon curves with R-carriers are given in polar coordinates by the equation (13) *in which F denotes a function defined in the interval* $(-\infty, \infty)$ *with the properties* (11) *and* (14).

Theory of general dispersions

In this Chapter a constructive theory of the functions known as general dispersions will be developed. In essence, these are solutions of the Kummer non-linear third order differential equation (11.1) for the case of oscillatory differential equations (Q), (q).

18 Introduction

18.1 Dispersions of the κ-th kind; $\kappa = 1, 2, 3, 4$

According to the theorem of § 13.11, all central dispersions of the first kind, ϕ_ν, of an oscillatory differential equation (q) in the interval $j = (a, b)$ satisfy the non-linear third order differential equation

$$-\{X, t\} + q(X) \cdot X'^2(t) = q(t). \qquad \text{(qq)}$$

Moreover, if the carrier q $(< 0) \in C_2$ all central dispersions of the second, third and fourth kinds, ψ_ν, χ_ρ, ω_ρ, satisfy the differential equations $(\hat{q}_1\hat{q}_1)$, $(\hat{q}_1 q)$, $(q\hat{q}_1)$ formed with the first associated carrier $\hat{q}_1$ of q.

One of the aims of our further study is to obtain *all* the regular (that is to say, satisfying everywhere the inequality $X' \neq 0$) integrals X of the non-linear third order differential equations (qq), $(\hat{q}_1\hat{q}_1)$, $(\hat{q}_1 q)$, $(q\hat{q}_1)$.

By the term dispersions of the first, second, third and fourth kinds of an oscillatory differential equation (q) in the interval $j = (a, b)$ we mean certain functions of a single variable defined constructively by means of the differential equations (q), $(\hat{q}_1)$. Their significance for the transformation theory under consideration lies in the fact that these functions represent all the regular integrals of the non-linear differential equations of the third order mentioned above.

Central dispersions of the 1st, 2nd, 3rd and 4th kinds are therefore special cases of the dispersions of the corresponding kinds. The theory of dispersions consists essentially in the description of properties of integrals of the above non-linear third order equations, and the connection of these integrals with the particular transformation problem relating to transformations of the differential equations (q), $(\hat{q}_1)$ into themselves and into each other.

In what follows we shall include the theory of dispersions in a broader problem as follows.

18.2 General dispersions

Let there be given two oscillatory differential equations (q), (Q) in the intervals $j = (a, b)$ and $J = (A, B)$ respectively:

$$y'' = q(t)y, \qquad \text{(q)}$$

$$\ddot{Y} = Q(T)Y. \qquad \text{(Q)}$$

By general dispersions of the differential equations (q), (Q) (in this order) we mean certain functions of a single variable defined constructively in the interval j by means of the differential equations (q), (Q). The significance of these general dispersions consists in the fact that they represent all the regular integrals X of the non-linear third order differential equation

$$-\{X, t\} + Q(X) \cdot X'^2(t) = q(t). \tag{Qq}$$

Obviously, from the general equation (Qq) we can specialize to the differential equation (qq) by setting $Q = q$, and similarly to the other differential equations $(\hat{q}_1\hat{q}_1)$, $(\hat{q}_1 q)$, $(q\hat{q}_1)$.

The theory of general dispersions consists essentially in the description of the properties of integrals of the differential equation (Qq), and the connection of these integrals with the general transformation problem relating to transformation of the differential equations (Q), (q).

It must be emphasized that the oscillatory character of the differential equations (Q), (q) is of crucial importance for this theory.

19 Linear mapping of the integral spaces of the differential equations (q), (Q) on each other

19.1 Fundamental concepts

We consider two oscillatory differential equations (q), (Q) in the intervals $j = (a, b)$ and $J = (A, B)$. We denote their integral spaces by r, R while (u, v), (U, V) will be arbitrary bases of r, R and w, W the Wronskians of these bases.

Every integral $y \in r$ of (q) has determinate constant coordinates c_1, c_2 with respect to the basis (u, v), such that $y = c_1 u + c_2 v$, and the same is true of every integral $Y \in R$ of (Q): $Y = C_1 U + C_2 V$. Conversely, given any ordered pair of constants c_1, c_2 or C_1, C_2 there corresponds precisely one integral $y \in r$ of (q) or precisely one integral $Y \in R$ of (Q), with the coordinates c_1, c_2 or C_1, C_2.

We now define a linear mapping $\boldsymbol{p}$ of the integral space r onto the integral space R in the following manner; given an integral $y \in r$ of (q).

$$y = \lambda u + \mu v,$$

we associate with this the integral $Y \in R$ of (Q) formed with the same constants λ, μ:

$$Y = \lambda U + \mu V.$$

The image Y of y under this mapping $\boldsymbol{p}$ will be denoted by $\boldsymbol{p}y$, consequently $Y = \boldsymbol{p}y$; it is convenient also to write $y \to Y(\boldsymbol{p})$, or more briefly $y \to Y$. The bases (u, v), (U, V) we call the *first* and *second* bases of the linear mapping $\boldsymbol{p}$. Obviously $u \to U(\boldsymbol{p})$, $v \to V(\boldsymbol{p})$. We say that the linear mapping p is determined by the bases (u, v), (U, V) (in this order), and represent this by writing $\boldsymbol{p} = [u \to U, v \to V]$. The ratio w/W is called the *characteristic* of the linear mapping $\boldsymbol{p}$, and is denoted by the symbol $\chi\boldsymbol{p}$.

The linear mapping $\boldsymbol{p}$ is also determined by an arbitrary first basis $(\bar{u}, \bar{v})$ and second basis $(\boldsymbol{p}\bar{u}, \boldsymbol{p}\bar{v})$, hence $\boldsymbol{p} = [\bar{u} \to \boldsymbol{p}\bar{u}, \bar{v} \to \boldsymbol{p}\bar{v}]$. The Wronskians of the bases $(\bar{u}, \bar{v})(\boldsymbol{p}\bar{u}, \boldsymbol{p}\bar{v})$ differ from w, W by the same (non-zero) multiplicative constant, so the characteristic of the linear mapping $\boldsymbol{p}$ is independent of the choice of bases of $\boldsymbol{p}$.

In the linear mapping $\boldsymbol{p}$, two linearly independent integrals of the differential equation (q) are mapped onto linearly independent integrals of (Q). The images of two dependent integrals of the differential equation (q) are also dependent, and differ from each other by the same multiplicative constant as their originals.

When speaking of two linear mappings of r onto R we say that they have the same or the opposite character according as their characteristics have the same sign or not.

Let $c \neq 0$ be an arbitrary number. The mapping determined by the bases (u, v), (cU, cV) we denote by $c\boldsymbol{p}$, that is to say $c\boldsymbol{p} = [u \to cU, v \to cV]$. This linear mapping maps every integral $y \in r$ of (q) onto the integral $c.\boldsymbol{p}y \in R$, and thus onto an integral of

(Q) which is linearly dependent upon $\boldsymbol{p}y$. We say that the linear mapping $c\boldsymbol{p}$ is linearly dependent or more briefly dependent upon $\boldsymbol{p}$; it is convenient to call this a *variation* of $\boldsymbol{p}$. The characteristic of $c\boldsymbol{p}$ is $(1/c^2)(w/W)$, so $\chi c\boldsymbol{p} = (1/c^2)\chi\boldsymbol{p}$. Consequently the linear mappings $\boldsymbol{p}$ and $c\boldsymbol{p}$ have the same character. The zeros of the images of each integral $y \in r$ of (q) in the linear mappings $\boldsymbol{p}$ and $c\boldsymbol{p}$ are obviously the same.

19.2 Composition of mappings

Alongside the linear mapping $\boldsymbol{p}$, we now consider a linear mapping $\boldsymbol{P}$ of the integral space R of (Q) onto the integral space $\bar{R}$ of another differential equation $(\bar{Q})$ in an interval $\bar{J} = (\bar{A}, \bar{B})$.

As the first basis of $\boldsymbol{P}$ we choose (U, V); let the second be $(\bar{U}, \bar{V})$; $\boldsymbol{P} = [U \to \bar{U}, V \to \bar{V}]$. We denote by $\bar{W}$ the Wronskian of $(\bar{U}, \bar{V})$, so we have $\chi\boldsymbol{P} = W/\bar{W}$.

Now we show that:

The composed mapping $\bar{\boldsymbol{P}} = \boldsymbol{P}\boldsymbol{p}$ of the integral space r onto the integral space $\bar{R}$ is the linear mapping $\bar{\boldsymbol{P}} = [u \to \bar{U}, v \to \bar{V}]$. Its characteristic is the product of $\chi\boldsymbol{P}$, $\chi\boldsymbol{p}$, thus

$$\chi\boldsymbol{P}\boldsymbol{p} = \chi\boldsymbol{P} \cdot \chi\boldsymbol{p}. \tag{19.1}$$

For, let $y \in r$ be an arbitrary integral of (q) and let $Y = \boldsymbol{p}y$, $\bar{Y} = \boldsymbol{P}Y$. Then with appropriate choice of the constants λ, μ we have the following relations

$$y = \lambda u + \mu v, \qquad Y = \lambda U + \mu V, \qquad \bar{Y} = \lambda\bar{U} + \mu\bar{V},$$

and the first part of our statement follows from this. The characteristic $\chi\bar{\boldsymbol{P}}$ is obviously $w/\bar{W} = (w/W)(W/\bar{W})$ which proves the second part.

The inverse mapping to $\boldsymbol{p}$, i.e. $\boldsymbol{p}^{-1}$, of the integral space R onto the integral space r is the linear mapping $\boldsymbol{p}^{-1} = [U \to u, V \to v]$. Its characteristic is the reciprocal of the characteristic of $\boldsymbol{p}$, that is

$$\chi\boldsymbol{p}^{-1} = (\chi\boldsymbol{p})^{-1}. \tag{19.2}$$

For the inverse of the mapping $\boldsymbol{p}$, which is the mapping $\boldsymbol{p}^{-1}$ of the integral space R to the integral space r, is so defined that $\boldsymbol{p}^{-1}\boldsymbol{p}$ is the identity mapping $\boldsymbol{e}$ of the integral space r onto itself: $\boldsymbol{p}^{-1}\boldsymbol{p} = \boldsymbol{e}$. From this the first part of our statement follows. Now obviously $\boldsymbol{e} = [u \to u, v \to v]$, $\chi\boldsymbol{e} = 1$, and when we use (1), with $\boldsymbol{P} = \boldsymbol{p}^{-1}$, the second part also follows.

19.3 Mapping of an integral space into itself

The above results naturally remain valid when two (or more) of the differential equations (q), (Q), $(\bar{Q})$ coincide, so that the corresponding integral spaces r, R, $\bar{R}$ coincide also.

Consider in particular the case $\bar{Q} = Q = q$, $\bar{R} = R = r$. We are then concerned with a linear mapping of the integral space r of the differential equation (q) onto itself.

Such a linear mapping $\boldsymbol{p}$ is determined by a first and second basis (u, v), (U, V) of the differential equation (q) in the above sense: $\boldsymbol{p} = [u \to U, v \to V]$. Given two linear mappings $\boldsymbol{p} = [u \to U, v \to V]$, $\boldsymbol{P} = [U \to \bar{U}, V \to \bar{V}]$ of the integral space r onto itself, their composed mapping $\boldsymbol{Pp}$ is the linear mapping $\boldsymbol{Pp} = [u \to \bar{U}, v \to \bar{V}]$ of r onto itself, and a formula such as (1) holds. The linear mapping $\boldsymbol{p}^{-1} = [U \to u, V \to v]$ is the linear mapping inverse to $\boldsymbol{p}$ of the integral space r onto itself, and there holds a formula such as (2). Moreover, $\boldsymbol{p}^{-1}\boldsymbol{p} = \boldsymbol{e}$, $\chi \boldsymbol{e} = 1$.

19.4 Determination of the linear mappings from the first phases

Let $\boldsymbol{p} = [u \to U, v \to V]$ be a linear mapping of the integral space r of (q) on the integral space R of (Q).

We choose arbitrary first phases α, $\mathbf{A}$ of the bases (u, v), (U, V) of the linear mapping $\boldsymbol{p}$. Then from (5.26) we have

$$\left.\begin{aligned} u &= \varepsilon \sqrt{|w|}\,\frac{\sin\alpha}{\sqrt{|\alpha'|}}, & v &= \varepsilon \sqrt{|w|}\,\frac{\cos\alpha}{\sqrt{|\alpha'|}}, \\ U &= \mathbf{E} \sqrt{|W|}\,\frac{\sin\mathbf{A}}{\sqrt{|\dot{\mathbf{A}}|}}, & V &= \mathbf{E} \sqrt{|W|}\,\frac{\cos\mathbf{A}}{\sqrt{|\dot{\mathbf{A}}|}}; \end{aligned}\right\} \tag{19.3}$$

in which ε, $\mathbf{E}$ take the values $+1$ or -1 according as the phases α, $\mathbf{A}$ are proper or not with respect to the bases (u, v), (U, V).

Let us take arbitrary coordinates $\lambda = \gamma \cos k_2$, $\mu = \gamma \sin k_2$ $(\gamma > 0, 0 \leqslant k_2 < 2\pi)$ and using these coordinates with respect to the bases (u, v), (U, V) let us form the integrals $y = \lambda u + \mu v$ $(\in r)$, $Y = \lambda U + \mu V$ $(\in R)$ of (q), (Q). Then y, Y may be expressed as follows:

$$y = k_1 \frac{\sin(\alpha + k_2)}{\sqrt{|\alpha'|}}, \qquad Y = \frac{\varepsilon E}{\sqrt{|\chi p|}} \cdot k_1 \frac{\sin(\mathbf{A} + k_2)}{\sqrt{|\dot{\mathbf{A}}|}} \qquad (k_1 = \varepsilon \sqrt{|w|}\gamma). \tag{19.4}$$

We see that for every choice of the phases α, $\mathbf{A}$ of the bases (u, v), (U, V) of $\boldsymbol{p}$, the linear mapping $\boldsymbol{p}$ is given by the formula $y \to Y$; here y, Y always represent two integrals of the differential equations (q), (Q) defined by the formulae (4) with arbitrary constants k_1, k_2, such that $k_1 \neq 0$, $0 \leqslant k_2 < 2\pi$. We call an ordered pair $(\alpha, \mathbf{A})$ of phases of the bases (u, v), (U, V) of the linear mapping $\boldsymbol{p}$, a *phase basis* of $\boldsymbol{p}$. For every choice of the phase basis $(\alpha, \mathbf{A})$ of the linear mapping $\boldsymbol{p}$, any two integrals $y \in r$, $Y = \boldsymbol{p}y \in R$ are expressible by the formulae (4) with the same constants k_1, k_2.

From (3) we obtain the relationship

$$\operatorname{sgn} \chi \boldsymbol{p} = \operatorname{sgn} \alpha' \cdot \operatorname{sgn} \dot{\mathbf{A}}, \tag{19.5}$$

that is, the characteristic $\chi \boldsymbol{p}$ is positive if both phases α, $\mathbf{A}$ increase or decrease; it is negative if one of these increases and the other decreases.

We have seen (§ 19.1) that the linear mapping $\boldsymbol{p}$ can also be determined from an arbitrary first basis $(\bar{u}, \bar{v})$ of (q) and the second basis $(\boldsymbol{p}\bar{u}, \boldsymbol{p}\bar{v})$. As first term of a phase basis $(\alpha, \mathbf{A})$ of the linear mapping $\boldsymbol{p}$ we can therefore choose an arbitrary phase α of (q); then the second term $\mathbf{A}$ is determined uniquely up to an integral multiple of π.

19.5 Phase bases of composed mappings

Once again we consider, in addition to the linear mapping $\boldsymbol{p}$, a linear mapping $\boldsymbol{P}$ of the integral space R of (Q) on to the integral space $\bar{R}$ of a further differential equation $(\bar{\mathrm{Q}})$ in an interval $\bar{J} = (\bar{A}, B)$.

Let $(\alpha, \mathbf{A})$ be a phase basis of $\boldsymbol{p}$ and $(\mathbf{A}, \bar{\mathbf{A}})$ a phase basis of $\boldsymbol{P}$. We let ε, $\mathbf{E}$, $\bar{\mathbf{E}}$ have their usual significance in respect of the phases α, $\mathbf{A}$, $\bar{\mathbf{A}}$.

It is easy to verify that:

The composed linear mapping $\boldsymbol{Pp}$ of the integral space r on the integral space $\bar{R}$ admits of the phase basis $(\alpha, \bar{\mathbf{A}})$ and for any two integrals $y \in r$, $\bar{Y} = \boldsymbol{Pp}y \in \bar{R}$ we have the formulae

$$y = k_1 \frac{\sin(\alpha + k_2)}{\sqrt{|\alpha'|}}, \qquad \bar{Y} = \frac{\varepsilon \bar{E}}{\sqrt{|\chi \boldsymbol{P}| \cdot |\chi \boldsymbol{p}|}} k_1 \frac{\sin(\bar{\mathbf{A}} + k_2)}{\sqrt{|\dot{\bar{\mathbf{A}}}|}}. \tag{19.6}$$

The mapping $\boldsymbol{p}^{-1}$ (inverse to the mapping $\boldsymbol{p}$) of the integral space R onto the integral space r admits of the phase basis $(\mathbf{A}, \alpha)$ and for any two integrals $Y \in R$, $y = \boldsymbol{p}^{-1}Y \in r$, we have

$$Y = k_1 \frac{\sin(\mathbf{A} + k_2)}{\sqrt{|\dot{\mathbf{A}}|}}, \qquad y = \varepsilon E \sqrt{|\chi \boldsymbol{p}|}\, k_1 \frac{\sin(\alpha + k_2)}{\sqrt{|\alpha'|}}. \tag{19.7}$$

Every phase basis of the identity mapping $\boldsymbol{e}$ of the integral space r onto itself is obviously of the form $(\alpha, \alpha + n\pi)$, where n is an integer and α is an arbitrary phase of the differential equation (q).

19.6 Determination of a mapping from arbitrary phases

In the above work we have associated with every linear mapping $\boldsymbol{p}$ of the integral space r onto the integral space R an ordered pair of phases α, $\mathbf{A}$ of the equations (q), (Q), namely the phase basis of $\boldsymbol{p}$, and this association has been done in such a way that it allows us to represent every integral $y \in r$ and its image $Y = \boldsymbol{p}y \in R$ by means of the formulae (4).

Conversely, it follows from the formulae (3), (4) that:

Arbitrary phases α, $\mathbf{A}$ of the differential equations (q), (Q) form a phase basis $(\alpha, \mathbf{A})$ of infinitely many linearly dependent mappings $c\boldsymbol{p}$ $(c \neq 0)$, each mapping the integral space r onto the integral space R. We obtain the bases (u, v), (U, V) of a linear mapping $\boldsymbol{p}$ of this system by means of the formulae (3) with arbitrary choice of the constants ε, $\mathrm{E} = \pm 1$; w, $W (\neq 0)$, and for every integral $y \in r$ and its image $Y = py \in R$ there hold formulae such as (4).

19.7 Normalized linear mappings

We shall continue to use the above notation.

Let $\boldsymbol{p} = [u \to U, v \to V]$, $\boldsymbol{P} = [U \to \bar{U}, V \to \bar{V}]$ be arbitrary linear mappings of the integral space r of (q) onto the integral space R of (Q) and of the integral space R onto the integral space $\bar{R}$ of ($\bar{\mathrm{Q}}$). Moreover, let (α, A), (A, $\bar{\mathrm{A}}$) be any phase bases of these linear mappings $\boldsymbol{p}$, $\boldsymbol{P}$.

Let $z \in j$, $Z \in J$ be arbitrary numbers.

We call the linear mapping $\boldsymbol{p}$ *normalized with respect to the numbers* z, Z (in this order) if it maps every integral $y \in r$ of (q) which vanishes at the point z onto an integral $Y \in R$ of (Q) which vanishes at the point Z; that is to say, when $y(z) = 0 \Rightarrow \boldsymbol{p}y(Z) = 0$.

Obviously, the linear mapping $\boldsymbol{p}$ is normalized with respect to the numbers z, Z if it maps a single integral $y \in r$ of (q) which vanishes at the point z onto an integral $Y \in R$ of (Q) which vanishes at the point Z. If therefore $y \in r$ is an arbitrary integral of (q) and $Y \in R$ its image under the linear mapping $\boldsymbol{p}$, then the linear mapping $\boldsymbol{p}$ is normalized with respect to every zero of y and every zero of Y. If, moreover, the linear mapping $\boldsymbol{p}$ is normalized with respect to the numbers z, Z then every mapping $c\boldsymbol{p}$ $(c \neq 0)$ which is dependent upon $\boldsymbol{p}$ has the same property.

We now have the important theorem:

Theorem. The linear mapping $\boldsymbol{p}$ *is normalized with respect to the numbers* z, Z *if and only if the values* $\alpha(z)$, $\mathrm{A}(Z)$ *of the terms of* (α, A) *at the points* z, Z *differ by an integral multiple of* π, *that is* $\alpha(z) - \mathrm{A}(Z) = n\pi$, n *integral.*

Proof. Let $y \in r$ be an integral of (q) which vanishes at the point z, and $Y = \boldsymbol{p}y \in R$ its image under the linear mapping $\boldsymbol{p}$. There hold therefore formulae such as (4) formed with appropriate constants k_1, k_2 and the number $\alpha(z) + k_2$ is an integral multiple of π.

If the linear mapping $\boldsymbol{p}$ is normalized with respect to the numbers z, Z, then $Y(Z) = 0$. Then the number $\mathrm{A}(Z) + k_2$ is an integral multiple of π, and the same is obviously true also of $\alpha(z) - \mathrm{A}(Z)$.

If, conversely, $\alpha(z) - \mathrm{A}(Z)$ is an integral multiple of π, then this is also true for $\mathrm{A}(Z) + k_2$, hence $Y(Z) = 0$. This completes the proof.

We can also show that: If the linear mapping $\boldsymbol{p}$ is normalized with respect to the numbers z, Z then it is also normalized with respect to every two numbers $\bar{z} \in j$, $\bar{Z} \in J$, such that $\bar{z}$ and $\bar{Z}$ are 1-conjugate with z and Z respectively. Conversely, let $\boldsymbol{p}$ be normalized with respect to z, Z and also with respect to two other numbers $\bar{z} \in j$, $\bar{Z} \in J$, where either $\bar{z}$ is conjugate with z or $\bar{Z}$ is conjugate with Z; then both these conjugacies hold, i.e. $\bar{z}$ is conjugate with z and $\bar{Z}$ with Z.

Proof. Let the linear mapping $\boldsymbol{p}$ be normalized with respect to the numbers z, Z, so that $\alpha(z) - \mathrm{A}(Z) = n\pi$, n being an integer.

(a) Let $\bar{z} \in j, \bar{Z} \in J$ be arbitrary numbers, with $\bar{z}$ conjugate to z and $\bar{Z}$ to Z. We have therefore $\bar{z} = \phi_\nu(z)$, $\bar{Z} = \Phi_N(Z)$, in which ϕ_ν, Φ_N are appropriate central dispersions of the differential equations (q) and (Q). Now the Abel functional equation (13.21) gives

$$\alpha(\bar{z}) = \alpha(z) + \nu\pi \cdot \operatorname{sgn} \alpha', \quad \mathbf{A}(\bar{Z}) = \mathbf{A}(Z) + N\pi \cdot \operatorname{sgn} \dot{\mathbf{A}},$$

and we see that the number $\alpha(\bar{z}) - \mathbf{A}(\bar{Z})$ is an integral multiple of π.

(b) Let $\boldsymbol{p}$ be normalized with respect to the numbers $\bar{z}$, $\bar{Z}$ and, for definiteness, assume that $\bar{z}$ is conjugate with z. Then $\alpha(\bar{z}) - \mathbf{A}(\bar{Z})$ is an integral multiple of π, and from the Abel functional equation it follows that the same is true for the number $\alpha(\bar{z}) - \alpha(z)$. We have therefore $\mathbf{A}(\bar{Z}) - \mathbf{A}(Z) = m\pi$, m an integer, and when we take account of the Abel functional equation, this gives $\bar{Z} = \Phi_M(Z)$ with $M = m \operatorname{sgn} \dot{\mathbf{A}}$. This completes the proof.

It is easy to verify the following statements:—

(i) Let the linear mappings $\boldsymbol{p}$, $\boldsymbol{P}$ be normalized with respect to z, Z and $Z, \bar{Z}$ respectively, then the composed linear mapping $\boldsymbol{Pp}$ is normalized with respect to $z, \bar{Z}$.

(ii) If the linear mapping $\boldsymbol{p}$ is normalized with respect to z, Z then the inverse linear mapping $\boldsymbol{p}^{-1}$ is normalized with respect to Z, z.

(iii) If the identity mapping $\boldsymbol{e}$ of the linear space r on itself is normalized with respect to the numbers z, Z then these are conjugate.

19.8 Canonical phase bases

We now show that:—

If the linear mapping $\boldsymbol{p}$ of the integral space r of (q) onto the integral space R of (Q) is normalized with respect to the numbers $z \in j$, $Z \in J$, then it possesses a phase basis $(\alpha, \mathbf{A})$ whose terms vanish at the point z and Z respectively, i.e. $\alpha(z) = 0$, $\mathbf{A}(Z) = 0$.

Proof. Assume that $\boldsymbol{p}$ is *normalized* with respect to the numbers z, Z.

We know that it is possible to choose an arbitrary phase of (q) as first term of a phase basis of $\boldsymbol{p}$, whereupon the second term is uniquely determined up to an integral multiple of π. Let us therefore choose as first term of a phase basis $(\alpha, \mathbf{A}_0)$ of $\boldsymbol{p}$ any phase α of (q) which vanishes at the point z: $\alpha(z) = 0$. Then, since $\boldsymbol{p}$ is *normalized* with respect to z, Z, we have $\mathbf{A}_0(Z) = n\pi$, n being an integer. If we now replace the phase $\mathbf{A}_0$ of (Q) by $\mathbf{A} \equiv \mathbf{A}_0 - n\pi$, then the phases α, $\mathbf{A}$ form a phase basis of $\boldsymbol{p}$ with the desired property.

We call a phase basis $(\alpha, \mathbf{A})$ of a linear mapping $\boldsymbol{p}$ which is normalized with respect to the numbers z, Z a *canonical phase basis with respect to the numbers* z, Z, if its terms α, $\mathbf{A}$ vanish at the points z, Z, i.e. $\alpha(z) = 0$, $\mathbf{A}(Z) = 0$.

The following facts are easily seen:

Let $(\alpha, \mathbf{A})$, $(\mathbf{A}, \bar{\mathbf{A}})$ be canonical phase bases of the linear mappings $\boldsymbol{p}$, $\boldsymbol{P}$ with respect to the numbers z, Z and Z, $\bar{Z}$ respectively. Then $(\alpha, \bar{\mathbf{A}})$ is a canonical phase basis of the linear mapping $\boldsymbol{Pp}$ with respect to the numbers z, $\bar{Z}$.

If $(\alpha, \mathbf{A})$ is a canonical phase basis of the linear mapping $\boldsymbol{p}$ with respect to the numbers z, Z then $(\mathbf{A}, \alpha)$ is a canonical phase basis of the inverse linear mapping $\boldsymbol{p}^{-1}$ with respect to the numbers Z, z.

Every canonical phase basis of the identity mapping $\boldsymbol{e}$ of the integral space $\boldsymbol{r}$ on itself with respect to the numbers z, $\phi_\nu(z)$ has the form $(\alpha, \alpha\phi_{-\nu})$, in which α is a phase of the differential equation (q) which vanishes at the point z: $\alpha(z) = 0$.

20 General dispersions of the differential equations (q), (Q)

We now come to the theory of general dispersions of differential equations (q), (Q). We shall continue to use the above concepts and notation.

20.1 Fundamental numbers and fundamental intervals of the differential equation (q)

Let $t_0 \in j$ be arbitrary and let t_ν be the ν-th 1-conjugate number with t_0, that is to say $t_\nu = \phi_\nu(t_0)$; $\nu = 0, \pm 1, \pm 2, \ldots$. The numbers t_ν are therefore the zeros of every integral v of (q) which vanishes at the point t_0: t_ν coincides with t_0 or represents the ν-th zero of v following or preceding this number, according as $\nu = 0$ or > 0 or < 0.

We call t_ν the ν-th *fundamental number of the differential equation* (q) *with respect to* t_0, more briefly the ν-th fundamental number.

The intervals $j_\nu = [t_\nu, t_{\nu+1})$ and $\bar{j}_\nu = (t_{\nu-1}, t_\nu]$ we call respectively the *right* and *left fundamental intervals of the differential equation* (q) *with respect to* t_0, more briefly the ν-th right and left fundamental intervals. Obviously, the fundamental intervals j_ν and $\bar{j}_\nu$ have only the number t_ν in common; the interior of j_ν coincides with the interior of $\bar{j}_{\nu+1}$.

Every number $t \in (a, b)$ lies in a determinate right fundamental interval j_ν and in a determinate left fundamental interval $\bar{j}_\mu$; if $t = t_\nu$ we have $\mu = \nu$, if $t \neq t_\nu$ we have $\mu = \nu + 1$. In particular, every zero of an arbitrary integral y of (q) lies in determinate fundamental intervals j_ν and $\bar{j}_\mu$; conversely, every fundamental interval j_ν or $\bar{j}_\mu$ contains precisely one zero of y.

20.2 The concept of general dispersion

We now associate with every normalized linear mapping $\boldsymbol{p}$ of the integral space r of (q) onto the integral space R of (Q) a certain function X defined in the interval $j = (a, b)$.

Let $\boldsymbol{p}$ be a normalized linear mapping of the integral space r onto the integral space R with respect to the (arbitrarily chosen) numbers $t_0 \in j$, $T_0 \in J$. We denote by t_ν, T_ν the fundamental numbers of the differential equations (q), (Q) with respect to t_0, T_0; we further denote by j_ν, J_ν the right and by $\bar{j}_\nu$, $\bar{J}_\nu$ the left fundamental intervals of the differential equations (q), (Q) with respect to t_0, T_0; $\nu = 0, \pm 1, \pm 2, \ldots$.

Let $t \in (a, b)$ be arbitrary and y an integral of the differential equation (q) vanishing at the point t; the number t lies in a determinate right ν-th fundamental interval j_ν.

Now we define the value $X(t)$ of the function X at the point t, according as $\chi\boldsymbol{p} > 0$ or $\chi\boldsymbol{p} < 0$, as follows:

In the case $\chi\boldsymbol{p} > 0$, $X(t)$ is the zero of the integral $\boldsymbol{p}y$ of (Q) which lies in the right ν-th fundamental interval J_ν.

In the case $\chi\boldsymbol{p} < 0$, $X(t)$ is the zero of the integral $\boldsymbol{p}y$ of (Q) which lies in the left $-\nu$-th fundamental interval $\bar{J}_{-\nu}$.

This function X is called the *general dispersion of the differential equations* (q), (Q) (in this order) *with respect to the numbers* t_0, T_0 *and the linear mapping* $\boldsymbol{p}$; more briefly: the general dispersion. The numbers t_ν, T_ν are the *fundamental numbers* and in particular t_0, T_0 are the *initial numbers of the general dispersion* X; the linear mapping $\boldsymbol{p}$ is called the *generator* of X. In the case $\chi\boldsymbol{p} > 0$ we call the general dispersion X *direct*, in the case $\chi\boldsymbol{p} < 0$ we call it *indirect*. In the case $Q = q$ we speak more briefly of the *general dispersion* X *of the differential equation* (q).

If we consider the general dispersion of the differential equations (q), (Q) with respect to a mapping $c\boldsymbol{p}$ ($c \neq 0$) (which is linearly dependent upon $\boldsymbol{p}$), and to the same numbers t_0, T_0 then this general dispersion coincides with X, since the linear mapping $c\boldsymbol{p}$ is still normalized with respect to t_0, T_0 and the zeros of the images $\boldsymbol{p}y$, $c\boldsymbol{p}y$ of every integral y of (q) are the same.

The general dispersion X is thus uniquely determined by its initial numbers t_0, T_0 and the generator $\boldsymbol{p}$.

Obviously we have, according as $\chi\boldsymbol{p} > 0$ or $\chi\boldsymbol{p} < 0$ the formulae

$$X(t_\nu) = T_\nu \qquad \text{or} \qquad X(t_\nu) = T_{-\nu}, \tag{20.1}$$

and moreover at every point $t \in (t_\nu, t_{\nu+1})$ we have

$$X(t) \in (T_\nu, T_{\nu+1}) \qquad \text{or} \qquad X(t) \in (T_{-\nu-1}, T_{-\nu}) \tag{20.2}$$
$$(\nu = 0, \pm 1, \pm 2, \ldots).$$

20.3 Properties of general dispersions

Fundamental for the theory of general dispersions is the following theorem

Theorem. Let X be the general dispersion with initial numbers t_0, T_0 and generator $\boldsymbol{p}$. Moreover, let (α, A) be a canonical phase basis of $\boldsymbol{p}$ with respect to the numbers t_0, T_0. Then the general dispersion X satisfies in the interval j the functional equation

$$\alpha(t) = \mathrm{A}(X(t)). \tag{20.3}$$

Proof. Let $x \in j$ be arbitrary. This must lie in a determinate interval j_ν and we thus have, since $\alpha(t_0) = 0$,

$$\nu\pi \leqslant \alpha(x) < (\nu + 1)\pi \qquad \text{or} \qquad -(\nu + 1)\pi < \alpha(x) \leqslant -\nu\pi, \tag{20.4}$$

according as $\operatorname{sgn} \alpha' = +1$ or -1.

Let $y \in r$ be an integral of the differential equation (q) vanishing at the point x; this is given by the formula (19.4) using appropriate constants k_1, k_2. Since $y(x) = 0$ we have

$$\alpha(x) + k_2 = n\pi, \qquad n \text{ integral}. \tag{20.5}$$

The integral $Y = py \in R$ of (Q) is given by (19.4) using the same constants k_1, k_2. Now, by our definition of X the number $X(x)$ is a zero of the integral Y, hence

$$\mathbf{A}(X(x)) + k_2 = N\pi, \qquad N \text{ integral.} \tag{20.6}$$

From the relations (5), (6) it follows that

$$\alpha(x) - \mathbf{A}(X(x)) = m\pi, \qquad m \text{ integral.} \tag{20.7}$$

We now distinguish two cases, according as $\chi \boldsymbol{p} > 0$ or $\chi \boldsymbol{p} < 0$.

In the case $\chi \boldsymbol{p} > 0$, the number $X(x)$ lies in the interval J_ν. We then have, taking account of the fact that $\mathbf{A}(T_0) = 0$,

$$\nu\pi \leqq \mathbf{A}(X(x)) < (\nu + 1)\pi \qquad \text{or} \qquad -(\nu + 1)\pi < \mathbf{A}(X(x)) \leqq -\nu\pi, \tag{20.8}$$

according as $\operatorname{sgn} \dot{\mathbf{A}} = +1$ or -1.

Now since (from (19.5)) $\operatorname{sgn} \alpha' \operatorname{sgn} \mathbf{A} = +1$, either the first or the second of the inequalities (4) and (8) holds. In both cases, we obtain from these inequalities

$$-\pi < \alpha(x) - \mathbf{A}(X(x)) < \pi. \tag{20.9}$$

This gives, together with (7), $\alpha(x) - \mathbf{A}(X(x)) = 0$.

In the case $\chi \boldsymbol{p} < 0$, the number $X(x)$ lies in $\bar{J}_{-\nu}$. We then have, by similar reasoning to that above,

$$-(\nu + 1)\pi < \mathbf{A}(X(x)) \leqslant -\nu\pi \qquad \text{or} \qquad \nu\pi \leqslant \mathbf{A}(X(x)) < (\nu + 1)\pi, \tag{20.10}$$

according as $\operatorname{sgn} \dot{\mathbf{A}} = +1$ or -1.

Now, from (19.5), $\operatorname{sgn} \alpha' \operatorname{sgn} \dot{\mathbf{A}} = -1$, and consequently there hold either the first inequality of (4) and the second inequality of (10), or the second inequality of (4) and the first of (10). In each case, we again obtain from these inequalities the formulae (9) and hence $\alpha(x) - \mathbf{A}(X(x)) = 0$. This completes the proof.

The importance of the functional equation (3) lies in the fact that we can apply it to obtain properties of general dispersions from properties of phases of the differential equations (q), (Q).

We consider a general dispersion X. Let t_0, T_0 be its initial numbers, and $\boldsymbol{p}$ its generator. Moreover, let $(\alpha, \mathbf{A})$ be a canonical phase basis of the linear mapping $\boldsymbol{p}$ with respect to the numbers t_0, T_0. From the above theorem, the functional equation (3) therefore holds in the interval j.

1. For $t \in j$ we have

$$X(t) = \mathbf{A}^{-1}\alpha(t); \tag{20.11}$$

where $\mathbf{A}^{-1}$ naturally denotes the function inverse to the phase A. This follows immediately from the functional equation (3).

2. The function X increases in the interval j from A to B or decreases in this interval from B to A, according as it is direct or indirect.

For if, for instance, the general dispersion X is direct, then (by (19.5)), we have $\operatorname{sgn} \alpha' \operatorname{sgn} \dot{\mathbf{A}} = +1$, so the functions α, $\mathbf{A}$ either increase in the intervals j, J from $-\infty$ to ∞ or decrease from ∞ to $-\infty$. Then (11) gives the corresponding part of our statement.

3. The function X^{-1} inverse to the general dispersion X is the general dispersion $x(T)$ of the differential equations (Q), (q) generated by the mapping $\boldsymbol{p}^{-1}$ inverse to $\boldsymbol{p}$, with the initial numbers T_0, t_0.

Proof. Formula (11) shows that the function X^{-1} inverse to X is the following:

$$X^{-1}(T) = \alpha^{-1}\mathbf{A}(T).$$

But $(\mathbf{A}, \alpha)$ is a canonical phase basis of the mapping $\boldsymbol{p}^{-1}$ inverse to $\boldsymbol{p}$ with respect to the numbers T_0, t_0 (§ 19.8) hence $X^{-1}(T) = x(T)$.

4. The general dispersion X is three times continuously differentiable in the interval j, and the following formulae hold at any two homologous points $t \in j$, $X \in J$; that is to say, two points linked by the relations $X = X(t)$, $t = X^{-1}(X)$:

$$\left.\begin{aligned} X'(t) \quad & \frac{\alpha'(t)}{\dot{\mathbf{A}}(X)}; \qquad X''(t) = \frac{1}{\dot{\mathbf{A}}^3(X)}[\alpha''(t)\dot{\mathbf{A}}^2(X) - \alpha'^2(t)\ddot{\mathbf{A}}(X)], \\ \dot{\mathbf{A}}(X) = & \frac{\alpha'(t)}{X'(t)}; \qquad \ddot{\mathbf{A}}(X) = \frac{1}{X'^3(t)}[\alpha''(t)X'(t) - X''(t)\alpha'(t)]. \end{aligned}\right\} \quad (20.12)$$

The first part of this statement follows immediately from (11), since the functions α, $\mathbf{A}$ are three times continuously differentiable in the intervals j, J and $\dot{\mathbf{A}}$ is always non-zero. The second part is obtained by differentiating the functional equation (3) twice.

5. The first derivative X' of the general dispersion X is always positive or always negative in the interval j according as the latter is direct or indirect: $\operatorname{sgn} X' = \operatorname{sgn} \chi\boldsymbol{p}$.

This follows from the above Theorem 2 and the first formula (12).

6. For $t \in j$ we have the formula

$$X[\phi_\nu(t)] = \Phi_{\nu \cdot \operatorname{sgn} X'}[X(t)] \quad (\nu = 0, \pm 1, \pm 2, \ldots); \qquad (20.13)$$

in which ϕ_ν, Φ_ν denote the ν-th central dispersions of the differential equations (q), (Q).

Proof. The functional equation (3) gives the following relation holding at the point $\phi_\nu(t)$:

$$\alpha\phi_\nu(t) = \mathbf{A}X\phi_\nu(t),$$

and on using (13.21), we have

$$\mathbf{A}X(t) + \nu\pi \cdot \operatorname{sgn} \alpha' = \mathbf{A}X\phi_\nu(t).$$

The function on the left hand side of this equation may obviously be written in the form

$$\mathbf{A}X(t) + (\nu \cdot \operatorname{sgn} \alpha' \cdot \operatorname{sgn} \dot{\mathbf{A}})\pi \cdot \operatorname{sgn} \dot{\mathbf{A}}$$

and since $\operatorname{sgn} \alpha' \operatorname{sgn} \dot{\mathbf{A}} = \operatorname{sgn} X'$, it may also be expressed as $\mathbf{A}\Phi_{\nu \cdot \operatorname{sgn} X'}[X(t)]$.

We have therefore

$$\mathbf{A}\Phi_{\nu \cdot \operatorname{sgn} X'}[X(t)] = \mathbf{A}X\phi_\nu(t),$$

from which the formula (13) follows.

7. The general dispersion X represents, in the interval j, a solution of the non-linear third order differential equation

$$-\{X, t\} + Q(X)X'^2 = q(t). \tag{Qq}$$

Proof. Let $t \in j$ be arbitrary. From the theorem of § 20.3, the functional equation (3) holds at the point t and in its neighbourhood. If we take the Schwarzian derivative of both sides, then use (1.17) and the first formula (12) we get

$$\{X, t\} + [\{\mathbf{A}, X\} + \dot{\mathbf{A}}^2(X)]X'^2 = \{\alpha, t\} + \alpha'^2(t).$$

and, by (5.16), this is precisely the equation (Qq).

The relationship (Qq) we call the *differential equation of the general dispersions of the differential equations* (q), (Q), more briefly the general dispersion equation.

8. The function x inverse to the general dispersion X represents in the interval J a solution of the non-linear third order differential equation

$$-\{x, T\} + q(x)\dot{x}^2 = Q(T). \tag{qQ}$$

This is an obvious consequence of the Theorems 3 and 7 above.

9. We now consider, alongside the linear mapping $\boldsymbol{p}$, a linear mapping $\boldsymbol{P}$ of the integral space R on the integral space $\bar{R}$, normalized with respect to the numbers T_0, Z_0 (Z_0 arbitrary); moreover, let $(\mathbf{A}, \bar{\mathbf{A}})$ be **a** canonical phase basis of $\boldsymbol{P}$ with respect to T_0, Z_0.

We know that the composed linear mapping $\boldsymbol{Pp}$ of the integral space r on the integral space $\bar{R}$ is normalized with respect to t_0, Z_0 and that it admits of the canonical phase basis $(\alpha, \bar{\mathbf{A}})$ with respect to t_0, Z_0 (§ 19.8).

Let X, $\bar{X}$ be two general dispersions of the differential equations (q), (Q) and (Q), ($\bar{\text{Q}}$) generated by the linear mappings $\boldsymbol{p}$, $\boldsymbol{P}$ with initial numbers t_0, T_0 and T_0, Z_0. We shall show that the composed function $\bar{X}X$ is the general dispersion $\bar{\bar{X}}$ of the differential equations (q), ($\bar{\text{Q}}$) generated by the linear mapping $\boldsymbol{Pp}$ with initial numbers t_0, Z_0.

For, from the formulae

$$X(t) = \mathbf{A}^{-1}\alpha(t), \qquad \bar{X}(T) = \bar{\mathbf{A}}^{-1}\mathbf{A}(T) \quad (t \in j,\ T \in J)$$

there follows

$$\bar{X}X(t) = \bar{\mathbf{A}}^{-1}\alpha(t).$$

Now $(\alpha, \bar{\mathbf{A}})$ is the canonical phase basis of the linear mapping $\boldsymbol{Pp}$ with respect to the numbers t_0, Z_0, and it follows, by (11), that $\bar{X}X(t) = \bar{\bar{X}}(t)$.

20.4 Determining elements of general dispersions

We know that given the initial numbers and a linear mapping $\boldsymbol{p}$ normalized with respect to them there is determined from these precisely one general dispersion. We now take up the question as to how far general dispersions are characterized by having a given linear mapping $\boldsymbol{p}$ as their generator. We continue to use the above notation.

1. Let $\boldsymbol{p}$ be a linear mapping of the integral space r of (q) onto the integral space R of (Q). By this linear mapping $\boldsymbol{p}$, there is determined precisely one countable system of general dispersions of the differential equations (q), (Q) with generator $\boldsymbol{p}$. If $X(t)$ is one general dispersion of this system, then the latter consists precisely of the functions $X\phi_\nu(t)$, $\nu = 0, \pm 1, \pm 2, \ldots$.

Proof. Let X be the general dispersion of the differential equations (q), (Q) with the initial numbers t_0, T_0 and generator $\boldsymbol{p}$. We consider a canonical phase basis (α, A) of $\boldsymbol{p}$ with respect to the numbers t_0, T_0 so that $\alpha(t_0) = 0$, $\mathrm{A}(T_0) = 0$, and formula (3) holds.

(a) Let Z be a general dispersion of the differential equations (q), (Q) with initial numbers t_0, Z_0 and generator p. Then we have $Z_0 = Z(t_0)$, and the linear mapping $\boldsymbol{p}$ is normalized with respect to the numbers t_0, Z_0 so $Z_0 = \Phi_\nu(T_0)$ with some appropriate index ν.

We now see, from the identity $\mathrm{A}(T) = \mathrm{A}\Phi_{-\nu}\Phi_\nu(T)$ that the function $\mathrm{A}\Phi_{-\nu}(T)$ vanishes at the point Z_0. Consequently $(\alpha, \mathrm{A}\Phi_{-\nu})$ is a canonical phase basis of the linear mapping $\boldsymbol{p}$ with respect to the numbers t_0, Z_0, and from (3) we have for $t \in j$

$$\mathrm{A}X(t) = \alpha(t) = \mathrm{A}\Phi_{-\nu}Z(t).$$

From this relation it follows that $X(t) = \Phi_{-\nu}Z(t)$ and that $Z(t) = \Phi_\nu X(t)$. This formula gives, on taking account of (13), $Z(t) = X\phi_{\pm\nu}(t)$.

(b) We now consider the function $X\phi_\nu$ formed from an arbitrary central dispersion ϕ_ν of the differential equation (q).

Since the linear mapping $\boldsymbol{p}$ is normalized with respect to the numbers t_0, T_0, it is also normalized with respect to the numbers $\phi_{-\nu}(t_0)$, T_0. From the identity $\alpha(t) = \alpha\phi_\nu\phi_{-\nu}(t)$ we see that the function $\alpha\phi_\nu$ vanishes at $\phi_{-\nu}(t_0)$ so $(\alpha\phi_\nu, \mathrm{A})$ is a canonical phase basis of the linear mapping $\boldsymbol{p}$ with respect to the numbers $\phi_{-\nu}(t_0)$, T_0. Let Z be the general dispersion of the differential equations (q), (Q) with initial numbers $\phi_{-\nu}(t_0)$, T_0 and generator $\boldsymbol{p}$. Then we have, from (3), for $t \in j$

$$\alpha\phi_\nu(t) = \mathrm{A}Z(t)$$

and moreover,

$$\mathrm{A}X(t) = \alpha(t) = \mathrm{A}Z\phi_{-\nu}(t).$$

From these relations it follows that $Z(t) = X\phi_\nu(t)$. This completes the proof.

In the second place we show that general dispersions of the differential equations (q), (Q) can be uniquely determined by means of initial conditions of the second order.

2. Let t_0; X_0, X_0' $(\neq 0)$, X_0'' be arbitrary numbers. There is precisely one general dispersion X of the differential equations (q), (Q) with the initial conditions

$$X(t_0) = X_0, \quad X'(t_0) = X_0', \quad X''(t_0) = X_0''. \tag{20.14}$$

This general dispersion X is direct or indirect according as $X_0' > 0$ or $X_0' < 0$.

Proof. We first assume that there exists a general dispersion X satisfying the above initial conditions. This is uniquely determined by the initial numbers t_0, X_0 and a linear mapping $\boldsymbol{p}$ of the integral space r on the integral space R which is normalized with respect to these numbers. We choose a canonical phase basis $(\alpha, \mathbf{A})$ of $\boldsymbol{p}$ with respect to the numbers t_0, X_0 in such a manner that

$$\alpha(t_0) = 0, \quad \alpha'(t_0) = 1, \quad \alpha''(t_0) = 0. \tag{20.15}$$

Then the general dispersion X satisfies in the interval j a functional equation such as (3) and the formulae (12) give the values of the functions $\dot{\mathbf{A}}$, $\ddot{\mathbf{A}}$ at the point X_0. In this way we obtain the values

$$\mathbf{A}(X_0) = 0, \quad \dot{\mathbf{A}}(X_0) = 1/X_0', \quad \ddot{\mathbf{A}}(X_0) = -X_0''/X_0'^3, \tag{20.16}$$

by which the first phase $\mathbf{A}$ of the differential equation (Q) is uniquely determined. (§ 7.1).

We see that every general dispersion X with the above initial values (14) coincides with that particular one which is determined by the initial numbers t_0, X_0 and the generator $\boldsymbol{p}$. The generator $\boldsymbol{p}$ is determined by the canonical phase basis $(\alpha, \mathbf{A})$ which is given uniquely by the initial values (15), (16). This completes the proof.

The general dispersions of the differential equations (q), (Q) thus form a system which is continuously dependent upon three parameters X_0, X_0', $(\neq 0)$, X_0''.

Finally we show that:

3. Given arbitrary phases α, $\mathbf{A}$ of the differential equations (q), (Q) there is determined precisely one general dispersion of these differential equations as a solution of the functional equation $\alpha(t) = \mathbf{A}(X(t))$. This dispersion X is the general dispersion of the differential equations (q), (Q) with respect to the zeros t_0, T_0 of the phases α, $\mathbf{A}$ and with respect to every linear mapping $\boldsymbol{p}$ with the canonical phase basis $(\alpha, \mathbf{A})$.

Proof. The general dispersion X of the differential equations (q), (Q) with zeros t_0, T_0 of the phases α, $\mathbf{A}$ as initial numbers and generator $\boldsymbol{p}$ with the canonical phase basis $(\alpha, \mathbf{A})$ satisfies the functional equation $\alpha(t) = \mathbf{A}(X(t))$ in the interval j (§ 20.3). At the same time, X is the unique solution of the latter: $X(t) = \mathbf{A}^{-1}\alpha(t)$, which proves this result.

20.5 Integration of the differential equation (Qq)

The above results open the way for us to determine all the regular integrals of the non-linear third order differential equation (Qq) in the interval j. By a regular integral X of the differential equation (Qq) we mean a solution whose derivative X' is always non-zero.

We shall prove the following theorem:—

Theorem. The set of all regular integrals of the differential equation (Qq) *defined in the interval j comprises precisely the general dispersions of the differential equations* (q), (Q).

Proof. (a) Let X be a general dispersion of the differential equations (q), (Q). From § 20.3, 5 and 7 this function represents a regular solution of the differential equation (Qq) in j.

(b) Now let X be a solution of the differential equation (Qq) defined in j.

We choose an arbitrary number t_0 and the first phases α, $\mathbf{A}$ of the differentia equations (q), (Q) determined by the initial values

$$\alpha(t_0) = 0, \quad \alpha'(t_0) = 1, \quad \alpha''(t_0) = 0,$$
$$\mathbf{A}(X_0) = 0, \quad \dot{\mathbf{A}}(X_0) = 1/X_0', \quad \ddot{\mathbf{A}}(X_0) = -X_0''/X_0'^3$$

where X_0, X_0', X_0'' are the values taken by X, X', X'' at the point t_0.

Then we have, in the interval j,

$$-\{\tan \alpha, t\} = q(t), \quad -\{\tan \mathbf{A}, X\} = Q(X)$$

and moreover, since the function X satisfies the differential equation (Qq),

$$-\{X, t\} - \{\tan \mathbf{A}, X\} \cdot X'^2 = -\{\tan \alpha, t\}.$$

It then follows, from (1.17), that

$$\{\tan \mathbf{A}(X), t\} = \{\tan \alpha, t\}$$

and further, on taking account of § 1.8, that

$$\tan \mathbf{A}(X) = \frac{c_{11} \tan \alpha(t) + c_{12}}{c_{21} \tan \alpha(t) + c_{22}},$$

where $c_{11}, \ldots, c_{22}$ denote appropriate constants.

Now the initial values of the phases α, $\mathbf{A}$ have the consequence that $c_{12} = 0$ $c_{11} = c_{22}$, $c_{21} = 0$ and moreover

$$\alpha(t) = \mathbf{A}(X).$$

Consequently X is the general dispersion of the differential equations (q), (Q determined by the initial numbers t_0, X_0 and the linear mapping p of the integra space r of (q) on the integral space R of (Q) determined by the phase basis (α, $\mathbf{A}$) This completes the proof.

20.6 Connection between general dispersions and the transformation problem

We consider a general dispersion X of the differential equations (q), (Q) with initia numbers t_0, T_0 and generator p: we have therefore $\chi p > 0$ or $\chi p < 0$ according a X is direct or indirect.

1. Let $Y \in R$ be an arbitrary integral of the differential equation (Q) and $y \in r$ it original in the linear mapping p, so that $y \to Y(p)$. Then the function $Y(X)/\sqrt{|X'}$ is an integral of the differential equation (q), and in the interval j we have the relation ship

$$\frac{YX(t)}{\sqrt{|X'(t)|}} = \pm \frac{1}{\sqrt{|\chi p|}} y(t); \tag{20.17}$$

in which the sign occurring on the right hand side is independent of the choice of th integral Y.

Proof. Let (α, $\mathbf{A}$) be a canonical phase basis of p with respect to the numbers t_0, T_0 Then the relationship $\mathbf{A}X(t) = \alpha(t)$ holds in the interval j (§ 20.3), and there hol also formulae such as (19.4) for the integrals $Y \in R, y \in r$. The relationship (17) follows immediately (since $\varepsilon E = \pm 1$).

2. For an appropriate variation $\boldsymbol{p}^*$ of $\boldsymbol{p}$, we have, for every integral $Y \in R$ and its original $y \in r$ ($\boldsymbol{p}^*$) in the interval j

$$\frac{YX(t)}{\sqrt{|X'(t)|}} = y(t); \tag{20.18}$$

where $\chi \boldsymbol{p}^* = \operatorname{sgn} X'$.

Proof. If in place of the linear mapping $\boldsymbol{p}$ we choose the linear mapping $\boldsymbol{p}^*$, dependent upon $\boldsymbol{p}$, given by the formula $\boldsymbol{p}^* = \varepsilon E \sqrt{|\chi \boldsymbol{p}|}\boldsymbol{p}$ then we obtain the formula (18). From (19.1), we have $\chi \boldsymbol{p}^* = \operatorname{sgn} \chi \boldsymbol{p}$.

3. Let U, V be independent integrals of the differential equation (Q), and W be the Wronskian of (U, V). Then the integrals $UX/\sqrt{|X'|}$ $(= u)$, $VX/\sqrt{|x'|}$ $(= v)$ of the differential equation (q) are also independent, and for the Wronskian w of (u, v) we have the formula $w = W \operatorname{sgn} X'$.

Proof. The first part of this statement follows from the fact that the images under a linear mapping $\boldsymbol{p}$ of two independent integrals of the differential equation (q) are themselves independent, (§ 19.1). The second part is obtained merely by a short calculation.

From Theorem 1 above we see that every general dispersion of the differential equations (q), (Q) represents a transforming function for these differential equations (q), (Q). From § 11.2 we know that every transforming function of the differential equations (q), (Q) in the interval j satisfies the differential equation (qQ), and consequently is a general dispersion of the differential equations (q), (Q) (§ 20.5). We thus have the following theorem:

Theorem. The transforming functions of the oscillatory differential equations (q), (Q) *in the interval j are precisely the general dispersions of these differential equations* (q), (Q).

The ordered pairs of functions $[\sqrt{|X'(t)|}, X(t)]$ formed from arbitrary general dispersions X of the differential equations (q), (Q) are therefore precisely the transformations of the differential equation (Q) into the differential equation (q).

20.7 Embedding of the general dispersions in the phase group

We consider in this section differential equations (q), (Q) whose intervals of definition j, J are assumed to coincide with $(-\infty, \infty)$: $j = J = (-\infty, \infty)$.

Let D be the set of general dispersions of the differential equations (q), (Q). We know (§ 20.3), that every general dispersion $X \in D$ is unbounded on both sides, belongs to the class C_3, and that its derivative X' never vanishes. Consequently, X is an unbounded phase function of the class C_3 (§ 5.7) so we conclude that D is a subset of the phase group $\mathfrak{G}$ (§ 10.1); $D \subset \mathfrak{G}$.

Let α, $\mathbf{A}$ be arbitrary (first) phases of the differential equations (q) and (Q). We know, (§§ 10.2, 10.3), that all the phases of the differential equations (q) and (Q) respectively form the right cosets $\mathfrak{E}\alpha$ and $\mathfrak{E}\mathbf{A}$ of $\mathfrak{E}$: $\mathfrak{E}$ denotes naturally the fundamental subgroup of $\mathfrak{G}$.

Now all the functions which are inverse to phases of the set $\mathfrak{E}\mathbf{A}$ form the left coset $\mathbf{A}^{-1}\mathfrak{E}$ of $\mathfrak{E}$ (see [81], page 141). We show that:

The set D of the general dispersions of (q), (Q) *is the product of the left coset* $\mathbf{A}^{-1}\mathfrak{E}$ *with the right coset* $\mathfrak{E}\alpha$. i.e.

$$D = \mathbf{A}^{-1}\mathfrak{E}\alpha.$$

Proof. (a) Let $X(t) \in D$. Then, from § 20.3, 1, for arbitrary choice of the phase $\bar{\mathbf{A}}$ of (Q) there holds the relationship

$$X(t) = \bar{\mathbf{A}}^{-1}\alpha(t). \tag{20.19}$$

Since $\mathbf{A}$, $\bar{\mathbf{A}}$ are phases of the same differential equation (Q) and consequently lie in the same right coset $\mathfrak{E}\mathbf{A} = \mathfrak{E}\bar{\mathbf{A}}$, we have $\bar{\mathbf{A}} = \xi\mathbf{A}$, $\xi \in \mathfrak{E}$. It follows that $\bar{\mathbf{A}}^{-1} = \mathbf{A}^{-1}\xi^{-1}$ and moreover, on taking account of (19)

$$X(t) = \mathbf{A}^{-1}\xi^{-1}\alpha \in \mathbf{A}^{-1}\mathfrak{E}\alpha.$$

We have therefore $D \subset \mathbf{A}^{-1}\mathfrak{E}\alpha$.

(b) Let $X(t) \in \mathbf{A}^{-1}\mathfrak{E}\alpha$. Then we have, for arbitrary choice of $\xi \in \mathfrak{E}$

$$X(t) = \mathbf{A}^{-1}\xi\alpha(t) = (\xi^{-1}\mathbf{A})^{-1}\alpha(t).$$

Now, $\xi^{-1} \in \mathfrak{E}$, and we see that $\bar{\mathbf{A}} = \xi^{-1}\mathbf{A} \in \mathfrak{E}\mathbf{A}$ is a phase of the differential equation (Q); consequently we have

$$X(t) = \bar{\mathbf{A}}^{-1}\alpha(t).$$

Then in view of § 20.4, 3, this relation gives $X(t) \in D$. We have, therefore, $\mathbf{A}^{-1}\mathfrak{E}\alpha \subset D$, and the proof is complete.

On this topic, see also § 29.1 on p. 237.

21 Dispersions of the $\varkappa$-th kind; $\varkappa = 1, 2, 3, 4$

We have already encountered, in § 18.1, the concept of a dispersion of the κ-th kind, $\kappa = 1, 2, 3, 4$.

The theory of general dispersions given in § 20 above contains, as a special case, a theory of dispersions of the κ-th kind with which we are now concerned and so gives additional information about the latter. We shall see, however, that besides this new information, there will also emerge some entirely fresh aspects as a consequence of the special character of these dispersions.

In the following we shall consider an oscillatory differential equation (q) in the interval $j = (a, b)$, assuming that $q < 0$ and $q \in C_2$ for $t \in j$.

21.1 Introduction

By dispersions of the first, second, third, or fourth kind of the differential equation (q) *we mean respectively a general dispersion of the differential equations* (q), (q); ($\hat{\text{q}}_1$), ($\hat{\text{q}}_1$); ($\hat{\text{q}}_1$), (q); (q), ($\hat{\text{q}}_1$).

Every dispersion ζ of the κ-th kind can be constructed as a general dispersion using appropriate initial values t_0, T_0 and generator $\boldsymbol{p}$ (§ 20.2). Depending on the value of κ, $\boldsymbol{p}$ is a linear mapping of the integral space r of (q) or of the integral space r_1 of ($\hat{\text{q}}_1$) onto r or r_1. According as $\kappa = 1, 2, 3, 4$, these mappings are $r \to r$, $r_1 \to r_1$, $r \to r_1$, $r_1 \to r$ respectively. We shall begin our further studies with some remarks about linear mappings in the above cases.

Let $\boldsymbol{p}$ be a linear mapping of the integral space r or r_1 onto the integral space r or r_1 and let $\boldsymbol{P}$ be the projection (§ 1.9) of the integral space r on r_1. Moreover, in each of the following cases

$$r \to r; \quad r_1 \to r_1; \quad r \to r_1; \quad r_1 \to r \quad (\boldsymbol{p}) \tag{21.1}$$

let $y \in r$ or $y_1 \in r_1$ be arbitrary integrals of the differential equation (q) or ($\hat{\text{q}}_1$) respectively and $Y \in r$ or $Y_1 \in r_1$ its image under $\boldsymbol{p}$, thus

$$y \to Y; \quad y_1 \to Y_1; \quad y \to Y_1; \quad y_1 \to Y \quad (\boldsymbol{p}).$$

Then in each of these cases we can define a linear mapping $\boldsymbol{p}_0$ of the integral space r onto itself by means of the appropriate one of the following formulae

$$y \to Y; \quad \boldsymbol{P}^{-1}y_1 \to \boldsymbol{P}^{-1}Y_1; \quad y \to \boldsymbol{P}^{-1}Y_1; \quad \boldsymbol{P}^{-1}y_1 \to Y \quad (\boldsymbol{p}_0).$$

This mapping $\boldsymbol{p}_0$ we call the *kernel* of $\boldsymbol{p}$ and we write $\boldsymbol{p}_0 = \mathrm{K}\boldsymbol{p}$.

Every linear mapping $\boldsymbol{p}$ of the integral space r or r_1 onto r or r_1 has therefore-precisely one kernel $\boldsymbol{p}_0$; this is a linear mapping of the integral space r onto itself. Conversely, every linear mapping $\boldsymbol{p}_0$ of the integral space r onto itself in each of the above cases (1) represents the kernel of precisely one linear mapping $\boldsymbol{p}$, and indeed the linear mapping $y \to Y(\boldsymbol{p}_0)$ is the kernel of the following linear mappings:

$$y \to Y; \quad \frac{y'}{\sqrt{-q}} \to \frac{Y'}{\sqrt{-q}}; \quad y \to \frac{Y'}{\sqrt{-q}}; \quad \frac{y'}{\sqrt{-q}} \to Y \quad (\boldsymbol{p}).$$

21.2 Determination of dispersions of the κ-th kind, $\kappa = 1, 2, 3, 4$

We have the following theorems:

1. All dispersions X_1, X_2, X_3, X_4 of the first, second, third, and fourth kinds of the differential equation (q) are determined by the following formulae:

$$\alpha(X_1) = \bar{\alpha}(t); \quad \beta(X_2) = \bar{\beta}(t); \quad \beta(X_3) = \bar{\alpha}(t); \quad \alpha(X_4) = \bar{\beta}(t); \qquad (21.2)$$

in which α, $\bar{\alpha}$ denote arbitrary or appropriate first phases and similarly β, $\bar{\beta}$ second phases of the differential equation (q).

Proof. We shall confine ourselves to the proof of the first two statements.

The relation (2) formed with two arbitrary first phases α, $\bar{\alpha}$ of (q) determines (using § 20.4) a general dispersion X_1 of the (coincident) differential equations (q), (q). This consequently represents a dispersion of the first kind of the differential equation (q). Conversely, every dispersion X_1 of the first kind of the differential equation (q), is a general dispersion of the differential equations (q), (q) (§ 21.1) and consequently satisfies the first relation (2) with some appropriate first phases α, $\bar{\alpha}$ of (q).

The second relation (2) formed with arbitrary first phases β, $\bar{\beta}$ of the differential equation ($\hat{q}_1$) determines (by § 20.4) a general dispersion X_2 of the differential equations ($\hat{q}_1$), ($\hat{q}_1$) (§ 5.11). Consequently this represents a dispersion of the second kind of the differential equation (q). Similarly it can be seen that every dispersion X_2 of the second kind of the differential equation (q), satisfies the second relation (2) with appropriate second phases β, $\bar{\beta}$ of (q).

2. Let X_κ be a dispersion of the κ-th kind of the differential equation (q) and $\boldsymbol{p}_\kappa$ its generator ($\kappa = 1, 2, 3, 4$). By means of suitable variation of the linear mapping $\boldsymbol{p}_\kappa$ we can establish the following relations in the interval j between the integrals y and $Y = \mathrm{K}\boldsymbol{p}_\kappa y$ of (q) and their derivatives y', Y':—

$$\left.\begin{aligned} &\frac{YX_1(t)}{\sqrt{|X_1'(t)|}} = y(t); \quad \frac{1}{\sqrt{|X_2'(t)|}} \cdot \frac{Y'X_2(t)}{\sqrt{-qX_2(t)}} = \frac{y'(t)}{\sqrt{-q(t)}}; \\ &\frac{1}{\sqrt{|X_3'(t)|}} \cdot \frac{Y'X_3(t)}{\sqrt{-qX_3(t)}} = y(t); \quad \frac{YX_4(t)}{\sqrt{|X_4'(t)|}} = \frac{y'(t)}{\sqrt{-q(t)}}. \end{aligned}\right\} \qquad (21.3)$$

This follows from the theorem of § 20.3, 9.

21.3 Determination of the central dispersion of the κ-th kind; $\kappa = 1, 2, 3, 4$

The dispersions of the κ-th kind of the differential equation (q) naturally include the central dispersions of the corresponding kind. The latter are, therefore, determined by formulae such as (2), in which the phases which occur satisfy appropriate conditions. On this topic we have the following theorems:

1. The dispersion X_1 of the first kind of the differential equation (q) determined by the first formula (2) is a central dispersion of the first kind of (q) if and only if the phases α, $\bar{\alpha}$ belong to the first phase system of the same basis of the differential equation (q), and in that case $X_1 = \phi_\nu$ ($\nu = 0, \pm 1, \pm 2, \ldots$) if and only if $\bar{\alpha} = \alpha + \nu\pi \operatorname{sgn} \alpha'$.

The dispersion of the second kind X_2 of the differential equation (q) determined by the second formula (2) is a central dispersion of the second kind of (q) if and only if the phases β, $\bar{\beta}$ belong to the second phase system of the same basis of the differential equation (q), and in this case $X_2 = \psi_\nu$ ($\nu = 0, \pm 1, \pm 2, \ldots$) if and only if $\bar{\beta} = \beta + \nu\pi \operatorname{sgn} \beta'$.

The dispersion of the third kind X_3 of the differential equation (q) determined by the third formula (2) is a central dispersion of the third kind of (q) if and only if the phases $\bar{\alpha}$, β belong to the first and second phase systems of the same basis of the differential equation (q), and in that case $X_3 = \chi_\rho$ ($\rho = \pm 1, \pm 2, \ldots$) if and only if

$$-\frac{1}{2}[1 - (2\rho - \operatorname{sgn} \rho)\varepsilon]\pi < \bar{\alpha} - \beta < \frac{1}{2}[1 + (2\rho - \operatorname{sgn} \rho)\varepsilon]\pi$$

$$(\varepsilon = \operatorname{sgn} \bar{\alpha}' = \operatorname{sgn} \beta').$$

The dispersion of the fourth kind, X_4 of the differential equation (q) determined by the fourth formula (2) is a central dispersion of the fourth kind of (q) if and only if the phases α, $\bar{\beta}$ belong to the first and second phase systems of the same basis of the differential equation (q) and in this case $X_\rho = \omega_\rho$ ($\rho = \pm 1, \pm 2, \ldots$) if and only if

$$-\frac{1}{2}[1 - (2\rho - \operatorname{sgn} \rho)\varepsilon]\pi < \bar{\beta} - \alpha < \frac{1}{2}[1 + (2\rho - \operatorname{sgn} \rho)\varepsilon]\pi$$

$$(\varepsilon = \operatorname{sgn} \alpha' = \operatorname{sgn} \bar{\beta}').$$

These theorems follow from the Abel functional equations (§ 13.7).

2. Let ζ be a dispersion of the κ-th kind of the differential equation (q) with initial numbers t_0, T_0 and generator $\boldsymbol{p}$; $\kappa = 1, 2, 3, 4$. Then, and only then, the following relations hold in the interval j:

(a) $\zeta = \phi_\nu$,	if	$T_0 = \phi_\nu(t_0)$	and	$\boldsymbol{p} = c\boldsymbol{e}$;
(b) $\zeta = \psi_\nu$,	if	$T_0 = \psi_\nu(t_0)$	and	$\boldsymbol{p} = c\boldsymbol{e}$;
(c) $\zeta = \chi_\rho$,	if	$T_0 = \chi_\rho(t_0)$	and	$\boldsymbol{p} = c\boldsymbol{P}$;
(d) $\zeta = \omega_\rho$,	if	$T_0 = \omega_\rho(t_0)$	and	$\boldsymbol{p} = c\boldsymbol{P}$;

where $\boldsymbol{e}$ denotes the identity mapping of the integral space r or r_1 onto itself, $\boldsymbol{P}$ the projection from r on r_1 and c ($\neq 0$) a constant; $\nu = 0, \pm 1, \pm 2, \ldots$; $\rho = \pm 1, \pm 2, \ldots$.

Proof. We shall prove, as an illustration, the result in case (c).

1. Let $\zeta = \chi_\rho$ for $t \in j$. First, we obviously have $T_0 = \zeta(t_0) = \chi_\rho(t_0)$. Now we consider an integral $u \in r$ of the differential equation (q) and let x be a zero of u; we have therefore $u(x) = 0$, $u'\zeta(x) = 0$. Moreover, let $u_1 = \boldsymbol{p}u \in r_1$, consequently $u_1 = \bar{u}'/\sqrt{(-q)}$, in which $\bar{u} \in r$. From the definition of ζ we have $u_1\zeta(x) = 0$, hence

$\bar{u}'\zeta(x) = 0$; the functions u', $\bar{u}'$ have therefore the common zero $\zeta(x)$ so the integrals u, $\bar{u} \in r$ of (q) are linearly dependent, i.e. $\bar{u} = cu$, where c ($\neq 0$) denotes a constant. We have therefore $\boldsymbol{p}u = c\boldsymbol{P}u$.

We next show that the value of the constant c does not depend on the choice of the integral $u \in r$. For, let $v \in r$ be another integral of (q). We first assume that v depends upon u, so that $v = ku$, $0 \neq k =$ constant; then we have $k\boldsymbol{p}u = \boldsymbol{p}(ku) = \boldsymbol{p}v = \bar{c}\boldsymbol{P}v = \bar{c}\boldsymbol{P}ku = \bar{c}k\boldsymbol{P}u = (\bar{c}k/c)\boldsymbol{p}u$ ($0 \neq \bar{c} =$ constant) and consequently $\bar{c} = c$. Secondly, let us assume that v is independent of u; then by the above reasoning, $\boldsymbol{p}v = \bar{c}\boldsymbol{P}v$ ($0 \neq \bar{c} =$ constant). Now consider the integral $u + v \in r$ of (q); on the one hand, we have $\boldsymbol{p}(u + v) = \boldsymbol{p}u + \boldsymbol{p}v = c\boldsymbol{P}u + \bar{c}\boldsymbol{P}v$, but we also have $\boldsymbol{p}(u + v) = C\boldsymbol{P}(u + v) = C(\boldsymbol{P}u + \boldsymbol{P}v)$ ($0 \neq C =$ constant). From these relations it follows that $(c - C)\boldsymbol{P}u + (\bar{c} - C)\boldsymbol{P}v = 0$ and hence $c = C = \bar{c}$; this shows that $\boldsymbol{p} = c\boldsymbol{P}$.

2. Let $T_0 = \chi_\rho(t_0)$ and $\boldsymbol{p} = c\boldsymbol{P}$.

The fundamental numbers t_ν of the differential equation (q) with respect to t_0 are $t_\nu = \phi_\nu(t_0)$ while the T_ν are those of the differential equation $(\hat{q}_1)$ with respect to T_0: $T_\nu = \chi_\rho\phi_\nu(t_0)$; $\nu = 0, \pm 1, \pm 2, \ldots$. Now let $t \in j$ be arbitrary and $u \in r$ an integral of (q) which vanishes at the point t, i.e. $u(t) = 0$. Then $\boldsymbol{p}u = c\boldsymbol{P}u = cu'/\sqrt{(-q)}$. From the definition of ζ, we have $\boldsymbol{p}u\zeta(t) = 0$ and hence $u'\zeta(t) = 0$ so we have $\zeta(t) = \chi_m(t)$ with some appropriate index m. We now show that $m = \rho$ independently of the choice of t. For, t lies in a certain right fundamental interval $[t_\nu, t_{\nu+1})$. From the relation $\operatorname{sgn} \chi\boldsymbol{p} = \operatorname{sgn} \chi(c\boldsymbol{P}) = +1$ we conclude that ζ is direct; consequently $\zeta(t)$ lies in the right fundamental interval $[T_\nu, T_{\nu+1})$. At the same time, from $t \in [\phi_\nu(t_0), \phi_{\nu+1}(t_0))$ it follows that $\zeta(t) \in [\chi_m\phi_\nu(t_0), \chi_m\phi_{\nu+1}(t_0))$. Thus $\chi_m\phi_\nu(t_0) = \chi_\rho\phi_\nu(t_0)$, and this relationship gives immediately $m = \rho$. We have therefore $\zeta(t) = \chi_\rho(t)$ for all $t \in j$, and this completes the proof.

21.4 The group of dispersions of the first kind of the differential equation (q)

In § 20.7 we have embedded the set D of general dispersions of two differential equations (q), (Q) over the intervals $j = J = (-\infty, \infty)$ in the phase group $\mathfrak{G}$. It was there proved that for arbitary choice of first phases α, $\mathbf{A}$ of (q) and (Q) respectively the set D can be represented as

$$D = \mathbf{A}^{-1}\mathfrak{E}\alpha; \tag{21.4}$$

where $\mathfrak{E}$ denotes the fundamental subgroup of $\mathfrak{G}$.

We now consider the case of coincident differential equations (q), (Q): $q = Q$ for all $t \in j = (-\infty, \infty)$; we here assume only that $q \in C_0$, not necessarily that $q \in C_2$. Then the set D represents the dispersions of the first kind of the differential equation (q). If we choose $\mathbf{A} = \alpha$, then formula (4) gives

$$D = \alpha^{-1}\mathfrak{E}\alpha. \tag{21.5}$$

The right side of this formula does not depend on the choice of the phase α, as can easily be seen: all phases of the differential equation (q) form precisely the right coset $\mathfrak{E}\alpha$ of $\mathfrak{E}$ (§ 10.2, 10.3). Consequently every phase $\bar{\alpha}$ of the differential equation (q) has the form $\bar{\alpha} = \xi\alpha$, where ξ is an appropriate element of $\mathfrak{E}$. We have therefore

$\bar{\alpha}^{-1} = \alpha^{-1}\xi^{-1}$ and moreover $\bar{\alpha}^{-1}\mathfrak{E}\bar{\alpha} = \alpha^{-1}(\xi^{-1}\mathfrak{E}\xi)\alpha = \alpha^{-1}\mathfrak{E}\alpha = D$ since obviously $\xi^{-1}\mathfrak{E}\xi = \mathfrak{E}$.

The fundamental subgroup $\mathfrak{E}$ of $\mathfrak{G}$, transformed by an arbitrary phase α of the differential equation (q) in accordance with the right side of (5), obviously represents precisely the set of dispersions of the first kind of the differential equation (q).

It is however known from group theory that under a transformation of this kind a subgroup of $\mathscr{R}$ goes over into another subgroup which is described as conjugate to the first.

The set of dispersions of the first kind of the differential equation (q) *forms a subgroup in the phase group* $\mathfrak{G}$ *conjugate with the fundamental subgroup.*

See also p. 237 for some recent results.

21.5 Group property of dispersions of the first kind

We now bring out the group property of the set of dispersions of the first kind from their definition, and then study the structure of this group.

To simplify our formulation, a *dispersion* will here always mean a dispersion of the first kind. Correspondingly, those linear mappings which come under consideration will always be linear mappings of the integral space r of the differential equation (q) onto itself, and by fundamental intervals we always mean the fundamental intervals of the differential equation (q); j naturally denotes the interval $(-\infty, \infty)$.

1. Let ζ be the dispersion determined by arbitrary equal initial numbers t_0, t_0 and the identically linear mapping $\boldsymbol{e}$. Then we have $\zeta(t) = t$ for $t \in j$.

For, let $t \in j$ be arbitrary. If $t = t_0$, then $\zeta(t) = t_0 = t$, so let us assume that $t \neq t_0$. Let $y \in r$ be an integral of the differential equation (q) vanishing at the point t; t lies in a certain ν-th right fundamental interval j_ν with respect to t_0. Since sgn $\chi\boldsymbol{e} = 1$ we conclude that ζ is direct; consequently $\zeta(t)$ is that zero of the integral $\boldsymbol{e}y = y$ which lies in the same fundamental interval j_ν. We have therefore $\zeta(t) = t$.

2. Let ζ be the dispersion determined by arbitrary initial numbers t_0, T_0 and an arbitrary generator $\boldsymbol{p}$. Then the inverse function ζ^{-1} represents the dispersion determined by the initial numbers T_0, t_0 and the inverse linear mapping $\boldsymbol{p}^{-1}$; this is direct or indirect according as ζ is direct or indirect.

Proof. We have $\chi\boldsymbol{p} > 0$ or < 0 according as ζ is direct or indirect. The mapping $\boldsymbol{p}^{-1}$ inverse to $\boldsymbol{p}$ is normalized with respect to the numbers T_0, t_0 (§ 19.7) and has characteristic $1/\chi\boldsymbol{p}$ (§ 19.2).

Let Z be the dispersion determined by the initial numbers T_0, t_0 and the generator $\boldsymbol{p}^{-1}$; Z is therefore direct or indirect according as ζ is direct or indirect.

Let $t \in j$ be an arbitrary number. If $t = T_0$, then $\zeta^{-1}(t) = \zeta^{-1}(T_0) = t_0 = Z(T_0) = Z(t)$, and consequently $\zeta^{-1}(t) = Z(t)$.

We now assume $t \neq T_0$. Let $y \in r$ be an integral of the differential equation (q) vanishing at the point t. Now t lies in a certain right fundamental interval with respect to T_0; let it be the ν-th such interval. It also lies in, say, the μ-th left fundamental interval. Consequently, $\zeta^{-1}(t)$ is the zero of the integral $\boldsymbol{p}^{-1}y \in r$ lying in the ν-th or $-\mu$-th right fundamental interval with respect to t_0, according as the dispersion ζ is direct or indirect. Hence, using the definition of Z, $\zeta^{-1}(t) = Z(t)$.

3. Let ζ_1, ζ_2 be the dispersions determined by arbitrary initial numbers $t_0, \bar{t}_0; \bar{t}_0, T_0$ and arbitrary generators $\boldsymbol{p}_1, \boldsymbol{p}_2$. Then the composite function $\zeta_2\zeta_1$ represents the dispersion determined by the initial numbers t_0, T_0 and the composite linear mapping $\boldsymbol{p}_2\boldsymbol{p}_1$. The latter is direct if both dispersions ζ_1, ζ_2 are direct or both are indirect; it is however indirect if one of these dispersions is direct and the other indirect.

Proof. We have $\chi\boldsymbol{p}_i > 0$ or < 0 according as ζ_i is direct or indirect ($i = 1, 2$). The linear mapping $\boldsymbol{p} = \boldsymbol{p}_2\boldsymbol{p}_1$ is normalized with respect to the numbers t_0, T_0 (§ 19.7) and has the characteristic $\chi\boldsymbol{p} = (\chi\boldsymbol{p}_2)\,(\chi\boldsymbol{p}_1)$ (§ 19.2).

Let ζ be the dispersion determined by the initial numbers t_0, T_0 and generator $\boldsymbol{p}$; ζ is therefore direct or indirect according as $(\chi\boldsymbol{p}_2)(\chi\boldsymbol{p}_1) > 0$ or < 0.

Let $t \in j$ be arbitrary. If $t = t_0$ then we have $\zeta_1(t) = \bar{t}_0$, $\zeta_2[\zeta_1(t)] = \zeta_2(\bar{t}_0) = T_0 = \zeta(t_0) = \zeta(t)$ and consequently we have $\zeta_2\zeta_1(t) = \zeta(t)$.

We assume now that $t \neq t_0$. Let $y \in r$ be an integral of the differential equation (q) vanishing at the point t; t lies in a certain right fundamental interval—say, the ν-th such interval—with respect to t_0. The number $\zeta_1(t)$ is therefore the zero of the integral $\boldsymbol{p}_1 y \in r$ lying in the ν-the right fundamental interval or the $-\nu$-th left fundamental interval with respect to $\bar{t}_0$ according as $\chi\boldsymbol{p}_1 > 0$ or < 0.

If $\chi\boldsymbol{p}_1 > 0$, then $\zeta_2[\zeta_1(t)]$ is that zero of the integral $\boldsymbol{p}_2(\boldsymbol{p}_1 y) = \boldsymbol{p}y \in r$ which is contained in the ν-th right or the $-\nu$-th left fundamental interval with respect to T_0, according as $\chi\boldsymbol{p}_2 > 0$ or < 0.

If $\chi\boldsymbol{p}_1 < 0$, then $\zeta_2[\zeta_1(t)]$ is that zero of $\boldsymbol{p}_2(\boldsymbol{p}_1 y) = \boldsymbol{p}y$ which lies in the $-\nu$-th left or ν-th right fundamental interval with respect to T_0, according as $\chi\boldsymbol{p}_2 > 0$ or < 0.

Thus the number $\zeta_2\zeta_1(t)$ is that zero of the integral $\boldsymbol{p}y \in r$ which lies in the ν-th right or $-\nu$-th left fundamental interval with respect to T_0 according as $(\chi\boldsymbol{p}_2)(\chi\boldsymbol{p}_1) > 0$ or < 0. It follows, using the definition of the function ζ, that

$$\zeta_2\zeta_1(t) = \zeta(t).$$

4. Let $t_0; \zeta_0, \zeta_0' (\neq 0), \zeta_0''$ be arbitrary numbers. There exists precisely one dispersion ζ satisfying the Cauchy initial conditions

$$\zeta(t_0) = \zeta_0, \quad \zeta'(t_0) = \zeta_0', \quad \zeta''(t_0) = \zeta_0''. \tag{21.6}$$

and this is direct or indirect according as $\zeta_0' > 0$ or $\zeta_0' < 0$.

This theorem is obviously a special case of that in § 20.4, 2. The dispersion ζ in question is determined uniquely by the first phases α, $\mathbf{A}$ of the differential equation (q), which are themselves determined by the initial values

$$\alpha(t_0) = 0, \quad \alpha'(t_0) = 1, \quad \alpha''(t_0) = 0; \quad \mathbf{A}(\zeta_0) = 0, \quad \mathbf{A}'(\zeta_0) = 1/\zeta_0',$$
$$\mathbf{A}''(\zeta_0) = -\zeta_0''/\zeta_0'^3$$

by use of the formula $\mathbf{A}\zeta(t) = \alpha(t)$; $t \in (-\infty, \infty)$.

We see that the dispersions form a system which is continuously dependent upon the parameters $\zeta_0, \zeta_0' (\neq 0), \zeta_0''$.

5. The dispersions form a 3-parameter group $\mathfrak{D}$ in which the operation of multiplication is defined as the composition of functions, with unit element ($\underline{1} =$) t. The direct, and consequently increasing, dispersions form in the group $\mathfrak{D}$ an invariant subgroup $\mathfrak{P}$ with index 2; the indirect, and consequently decreasing, dispersions form in

the group $\mathfrak{D}$ the coset of $\mathfrak{P}$, and consequently the second element of the factor group $\mathfrak{D}/\mathfrak{P}$.

Proof. The first part of this theorem follows immediately from the above results 1–4. Obviously, $\underline{1} \in \mathfrak{P}$ and for arbitrary elements $\zeta_1, \zeta_2 \in \mathfrak{P}$, we have $\zeta_1^{-1} \in \mathfrak{P}$, $\zeta_2\zeta_1 \in \mathfrak{P}$ is a subgroup of $\mathfrak{D}$.

Now let $A \subset \mathfrak{D}$ be the set comprising the indirect dispersions, and consider an arbitrary dispersion $\zeta \in \mathfrak{D}$. From the definition of the left (right) coset $\zeta\mathfrak{P}$ ($\mathfrak{P}\zeta$) of the elements $\zeta \in \mathfrak{D}$ with respect to the subgroup $\mathfrak{P}$, the former constitutes the set comprising the dispersions $\zeta X (X\zeta)$ in which X ranges over the elements of $\mathfrak{P}$. But, from 3, $\zeta X \in \mathfrak{P}$ or $\zeta X \in A$ ($X\zeta \in \mathfrak{P}$ or $X\zeta \in A$), according as $\zeta \in \mathfrak{P}$ or $\zeta \in A$. We have therefore $\xi\mathfrak{P} = \mathfrak{P} = \mathfrak{P}\xi$ in the case $\xi \in \mathfrak{P}$ and $\xi\mathfrak{P} = A = \mathfrak{P}\xi$ in the case $\xi \in A$, so that in both cases $\zeta\mathfrak{P} = \mathfrak{P}\zeta$. This shows that the subgroup $\mathfrak{P}$ in $\mathfrak{D}$ is invariant. The factor group $\mathfrak{D}/\mathfrak{P}$ is obviously formed from the two elements $\mathfrak{P}$, A. This completes the proof.

We call $\mathfrak{D}$ the *dispersion group of the first kind* of the differential equation (q), or more briefly the dispersion group.

21.6 Representation of the dispersion group

1. In the following we shall continue to denote the dispersion group of the first kind of the differential equation (q) by $\mathfrak{D}$. We shall introduce also the following notation:

$\mathfrak{P}$: The invariant subgroup of $\mathfrak{D}$ comprising the direct dispersions.

$\mathfrak{C}$: The infinite cyclic group formed from the central dispersions of the first kind of the differential equation (q) (§ 12.5).

$\mathfrak{S}$: The infinite cyclic group formed from the central dispersions of the first kind with even indices of the differential equation (q).

We have therefore the following inclusion relationships

$$\mathfrak{D} \supset \mathfrak{P} \supset \mathfrak{C} \supset \mathfrak{S} \supset \{\underline{1}\}. \tag{21.7}$$

We now choose a basis (u, v) of the differential equation (q) and denote its Wronskian by w. Let $\zeta \in \mathfrak{D}$ be an arbitrary dispersion, and U, V be the functions

$$U = \frac{u(\zeta)}{\sqrt{|\zeta'|}}, \qquad V = \frac{v(\zeta)}{\sqrt{|\zeta'|}}. \tag{21.8}$$

From § 20.6, 3, U, V are independent integrals of (q) and the Wronskian of the basis of (q) formed from these is $W = w \operatorname{sgn} \zeta'$.

Obviously, the following formulae hold in the interval j:

$$\left.\begin{aligned} \frac{u(\zeta)}{\sqrt{|\zeta'|}} &= c_{11}u + c_{12}v, \\ \frac{v(\zeta)}{\sqrt{|\zeta'|}} &= c_{21}u + c_{22}v; \end{aligned}\right\} \tag{21.9}$$

in which c_{11}, c_{12}, c_{21}, c_{22} are appropriate constants. The 2×2 matrix $C = \|c_{ik}\|$ of these constants is determined uniquely by the dispersion ζ.

Now it follows from (9) that $W = |C|w$, and hence

$$|C| = \operatorname{sgn} \zeta';$$

where $|C|$ naturally denotes the determinant of C. C is thus a unimodular matrix with determinant sgn ζ'. From § 20.3, 2 it follows that $|C|$ has the value 1 or -1 according as the dispersion ζ is direct or indirect.

To simplify our notation we shall write the formula (9) in the vector form

$$\frac{\boldsymbol{u}(\zeta)}{\sqrt{|\zeta'|}} = C\boldsymbol{u}, \tag{21.10}$$

where $\boldsymbol{u}$ denotes the vector formed from the components u, v.

2. We now associate with every dispersion $\zeta \in \mathfrak{D}$ the matrix C defined by means of the formula (10). In this way we obtain a mapping $\boldsymbol{d}$ of the group $\mathfrak{D}$ onto the group $\mathfrak{L}$ formed from all unimodular 2×2 matrices. We are in fact concerned with a mapping *onto* the group $\mathfrak{L}$ as can be seen as follows:

Let $C = ||c_{ik}||$ be an arbitrary element of $\mathfrak{L}$; we have to show the existence of an original $\zeta \in \mathfrak{D}$ of C under the mapping $\boldsymbol{d}$. To accomplish this, we choose an arbitrary zero t_0 of the integral $c_{21}u + c_{22}v \in r$ and a zero T_0 of $v \in r$, doing this in such a way that

$$\operatorname{sgn} u(T_0) = \operatorname{sgn} (c_{11}u(t_0) + c_{12}v(t_0)) \tag{21.11}$$

(it is easy to show that such a choice is always possible). Let ζ be the dispersion determined by the initial numbers t_0, T_0 and the generator $\boldsymbol{p} = [c_{11}u + c_{12}v \to u, c_{21}u + c_{22}v \to v]$. Since $|C| = \operatorname{sgn} |C|$, we have $\chi\boldsymbol{p} = \operatorname{sgn} |C|$. Now the formula (20.17) applied to the integrals $(Y =)$ u, v and the dispersion ζ gives, when we take account of (11), relationships such as (9). Consequently C is the original of the dispersion ζ in the mapping $\boldsymbol{d}$.

We now wish to show that $\boldsymbol{d}$ is a homomorphic mapping (deformation) of the group $\mathfrak{D}$ onto the matrix group $\mathfrak{L}$.

We therefore consider arbitrary dispersions ζ_1, $\zeta_2 \in \mathfrak{D}$ and their $\boldsymbol{d}$-images C_1, $C_2 \in \mathfrak{L}$. From the formulae

$$\frac{\boldsymbol{u}(\zeta_1)}{\sqrt{|\zeta_1'|}} = C_1\boldsymbol{u}, \qquad \frac{\boldsymbol{u}(\zeta_2)}{\sqrt{|\zeta_2'|}} = C_2\boldsymbol{u}$$

there follows the relationships

$$\frac{\boldsymbol{u}(\zeta_2\zeta_1)}{\sqrt{|\zeta_2(\zeta_1)|}} \cdot \frac{1}{\sqrt{|\zeta_1'|}} = C_2 \frac{\boldsymbol{u}(\zeta_1)}{\sqrt{|\zeta_1'|}} = C_2C_1\boldsymbol{u},$$

and moreover

$$\frac{\boldsymbol{u}(\zeta_2\zeta_1)}{\sqrt{|(\zeta_2\zeta_1)'|}} = C_2C_1\boldsymbol{u},$$

so that we have, indeed $\boldsymbol{d}(\zeta_2\zeta_1) = C_2C_1$.

3. The unit element of the group $\mathfrak{L}$ is naturally the unit matrix $E = ||\delta_{ik}||$ ($\delta_{11} = \delta_{22} = 1$; $\delta_{12} = \delta_{21} = 0$). We now prove the theorem:

The deformation $\boldsymbol{d}$ of the group $\mathfrak{D}$ onto the group $\mathfrak{L}$ maps onto the unit element $E \in \mathfrak{L}$ precisely the central dispersions of the first kind of the differential equation (q) with even indices while onto the element $-E \in \mathfrak{L}$ it maps precisely the central dispersions of the first kind of the differential equation (q) with odd indices.

Proof. Let ζ be a central dispersion of the first kind of the differential equation (q), so that from (13.10) we have

$$\frac{u(\zeta)}{\sqrt{|\zeta'|}} = u, \quad \text{or} \quad \frac{u(\zeta)}{\sqrt{|\zeta'|}} = -u, \tag{21.12}$$

according as the index of ζ is even or odd. It follows that $\boldsymbol{d}\zeta = E$ or $\boldsymbol{d}\zeta = -E$. Conversely, let any dispersion ζ be such that $\boldsymbol{d}\zeta = E$ or $\boldsymbol{d}\zeta = -E$; then the formulae (12) hold and from that and § 3.12 we conclude that ζ is a central dispersion of the first kind of (q) with an even or an odd index.

If we now apply the first isomorphism theorem for groups (see [81], p. 178), then we obtain the following result:

The group $\mathfrak{S}$ formed from the central dispersions of the first kind of the differential equation (q) with even indices is invariant in the group $\mathfrak{D}$, and the factor group $\mathfrak{D}/\mathfrak{S}$ is isomorphic to the matrix group $\mathfrak{L}$. All the dispersions included in the same element of $\mathfrak{D}/\mathfrak{S}$ are mapped by the deformation $\boldsymbol{d}$ onto the same element of $\mathfrak{L}$.

4. From the formula (20.13) we conclude that every central dispersion of the first kind of the differential equation (q) commutes with each direct dispersion of (q). This implies that the group $\mathfrak{C}$ is a subgroup of the centre of $\mathfrak{P}$. We now show that $\mathfrak{C}$ coincides with the centre, in other words we have the following theorem:

Theorem. The central dispersions of the first kind of the differential equation (q) *form the centre of the group* $\mathfrak{P}$ *of the direct dispersions of* (q).

Proof. It is clearly sufficient to show that every direct dispersion which commutes with all direct dispersions is a central dispersion of the first kind of (q).

Let ζ_0 be a direct dispersion which commutes with all direct dispersions, and consequently with all elements of $\mathfrak{P}$. Moreover let ζ be an arbitrary element of $\mathfrak{P}$. Further, let $C_0 = ||c_{ik}^0||$, $C = ||c_{ik}||$ be the $\boldsymbol{d}$-images of ζ_0 and ζ respectively. We have therefore $C_0 = \boldsymbol{d}\xi_0$, $C = \boldsymbol{d}\xi$; $|C_0| = 1$, $|C| = 1$. From $\zeta_0\zeta = \zeta\zeta_0$ there follows the relationship

$$C_0 C = C C_0. \tag{21.13}$$

Since every element of $\mathfrak{L}$ possesses a $\boldsymbol{d}$-original, we conclude from (13) that the matrix C_0 commutes with every matrix $C \in \mathfrak{L}$, $(|c| = 1)$. Let us now first choose $c_{12} = c_{21} = 0$, $c_{11} \neq c_{22}$, $c_{11}c_{22} = 1$ and then $c_{11} = c_{22} = 0$, $c_{12}c_{21} = -1$; then we obtain $c_{12}^0 = c_{21}^0 = 0$, $c_{11}^0 = c_{22}^0$; $c_{11}^0 c_{22}^0 = 1$. We have therefore $C_0 = E$ or $C_0 = -E$ and we see that ζ_0 is a central dispersion of the first kind of the differential equation (q). This completes the proof.

The above theorem was the motive for our choice of the adjective "central" in the term central dispersion (§ 12.2). We sum up as follows:

The dispersions of the first kind of the differential equation (q) *form a 3-parameter continuous group* $\mathfrak{D}$. *The direct* (*increasing*) *dispersions form in* $\mathfrak{D}$ *an invariant subgroup* $\mathfrak{P}$ *with index 2; the indirect* (*decreasing*) *dispersions form the coset of* $\mathfrak{P}$. *The infinite cyclic group* $\mathfrak{C}$ *formed from the central dispersions of the first kind of the differential equation* (q) *is the centre of* $\mathfrak{P}$. *The central dispersions of the first kind of the differential equation* (q) *with even indices form an invariant subgroup* $\mathfrak{S}$ *in* $\mathfrak{D}$, *and the factor group* $\mathfrak{D}/\mathfrak{S}$ *is isomorphic to the group formed from all* 2×2 *unimodular matrices.*

21.7 The group of dispersions of the second kind of the differential equation (q)

We now assume that the differential equation (q) admits of the first associated differential equation ($\hat{q}_1$) (§ 1.9). Then the differential equation (q) possesses dispersions of the second kind, and these coincide with those of the first kind of ($\hat{q}_1$). Consequently, the dispersions of the second kind of the differential equation (q) form a continuous 3-parameter group $\mathfrak{D}_1$, whose structure is naturally analogous to that of the group $\mathfrak{D}$, with, of course, central dispersions of the second kind replacing those of the first kind.

21.8 The semigroupoid of general dispersions of the differential equations (q), (Q)

Let (q), (Q) be arbitrary oscillatory differential equations in the interval $j = (-\infty, \infty)$.

We consider the non-linear third order differential equations

$$(qq), \quad (qQ), \quad (Qq), \quad (QQ)$$

and denote by

$$G_{11}, \quad G_{12}, \quad G_{21}, \quad G_{22}$$

respectively the sets consisting of all regular integrals of these differential equations in the interval $(-\infty, \infty)$. Thus, for instance, G_{11} is the set of all dispersions of the first kind of the differential equation (q), G_{12} those of the general dispersions of the differential equations (Q), (q), etc.

We know that the function $X(t) = t$ lies in the sets G_{11} and G_{22}. We also know that the function x_{ik}^{-1} inverse to any general dispersion $x_{ik} \in G_{ik}$ is a member of the set G_{ki}: $x_{ik}^{-1} \in G_{ki}$ $(i, k = 1, 2)$.

Let $x_{ik} \in G_{ik}$, $y_{km} \in G_{km}$ $(i, k, m = 1, 2)$ be arbitrary general dispersions of the corresponding differential equations (q) and (Q). It is easily seen that the function $x_{ik}y_{km}$ formed by composition of these general dispersions (the order is significant) gives a general dispersion contained in the set G_{im}; $x_{ik}y_{km} \in G_{im}$; conversely every element $z_{im} \in G_{im}$ is the composition of appropriate general dispersions $x_{ik} \in G_{ik}$, $y_{km} \in G_{km}$: i.e. $x_{ik}y_{km} = z_{im}$. These facts can conveniently be expressed by means of the formula

$$G_{ik}G_{km} = G_{im} \quad (i, k, m = 1, 2)$$

or by the following multiplication table

	G_{11}	G_{12}	G_{21}	G_{22}
G_{11}	G_{11}	G_{12}	—	—
G_{12}	—	—	G_{11}	G_{12}
G_{21}	G_{21}	G_{22}	—	—
G_{22}	—	—	G_{21}	G_{22}

We now consider the semigroupoid Γ constructed from the union set $G_{11} \cup G_{12} \cup G_{21} \cup G_{22}$ with the operation of multiplication defined as composition of functions.

From the above we see that:

The sets G_{11}, G_{12} are groups with the common unit element $(\underline{1} =)$ $X(t) = t$. The sets G_{12}, G_{21} are formed from pairwise inverse elements. Since $G_{11}G_{12} = G_{12}$, $G_{12}G_{22} = G_{12}$, the group G_{11} is a left operator region and the group G_{22} is a right operator region of the set G_{12}. Since $G_{22}G_{21} = G_{21}$, $G_{21}G_{11} = G_{21}$ the group G_{22} is a left and the group G_{11} is a right operator region of the set G_{21}.

In this way we arrive at the following structure of the semigroupoid Γ of the general dispersions of the differential equations (q), (Q):

The semigroupoid Γ consists of two groups G_{11}, G_{22} with the common unit element $\underline{1}$ and of two further, equivalent, sets G_{12}, G_{21}. These sets have the two products $G_{12}G_{21}$ and $G_{21}G_{12}$, which coincide with the groups G_{11} and G_{22}, and consist of pairwise inverse elements whose product is always the unit element of the groups G_{11}, G_{22}. The group G_{11} is a left and the group G_{22} is a right operator region of the set G_{12}; similarly the group G_{11} is a right and the group G_{22} is a left operator region of the set G_{21}.

We conclude this discussion with one remark: the groups G_{11}, G_{22} always have in common the group $\{\underline{1}\}$ consisting only of the unit element; in special cases, however their intersection can be larger. This occurs, for instance, when the differential equations (q), (Q) have the same fundamental dispersion ϕ of the first kind, for in this case their intersection $G_{11} \cap G_{22}$ contains the cyclic group consisting of all the central dispersions ϕ_ν of the first kind of the differential equations (q), (Q).

III General transformation theory

Up till now we have encountered the transformation problem exclusively with reference to oscillatory differential equations (q), (Q) (§§ 13.5, 20.6).

In this third part we shall be concerned with the transformation problem for arbitrary differential equations (q), (Q). Our study will be divided into two parts, according to the assumptions which are made with regard to the transforming function. In the first chapter we shall be concerned with general transformations in which the transforming function is in general not restricted by any conditions. The second chapter will contain the theory of so-called complete transformations. These are characterized by the fact that the intervals of definition and the ranges of their kernels coincide with the intervals of definition of the differential equations (q), (Q).

General transformations

In this chapter we shall first establish the special form of the transformation formula (11.11). We shall then build up a transformation theory at the heart of which lie problems of existence and uniqueness relating to solutions of the differential equation (Qq), for general differential equations (q), (Q).

22 Establishment of the special form of the transformation formula

22.1 A theorem on transformations of second order differential equations

The special form of the transformation formula (11.11), which is linear with respect to the solutions Y, y of the equations (Q), (q), may perhaps appear to be arbitrary and conditioned by the methods applied to the solution of the transformation problem. The question now arises whether this transformation formula may not be replaced by a more general relationship constructed with an appropriate function f, of the form

$$y(t) = f(t, Y[X(t)])$$

We shall show that the answer to this question is, in general, negative.

P. Stäckel [J. reine angew. Math. 111 (1893)], S. Lie (Leipziger Ber. 1894) and E. J. Wilczynski [Amer. J. Math. 23 (1901)] have shown by various methods that generally the linear form of the transformation formula (11.11) is the only one possible.

We can express this result in the following theorem. The formulation of the theorem in a manner suitable for application to the Jacobian form of the differential equation, and also its proof, were kindly supplied by Frau Z. Mikolajska-Młak.

Theorem. Let s, S *be three-dimensional spaces with point coordinates* (t, y, z), (T, Y, Z), *in which*

$$a < t < b; \qquad -\infty < y, z < \infty; \qquad A < T < B; \qquad -\infty < Y, Z < \infty.$$

Moreover, let

$$\left.\begin{aligned} t &= x(T), \\ y &= f(t, Y, Z), \\ z &= g(t, Y, Z), \end{aligned}\right\} \qquad (t)$$

$$\left.\begin{aligned} T &= X(t), \\ Y &= F(t, y, z), \\ Z &= G(t, y, z) \end{aligned}\right\} \qquad (T)$$

be simple, mutually inverse, mappings of the spaces S *on* s *and of the space* s *on* S. *We assume:*

1°. *The mappings* **t**, **T** *are of class* C_2 *and the following relations hold for all values* $t \in (a, b)$, $T \in (A, B)$; $-\infty < Y, Z < \infty$

$$\dot{x}(T) \neq 0; \quad f(t, 0, 0) = 0; \quad f_Y(t, Y, Z) \neq 0; \quad g_Z(t, Y, Z) \neq 0.$$

2°. *The mapping* $\boldsymbol{t}$ *carries over the solutions of each system of differential equations with continuous coefficients* Q *of the form*

$$\left.\begin{aligned}\dot{Y} &= Z,\\ \dot{Z} &= Q(T)Y\end{aligned}\right\} \qquad \text{(Q)}$$

into solutions of an analogous system

$$\left.\begin{aligned}y' &= z,\\ z' &= q(t)y.\end{aligned}\right\} \qquad \text{(q)}$$

On these hypotheses, the mapping $\boldsymbol{t}$ *has the following linear form*

$$\begin{aligned}t &= x(T),\\ y &= w(t)Y,\\ z &= w'(t)Y + w(t)X'(t)Z.\end{aligned}$$

Proof. Consider a system (Q) and one of its solution curves $\mathfrak{K}$: $Y(T)$, $Z(T)$. The image $\mathfrak{K}$ of the latter in the mapping $\boldsymbol{t}$ has the parametric coordinates

$$\mathfrak{k}: \ y(t) = f(t, Y[X(t)], Z[X(t)]); \ \ z(t) = g(t, Y[X(t)], Z[X(t)]).$$

From 2°, $\mathfrak{k}$ is a solution curve of the system (q). We have therefore

$$z(t) = y'(t) = f_t(t, Y, Z) + f_Y(t, Y, Z)\dot{Y}[X(t)]X'(t) + f_Z(t, Y, Z)\dot{Z}[X(t)]X'(t),$$
$$q(t)\cdot y(t) = z'(t) = g_t(t, Y, Z) + g_Y(t, Y, Z)\dot{Y}[X(t)]X'(t) + g_Z(t, Y, Z)\dot{Z}[X(t)]X'(t)$$

and moreover, since Y, Z satisfy the system (Q)

$$z(t) = f_t(t, Y, Z) + f_Y(t, Y, Z)Z[X(t)]X'(t) + f_Z(t, Y, Z)Q[X(t)]Y[X(t)]X'(t),$$
$$\begin{aligned}q(t)\cdot y(t) = g_t(t, Y, Z) &+ g_Y(t, Y, Z)Z[X(t)]X'(t)\\ &+ g_Z(t, Y, Z)Q[X(t)]Y[X(t)]X'(t).\end{aligned}$$

These relations hold for every solution curve $\mathfrak{K}$ of the system (Q). Since one solution curve of (Q) passes through each point $(T, Y, Z) \in S$, we have for $t \in (a, b)$ and for all Y, Z:

$$g(t, Y, Z) = f_t(t, Y, Z) + f_Y(t, Y, Z)Z\cdot X'(t) + f_Z(t, Y, Z)Q[X(t)]Y\cdot X'(t), \tag{22.1}$$

$$q(t)\cdot f(t, Y, Z) = g_t(t, Y, Z) + g_Y(t, Y, Z)Z\cdot X'(t) + g_Z(t, Y, Z)Q[X(t)]Y\cdot X'(t). \tag{22.2}$$

In these formulae the continuous function Q can (from 2°) be chosen arbitrarily. It follows that

$$f_Z(t, Y, Z)Y\cdot X'(t) = 0$$

and moreover, since (from 1°) $X'(t) \neq 0$,

$$f_Z(t, Y, Z) = 0.$$

The function f is therefore independent of Z:

$$f(t, Y, Z) = h(t, Y). \tag{22.3}$$

We now have, from (1) and (3),

$$g(t, Y, Z) = h_t(t, Y) + h_Y(t, Y)Z \cdot X'(t)$$

and moreover

$$g_t(t, Y, Z) = h_{tt}(t, Y) + h_{tY}(t, Y)Z \cdot X'(t) + h_Y(t, Y)Z \cdot X''(t),$$
$$g_Y(t, Y, Z) = h_{tY}(t, Y) + h_{YY}(t, Y)Z \cdot X'(t),$$
$$g_Z(t, Y, Z) = h_Y(t, Y) \cdot X'(t).$$

These formulae show, on taking account of (2), that

$$q(t) \cdot h(t, Y) = h_{tt}(t, Y) + 2h_{tY}(t, Y)Z \cdot X'(t) + h_{YY}(t, Y)Z^2 \cdot X'^2(t)$$
$$+ h_Y(t, Y)[Z \cdot X''(t) + X'^2(t)Y \cdot Q[X(t)]].$$

Thence we obtain, by differentiating twice partially with respect to Z,

$$2h_{tY}(t, Y) \cdot X'(t) + 2h_{YY}(t, Y)Z \cdot X'^2(t) + h_Y(t, Y) \cdot X''(t) = 0,$$
$$h_{YY}(t, Y) = 0.$$

This shows that the function h is linear with respect to Y:

$$h(t, Y) = w(t)Y + a(t).$$

From 1° it follows that $a(t) = 0$, and hence we obtain the linear form of the function f:

$$f(t, Y, Z) = w(t) \cdot Y.$$

Finally, from (1) we obtain

$$g(t, Y, Z) = w'(t)Y + w(t)X'(t)Z,$$

and this completes the proof.

22.2 Introduction of the differential equation (Qq)

We consider two differential equations (q), (Q) over arbitrary intervals of definition $j = (a, b)$, $J = (A, B)$:

$$y'' = q(t)y; \tag{q}$$
$$\ddot{Y} = Q(T)Y. \tag{Q}$$

Let ($[w, X] =$) $w(t)$, $X(t)$ be a transformation of the differential equation (Q) into the differential equation (q). The functions w, X are therefore defined in a sub-interval

$i \subset j$ and have the properties 1–3 set out in § 11.2. In particular, the range of X, $I = X(i)$ is therefore a sub-interval of J: $I \subset J$.

We know (§ 11.2) that the transforming function X satisfies the non-linear differential equation of the third order:

$$-\{X, t\} + Q(X)X'^2 = q(t) \tag{Qq}$$

in its interval of definition i. We also know that the multiplier w of $[w, X]$ is determined from the function X uniquely except for a multiplicative constant k ($\neq 0$), by means of the formula (11.12).

23 Transformation properties of solutions of the differential equation (Qq)

23.1 Relations between solutions of the differential equations (Qq), (qQ), (qq), (QQ)

We are only interested in regular solutions of these differential equations, that is to say in solutions $X \in C_3$ with non-vanishing derivative X'. If therefore X is such a solution of one of these differential equations in a partial interval k of j or J, and $K = X(k)$ is its range, then in this interval K there exists the inverse function $x \in C_3$ of X. This has a derivative $\dot{x}$ which is always non-zero in this interval. The range of x is naturally the interval k; $x(K) = k$. We use the term homologous (with respect to the relevant differential equation) to describe any two numbers $t \in k$, $T \in K$ which are linked by the relationships $T = X(t)$, $t = x(T)$.

1. Let $X(t)$, $t \in i$ ($\subset j$) be a solution of the differential equation (Qq). Then the function inverse to X, $x(T)$, $T \in I (= X(i) \subset J)$ is a solution of the differential equation (qQ).

Proof. Let $t \in i$, $T \in I$ be two homologous numbers. Since X is a solution of (Qq), at the point t we have the relation

$$-\frac{\{X, t\}}{X'} + Q(X)X' = \frac{q(t)}{X'}. \tag{23.1}$$

From this, taking account of formulae (1.10), (1.6) we have

$$\frac{\{x, T\}}{\dot{x}} + Q(T)\frac{1}{\dot{x}} = q(x)\dot{x}$$

and further

$$-\{x, T\} + q(x)\dot{x}^2 = Q(T).$$

This completes the proof.

2. Let X, x be inverse solutions of the differential equations (Qq), (qQ) with intervals of definition i ($\subset j$), I ($\subset J$). Then at any two homologous points $t \in i$, $T \in I$ there hold the symmetric relations

$$Q(X)X' - \frac{1}{2}\frac{\{X, t\}}{X'} = q(x)\dot{x} - \frac{1}{2}\frac{\{x, T\}}{\dot{x}}, \tag{23.2}$$

$$Q(X)X' + \frac{1}{4}\left(\frac{1}{X'}\right)'' = q(x)\dot{x} + \frac{1}{4}\left(\frac{1}{\dot{x}}\right)^{\cdot\cdot}. \tag{23.3}$$

To see this, we start from the formula (1); from this and (1.6) it follows that

$$Q(X)X' - \frac{1}{2}\frac{\{X, t\}}{X'} = q(x)\dot{x} + \frac{1}{2}\frac{\{X, t\}}{X'} \tag{23.4}$$

and thence, using (1.10), we get the relation (2). Formula (3) is obtained from (4) and (1.16).

3. We continue to employ the symbols X, x with the above meaning. Let f, F be two functions constructed in the intervals i, I with arbitrary constants a_0, a_1; A_0, A_1 as follows

$$\left.\begin{aligned} f(t) &= a_0 + a_1 t + \frac{1}{4}\cdot\frac{1}{X'(t)} + \int_{X(t_0)}^{X(t)} [t - x(\eta)]Q(\eta)\, d\eta, \\ F(T) &= A_0 + A_1 T + \frac{1}{4}\cdot\frac{1}{\dot{x}(T)} + \int_{x(T_0)}^{x(T)} [T - X(H)]q(H)\, dH; \end{aligned}\right\} \tag{23.5}$$

where $t_0 \in i$, $T_0 \in I$ denote arbitrarily chosen homologous numbers.

Then, at any two homologous points $t \in i$, $T \in I$ we have the relationship

$$f''(t) = \ddot{F}(T), \tag{23.6}$$

the proof of which follows from (3) above.

In order to formulate the following theorems more simply, we shall denote the functions Q, q by Q_1, Q_2 and the differential equations (QQ), (Qq), (qQ), (qq) respectively by (Q_{11}), (Q_{12}), (Q_{21}), (Q_{22}).

4. Let X, Y be arbitrary solutions of the differential equations ($Q_{\alpha\beta}$), ($Q_{\beta\gamma}$) ($\alpha, \beta, \gamma = 1, 2$). Let i, k be the intervals of existence of the functions X, Y and let I, K be the ranges of the latter. Moreover let $i \cap K \neq \varnothing$, so that the composite function $Z = XY$ is defined in a certain interval $\bar{k}$ ($\subset k$).

We can show that the function Z is a solution of the equation ($Q_{\alpha\gamma}$) in the interval k. For, by our assumptions, in the interval $\bar{k}$ we have:

$$-\{Y, t\} + Q_\beta(Y)Y'^2 = Q_\gamma(t),$$

$$-\{X, Y\} + Q_\alpha(Z)X'^2(Y) = Q_\beta(Y),$$

and at the same time from (1.17) we have

$$\{Z, t\} = \{X, Y\}Y'^2(t) + \{Y, t\}.$$

From these relationships it follows that

$$-\{Z, t\} + Q_\alpha(Z)Z'^2 = Q_\gamma(t),$$

and the proof is complete.

23.2 Reciprocal transformations of integrals of the differential equations (q), (Q)

We now return to the situation considered in § 22.2 and concern ourselves with the question of how far the transformations of the equation (Q) into the equation (q) are determined by the solutions of the equation (Qq).

Let X be a solution of (Qq) with the interval of definition i ($\subset j$). We know that the function x inverse to X, with the definition interval ($X(i) =$) I ($\subset J$), satisfies the differential equation (qQ) (§ 23.1).

We choose an arbitrary number $t_0 \in i$, and denote the values of the functions X, X', X'' at the point t_0 by X_0, X_0' ($\neq 0$), X_0''; analogously, x_0, $\dot{x}_0$ ($\neq 0$), $\ddot{x}_0$ denote the values of x, $\dot{x}$, $\ddot{x}$ at the point $T_0 \in I$ homologous to t_0. The numbers X_0, X_0', X_0'' are inter-related, since $X_0 = T_0$, $x_0 = t_0$, and the formulae (1.6) hold.

1. If Y is an integral of the differential equation (Q), then the function $\bar{y}$, defined in the interval i by means of the formula

$$\bar{y}(t) = \frac{Y[X(t)]}{\sqrt{|X'(t)|}} \tag{23.7}$$

satisfies the differential equation (q), and this solution $\bar{y}$ is that portion lying in the interval i of the integral y of (q) which is determined by the Cauchy initial conditions

$$\left.\begin{aligned} y(t_0) &= \frac{Y(X_0)}{\sqrt{|X_0'|}}, \\ y'(t_0) &= \frac{\dot{Y}(X_0)}{\sqrt{|X_0'|}} \cdot X_0' - \frac{1}{2}\frac{Y(X_0)}{\sqrt{|X_0'|}} \cdot \frac{X_0''}{X_0'}. \end{aligned}\right\} \tag{23.8}$$

Proof. Clearly, the function $\bar{y}$ is everywhere twice differentiable in the interval i, and it is easy to verify that the following formulae hold:

$$\left.\begin{aligned} \bar{y}'(t) &= \frac{\dot{Y}(X)}{\sqrt{|X'|}} \cdot X' + Y(X)\left(\frac{1}{\sqrt{|X'|}}\right)', \\ \bar{y}''(t) &= \frac{\ddot{Y}(X)}{\sqrt{|X'|}} \cdot X'^2 - \frac{Y(X)}{\sqrt{|X'|}} \cdot \{X, t\}. \end{aligned}\right\} \tag{23.9}$$

Since the functions Y, X satisfy respectively the differential equations (Q) and (Qq), at every point $t \in i$ we have

$$\begin{aligned} \ddot{Y}(X) &= Q(X)\,Y(X), \\ -\{X, t\} &= -Q(X)X'^2 + q(t). \end{aligned}$$

We have therefore

$$\bar{y}''(t) = \frac{Y(X)}{\sqrt{|X'|}}\,Q(X)X'^2 + \frac{Y(X)}{\sqrt{|X'|}}\,[-Q(X)X'^2 + q(t)]$$

and consequently

$$\bar{y}''(t) = q(t)\bar{y}.$$

so the function $\bar{y}$ is a solution of the equation (q). The values $\bar{y}(t_0)$, $\bar{y}'(t_0)$ are given from (7) and (9) by means of (8).

2. Let Y, y be the integrals of the differential equations (Q), (q) considered in Theorem 1 above. Then the portion $\bar{Y}$ of Y defined in the interval I is given by the inverse formula to (7), namely

$$\bar{Y}(T) = \frac{y[x(T)]}{\sqrt{|\dot{x}(T)|}} \tag{23.10}$$

and the Cauchy initial values $Y(T_0)$, $Y'(T_0)$ are

$$\left.\begin{aligned} Y(T_0) &= \frac{y(x_0)}{\sqrt{|\dot{x}_0|}}, \\ \dot{Y}(T_0) &= \frac{y'(x_0)}{\sqrt{|\dot{x}_0|}} \cdot \dot{x}_0 - \frac{1}{2}\frac{y(x_0)}{\sqrt{|\dot{x}_0|}} \cdot \frac{\ddot{x}_0}{\dot{x}_0}. \end{aligned}\right\} \tag{23.11}$$

Proof. Since y is an integral of (q) and x a solution of (qQ), the theorem above shows that the function

$$\tilde{Y}(T) = \frac{y[x(T)]}{\sqrt{|\dot{x}(T)|}}$$

is a solution of the differential equation (Q) in the interval I.

Now, at two homologous points $T \in I$, $t \in i$ there hold the relations

$$\tilde{Y}(T) = \frac{\bar{y}(t)}{\sqrt{|\dot{x}(T)|}} = \frac{Y[X(t)]}{\sqrt{|\dot{x}(T) \cdot X'(t)|}} = Y(T) = \bar{Y}(T).$$

Consequently, $\tilde{Y}$ is the portion $\bar{Y}$ of the integral Y defined in I. From formula (8) the Cauchy initial conditions for Y are given by the formulae (11).

The above study thus yields the following theorem.

Theorem. The ordered pair of functions, $w(t) = k/\sqrt{|X'(t)|}$, $X(t)$ *constructed with an arbitrary constant* k ($\neq 0$), *represents a transformation* $[w, X]$ *of the differential equation* (Q) *into the differential equation* (q). *At the same time, the ordered pair of functions* $W(T) = k^{-1}/\sqrt{|\dot{x}(T)|}$, $x(T)$ *represents a transformation of the differential equation* (q) *into* (Q).

Every integral Y *of the differential equation* (Q) *is transformed by means of the transformation* $[w, X]$ *into its image*

$$\bar{y}(t) = k\,\frac{Y[X(t)]}{\sqrt{|X'(t)|}}, \tag{23.12}$$

which forms a portion of the image integral y *of* Y *determined by the initial values*

$$\left.\begin{aligned} y(t_0) &= k\,\frac{Y(X_0)}{\sqrt{|X_0'|}}, \\ y'(t_0) &= k\left[\frac{\dot{Y}(X_0)}{\sqrt{|X_0'|}} \cdot X_0' - \frac{1}{2}\frac{Y(X_0)}{\sqrt{|X_0'|}} \cdot \frac{X_0''}{X_0'}\right]. \end{aligned}\right\} \tag{23.13}$$

At the same time, the integral y of the differential equation (q) *is transformed by means of the transformation* $[W, x]$ *into its image*

$$\bar{Y}(T) = \frac{1}{k}\frac{y[x(T)]}{\sqrt{|\dot{x}(T)|}}, \tag{23.14}$$

which forms a portion of the image integral Y of y determined by the initial values

$$\left.\begin{aligned} Y(T_0) &= \frac{1}{k}\frac{y(x_0)}{\sqrt{|\dot{x}_0|}}, \\ \dot{Y}(T_0) &= \frac{1}{k}\left[\frac{y'(x_0)}{\sqrt{|\dot{x}_0|}}\cdot \dot{x}_0 - \frac{1}{2}\frac{y(x_0)}{\sqrt{|\dot{x}_0|}}\cdot\frac{\ddot{x}_0}{\dot{x}_0}\right]. \end{aligned}\right\} \tag{23.15}$$

3. Let Y, y be the integrals of the differential equations (Q), (q), considered in the above theorem. Then at every two homologous points $T \in I$, $t \in i$ there hold the relations

$$\left.\begin{aligned} \sqrt[4]{|\dot{x}(T)|}\cdot kY(T) &= \sqrt[4]{|X'(t)|}\cdot y(t), \\ \frac{k}{\sqrt[4]{|\dot{x}(T)|}}\left[\dot{Y}(T) + \frac{1}{4}Y(T)\cdot\frac{\ddot{x}(T)}{\dot{x}(T)}\right] &= \frac{\varepsilon}{\sqrt[4]{|X'(t)|}}\left[y'(t) + \frac{1}{4}y(t)\frac{X''(t)}{X'(t)}\right] \end{aligned}\right\} \tag{23.16}$$

with $\varepsilon = \operatorname{sgn} X'_0 = \operatorname{sgn} \dot{x}_0$.

These relations can be obtained from the formulae (12), (14) and their derivatives, by application of (1.6).

4. The image integrals u, v of two independent integrals U, V of the differential equation (Q) formed by the transformation $[w, X]$ are independent, and an analogous statement holds for the transformation $[W, x]$. This follows immediately from the formulae (13) and (15).

23.3 Transformations of the derivatives of integrals of the differential equations (q), (Q)

The above results can be used to determine transformations of the integrals (or of their first derivatives) of one of the differential equations (q), (Q) into portions of integrals (or their derivatives) of the other equation.

We assume that the carriers q, Q of the equations (q), (Q) $\in C_2$ and are always non-zero in their intervals of definition j, J. Then the differential equations (q), (Q) admit of associated differential equations ($\hat{q}_1$), ($\hat{Q}_1$), as in § 1.9. Their carriers $\hat{q}_1$, $\hat{Q}_1$ are determined by means of (1.18) and (1.20) while the relation between the derivatives y', $\dot{Y}$ of the integrals y, Y of the differential equations (q), (Q) and the integrals y_1, Y_1 of ($\hat{q}_1$), ($\hat{Q}_1$) is that of (1.21).

When we apply the above results to the differential equations ($\hat{q}_1$), (Q) and ($\hat{q}_1$), ($\hat{Q}_1$) we obtain information about transformations of integrals y, Y of the differential equations (q), (Q) and their derivatives y', $\dot{Y}$. The transformations corresponding to the relations (12), (14) are

$$\left.\begin{aligned} \bar{y}'(t) &= k\sqrt{|q(t)|}\,\frac{Y[X_1(t)]}{\sqrt{|X_1'(t)|}}, & \bar{Y}(T) &= \frac{1}{k}\,\frac{1}{\sqrt{|q[x_1(T)]|}}\,\frac{y'[x_1(T)]}{\sqrt{|\dot{x}_1(T)|}},\\ \bar{y}'(t) &= k\sqrt{\left|\frac{q(t)}{Q[X_2(t)]}\right|}\,\frac{\dot{Y}[X_2(t)]}{\sqrt{|X_2'(t)|}}, & \dot{\bar{Y}}(T) &= \frac{1}{k}\sqrt{\left|\frac{Q(T)}{q[x_2(T)]}\right|}\,\frac{y'[x_2(T)]}{\sqrt{|\dot{x}_2(T)|}}; \end{aligned}\right\} \tag{23.17}$$

X_1, x_1 here represent mutually inverse solutions of the differential equations ($Q\hat{q}_1$), ($\hat{q}_1 Q$) and X_2, x_2 are mutually inverse solutions of ($\hat{Q}_1\hat{q}_1$), ($\hat{q}_1\hat{Q}_1$).

23.4 Relations between solutions of the differential equation (Qq) and first phases of the differential equations (q), (Q)

The phases of the differential equations (q), (Q) considered in this paragraph are always first phases so we shall speak in what follows simply of phases instead of first phases.

We continue to use the symbols X, x, etc. as in § 23.2.

1. Let $\mathbf{A}$ be a phase of the differential equation (Q). Then the function $\bar{\alpha}$ defined in the interval i by means of the formula

$$\bar{\alpha}(t) = \mathbf{A}[X(t)] \tag{23.18}$$

is a portion of a phase α of the differential equation (q) and this phase α is determined by the Cauchy initial conditions

$$\alpha(t_0) = \mathbf{A}(X_0); \quad \alpha'(t_0) = \dot{\mathbf{A}}(X_0)X_0'; \quad \alpha''(t_0) = \ddot{\mathbf{A}}(X_0)X_0'^2 + \dot{\mathbf{A}}(X_0)X_0''. \tag{23.19}$$

Obviously, the phases α, $\mathbf{A}$ are linked (§ 9.2).

Proof. The phase $\mathbf{A}$ is contained in the phase system of a basis (U, V) of the differential equation (Q), (§ 5.6). Consequently, we have the relation $\tan \mathbf{A} = U/V$ holding in the interval J, except at the zeros of V.

We consider the transformation $w(t) = 1/\sqrt{|X'(t)|}$, $X(t)$ of the differential equation (Q) into the differential equation (q). Let u, v be the image integrals of U, V under this transformation $[w, X]$. From § 23.2, 4 (u, v) is a basis of (q); let α_0 be a phase of this basis. Then we have, for $t \in i$ (apart from the singular points),

$$\tan \alpha_0(t) = \frac{u(t)}{v(t)} = \frac{U[X(t)]}{V[X(t)]} = \tan \mathbf{A}[X(t)],$$

and hence $\alpha_0(t) + m\pi = \mathbf{A}[X(t)]$, m being an appropriate integer. Now, the function $\alpha = \alpha_0 + m\pi$ represents a phase of the basis (u, v) and $\bar{\alpha}$ is the portion of α defined in i. By differentiating (18) we obtain the initial values $\alpha'(t_0)$, $\alpha''(t_0)$ as stated in (19). This completes the proof.

Naturally, the solution x inverse to X of the differential equation (qQ) transforms the phase α into a portion $\bar{\mathbf{A}}$ of $\mathbf{A}$:

$$\bar{\mathbf{A}}(T) = \alpha[x(T)] \quad (T \in I = X(i)). \tag{23.20}$$

2. Let α, $\mathbf{A}$ be arbitrary linked phases of the differential equations (q), (Q) and let

$$L = a(j) \cap \mathbf{A}(J); \quad k = \alpha^{-1}(L), \quad K = \mathbf{A}^{-1}(L). \tag{23.21}$$

Then corresponding to every number $t \in k$ or $T \in K$ there is precisely one number $Z(t) \in K$ or $z(T) \in k$ satisfying respectively the equation

$$\alpha(t) = \mathbf{A}[Z(t)] \qquad \text{or} \qquad \alpha(z(T)] = \mathbf{A}(T). \tag{23.22}$$

The functions $Z(t) = \mathbf{A}^{-1}\alpha(t)$, $z(T) = \alpha^{-1}\mathbf{A}(T)$, which are defined by (22) in the intervals k, K and are obviously inverse functions, belong to the class C_3 and represent regular solutions of the differential equations (Qq), (qQ) respectively. The curves defined by the functions Z, z go from boundary to boundary of the rectangular region $(a, b) \times (A, B)$.

Proof. (a) Let $t \in k$ be arbitrary. Then $\alpha(t) \in L = \mathbf{A}(K)$, and since $\mathbf{A}$ increases or decreases, there is precisely one number $Z(t) \in K$ satisfying the first equation (22). A similar result holds for the second equation (22).

(b) From $Z(t) = \mathbf{A}^{-1}\alpha(t)$, $z(T) = \alpha^{-1}\mathbf{A}(T)$ it follows that the functions Z, z belong to the class C_3 and their derivatives Z', $\dot{z}$ are always non-zero. If we take the Schwarzian derivative of (22) it is clear that the functions Z, z satisfy the differential equations (Qq), (qQ).

(c) The validity of the last statement follows from the result of § 9.2 relating to the intervals k, K. This completes the proof.

We call Z, z the solutions of the differential equations (Qq), (qQ) *generated by* the phases α, $\mathbf{A}$.

The solution X of the differential equation (Qq) considered in 1 above is obviously that portion with domain of definition i of the solution Z of the differential equation (Qq) generated by the phases α, $\mathbf{A}$.

23.5 Reciprocal transformations of first and second phases of the differential equations (q), (Q)

From § 23.4, 1, a solution X of the differential equation (Qq) transforms each first phase $\mathbf{A}$ of the equation (Q) into a portion $\bar{\alpha}$ of a first phase α of the equation (q), according to the formula (18). An analogous statement holds for a solution x of the equation (qQ) and each first phase α of (q): the function x similarly transforms the phase α into a portion $\bar{\mathbf{A}}$ of a first phase of (Q). From § 23.4, 2 any two linked first phases α, $\mathbf{A}$ of the equations (q), (Q) generate inverse solutions of the equations (Qq), (qQ).

Now we assume that the functions q, $Q \in C_2$ and are always non-zero in their intervals of definition j, J, so that the differential equations (q), (Q) admit of associated differential equations ($\hat{q}_1$), ($\hat{Q}_1$). Then every first phase α_1 of ($\hat{q}_1$) represents a second phase β of (q), and similarly every first phase $\mathbf{A}_1$ of $(\hat{\mathrm{Q}}_1)$ represents a second phase **B** of (Q): $\alpha_1 = \beta$, $\mathbf{A}_1 = \mathbf{B}$. If we apply the results of § 23.4, 1 and 2, to the differential equations $(\hat{\mathrm{q}}_1)$, (Q) and $(\hat{\mathrm{q}}_1)$, $(\hat{\mathrm{Q}}_1)$, we obtain results relating to transformations of first and second phases α, **A** or β, **B** of the differential equations (q), (Q) into each other. The transformation formulae corresponding to the relations (18), (20) are

$$\bar{\beta}(t) = \mathbf{A}[X_1(t)]; \qquad \bar{\mathbf{A}}(T) = \beta[x_1(T)];$$
$$\bar{\beta}(t) = \mathbf{B}[X_2(t)]; \qquad \bar{\mathbf{B}}(T) = \beta[x_2(T)];$$

in which X_1, x_1, represent mutually inverse solutions of the equations $(\mathrm{Q}\hat{\mathrm{q}}_1)$, $(\hat{\mathrm{q}}_1\mathrm{Q})$ and X_2, x_2 are similarly mutually inverse solutions of $(\hat{\mathrm{Q}}_1\hat{\mathrm{q}}_1)$, $(\hat{\mathrm{q}}_1\hat{\mathrm{Q}}_1)$.

24 Existence and uniqueness problems for solutions of the differential equation (Qq)

24.1 The existence and uniqueness theorem for solutions of the differential equation (Qq)

At the basis of general transformation theory lies the following theorem:

Theorem. Let $t_0 \in j$, $X_0 \in J$, X_0' $(\neq 0)$, X_0'' *be arbitrary. Then there is precisely one "broadest" solution* $Z(t)$ *of the differential equation* (Qq) *in a certain interval* k $(\subset j)$ *with the Cauchy initial conditions*

$$Z(t_0) = X_0, \qquad Z'(t_0) = X_0', \qquad Z''(t_0) = X_0''; \tag{24.1}$$

where "broadest" is used in the sense that every solution of (Qq) *satisfying the same initial conditions is a portion of* $Z(t)$.

Let α, $\mathbf{A}$ *be arbitrary phases of the differential equations* (q), (Q), *whose values at the points* t_0, X_0 *are linked as follows:*

$$\alpha(t_0) = \mathbf{A}(X_0); \quad \alpha'(t_0) = \dot{\mathbf{A}}(X_0)X_0'; \quad \alpha''(t_0) = \ddot{\mathbf{A}}(X_0)X_0'^2 + \dot{\mathbf{A}}(X_0)X_0''. \tag{24.2}$$

Then $Z(t)$ *is the solution of the differential equation* (Qq) *generated by the linked phases* α, $\mathbf{A}$:

$$Z(t) = \mathbf{A}^{-1}\alpha(t). \tag{24.3}$$

Proof. We choose one of the phases α, $\mathbf{A}$, for instance the phase α, arbitrarily; then the other, $\mathbf{A}$, is determined uniquely as in § 7.1 by the values $\mathbf{A}(X_0)$, $\dot{\mathbf{A}}(X_0)$, $\ddot{\mathbf{A}}(X_0)$ given by the formulae (2), (§ 7.1).

The solution $Z(t)$ generated by the phases α, $\mathbf{A}$ obviously satisfies the initial conditions (1). We have therefore to show that every solution $X(t)$ of (Qq) defined in an interval i $(\subset j)$ with the initial values (1) is a portion of $Z(t)$. From § 23.4, 1, the function $\bar{\alpha}(t) = \mathbf{A}[X(t)]$, which is defined in the interval i, is a portion of a phase α_0 of (q); more precisely, of that phase α_0 which is determined by the same initial values (2) as for α. It follows that $\alpha_0(t) = \alpha(t)$ for $t \in j$ and further that $\alpha(t) = \mathbf{A}[X(t)]$ for $t \in i$, thus $X(t)$ is the portion of $Z(t)$ which exists in the interval i. This completes the proof.

From § 23.4, 2 the curve defined by the function $Z(t)$ passes from boundary to boundary of the rectangular region $j \times J$.

24.2 Transformations of given integrals of the differential equations (q), (Q) into each other

We now concern ourselves with the following question; if two integrals y, Y of the differential equations (q), (Q) are given *arbitrarily*, can we transform one of them (say, Y) into a portion $\bar{y}$ of the other integral y, by means of (23.7), using a suitable solution $X(t)$ of the differential equation (Qq), $t \in i$ $(\subset j)$? If the answer is yes, then naturally the integral y is transformed by the solution x of the differential equation (qQ), inverse to X, into a portion $\bar{Y}$ of Y as in (23.10).

The answer to this question is in the affirmative, provided only that we be allowed, if necessary, to change the sign of one of the two integrals y, Y. We can even prescribe arbitrarily the value X_0 taken by the function X at an arbitrary point $t_0 \in j$, $X_0 = X(t_0)$. However, it must be emphasized that the data mentioned above cannot be chosen completely arbitrarily, since at two homologous points $T = X(t)$ $(\in I = X(i))$, and $t = x(T)$ $(\in i)$ the transformation formulae (23.7), (23.10) show that the two integrals y, Y must have the same sign or must both vanish.

We set out the principal result more precisely in the following theorem:

Theorem. Let y, Y be arbitrary integrals of the differential equations (q), (Q). *Moreover, let $t_0 \in j$, $X_0 \in J$ be arbitrary numbers, which satisfy one or other of the following conditions* (a), (b):

(a) $y(t_0) \neq 0 \neq Y(X_0)$, (b) $y(t_0) = 0 = Y(X_0)$.

Then there exist broadest solutions X of the differential equation (Qq), *which take the value X_0 at the point t_0, i.e. $X_0 = X(t_0)$, and in their intervals of definition transform the integral Y into a portion $\bar{y}$ of y:*

$$\bar{y}(t) = \eta \frac{Y[X(t)]}{\sqrt{|X'(t)|}}. \tag{24.4}$$

In case (a) *there is precisely one increasing and precisely one decreasing broadest solution X of the differential equation* (Qq); *in the case* (b) *there are ∞^1 increasing and the same number of decreasing broadest solutions X.*

In both cases (a), (b) *the symbol η denotes the number ± 1, as follows:*

$$\text{(a)}\ \eta = \operatorname{sgn} y(t_0)\, Y(X_0)$$

$$\text{(b)}\ \eta = \begin{cases} \operatorname{sgn} y'(t_0)\, \dot{Y}(X_0) & \text{for increasing solutions,} \\ -\operatorname{sgn} y'(t_0)\, \dot{Y}(X_0) & \text{for decreasing solutions.} \end{cases}$$

Proof. We first assume that there is a solution X of the differential equation (Qq) defined in an interval k $(\subset j)$ and which is broadest in the sense of this theorem. Then the following relations hold in the interval k

$$\left.\begin{aligned} \bar{y}(t) &= \eta \frac{Y[X(t)]}{\sqrt{|X'(t)|}}, \\ \bar{y}'(t) &= \eta \left[\frac{\dot{Y}[X(t)]}{\sqrt{|X'(t)|}} X'(t) - \frac{1}{2} \frac{Y[X(t)]}{\sqrt{|X'(t)|}} \cdot \frac{X''(t)}{X'(t)}\right]. \end{aligned}\right\} \tag{24.5}$$

It is easy to verify that the functions X, X', X'' take the following values at the point t_0 in the two cases (a), (b):

$$\left.\begin{aligned} &\text{(a)}\quad X(t_0) = X_0, \qquad X'(t_0) = \varepsilon \frac{Y^2(X_0)}{y^2(t_0)}, \\ &X''(t_0) = 2 \frac{Y^2(X_0)}{y^4(t_0)} [Y(X_0)\dot{Y}(X_0) - \varepsilon y(t_0) y'(t_0)]; \\ &\text{(b)}\quad X(t_0) = X_0, \qquad X'(t_0) = \varepsilon \frac{y'^2(t_0)}{\dot{Y}^2(X_0)}; \end{aligned}\right\} \tag{24.6}$$

where $\varepsilon = \pm 1$. In case (b) the value $X''(t_0)$ is not determined by the conditions (5). Obviously, $\varepsilon = 1$ or $\varepsilon = -1$ according as X is increasing or decreasing in the interval k.

In case (a), therefore, the initial values $X(t_0) = X_0$, $X'(t_0)$ $(\neq 0)$ and $X''(t_0)$ are uniquely determined by (i) the integrals y, Y (ii) the choice of the values $t_0 \in j$, $X_0 \in J$ and (iii) whether the function X is increasing or decreasing. In case (b) this holds only for the initial values $X(t_0)$, $X'(t_0)$. From the theorem of § 24.1, it follows that the number of broadest solutions X of the differential equation (Qq) satisfying the condition of the theorem cannot exceed the number stated in this theorem.

Now let X be the broadest solution of the differential equation (Qq) determined by the initial conditions (6) (a) or (b); in the case (b) let X_0'' be arbitrary. The existence of this solution X is ensured by the theorem of § 24.1; let the interval of definition of X be k $(\subset j)$.

According to § 23.2, 1, the function

$$\bar{y}(t) = \frac{Y[X(t)]}{\sqrt{|X'(t)|}}, \tag{24.7}$$

which is defined in the interval k, is a solution of the differential equation (q) and it is in fact the portion contained in k of the integral $\bar{y}$ of (q) determined by the Cauchy initial conditions

$$\bar{y}(t_0) = \frac{Y(X_0)}{\sqrt{|X'(t_0)|}},$$

$$\bar{y}'(t_0) = \frac{\dot{Y}(X_0)}{\sqrt{|X'(t_0)|}} X'(t_0) - \frac{1}{2}\frac{Y(X_0)}{\sqrt{|X'(t_0)|}} \cdot \frac{X''(t_0)}{X'(t_0)}.$$

If we replace $X'(t_0)$, $X''(t_0)$ by the values given in the formulae (6), then in both cases (a), (b) we have

$$\bar{y}(t_0) = \eta y(t_0); \quad \bar{y}'(t_0) = \eta y'(t_0),$$

and it follows that for $t \in k$

$$\bar{y}(t) = \eta y(t).$$

Consequently the solution X of the differential equation (Qq) transforms (by (7)) the integral ηY into the portion of the integral y defined in the interval k. This completes the proof.

One remark needs to be added. The formula (7) can also be expressed as:

$$y(t) = \eta \frac{Y[X(t)]}{\sqrt{|X'(t)|}}, \tag{24.8}$$

where, however, validity is limited to the interval k $(\subset j)$. In special cases it can happen that (8) is valid in the whole interval j and at the same time the range of the function X coincides with the interval J. Then the function X transforms (by (8)) the integral ηY in its whole domain into the integral y. Naturally, this situation only occurs if the interval of definition k of X is identical with j and also the interval of definition, K, of the function x inverse to X is identical with J. This occurs, in particular, if the differential equations (q), (Q) are oscillatory. Then any two arbitrary phases

α, A of these differential equations are similar to each other; consequently the intervals k and j coincide and the intervals K, J coincide also (§ 9.2).

For example, the function $\sin t$ (arising from the carrier $q = -1$), is transformed into the integral $\sqrt{T}J_\nu(T)$ of the Bessel differential equation (1.24) over the whole range $t \in (-\infty, \infty)$, by means of a suitable increasing function $x_\nu(T)$ ($\in C_3$), $T \in (0, \infty)$. Hence we have the following representation of the Bessel function $J_\nu(T)$:

$$J_\nu(T) = \frac{\sin x_\nu(T)}{\sqrt{T \cdot \dot{x}_\nu(T)}}.$$

25 Physical application of general transformation theory

25.1 Straight line motion in physical space

We consider two physical spaces I and II. In these spaces let time be measured during certain (open) time intervals k, K respectively on clocks [I] and [II]. We assume that the clocks [I], [II] are coordinated by means of two inverse functions $X(t)$, $t \in k$ and $x(T)$, $T \in K$, defined in the intervals k, K: that is, at any instant $t \in k$ measured on clock [I], clock [II] shows the time $T = X(t)$ $(\in K)$, and at any instant when clock [II] indicates the time $T \in K$ clock [I] shows the time $t = x(T)$ $(\in k)$. We call X and x the *time functions* for the spaces II and I respectively. With a view to the following application we assume that these functions belong to the class C_3 and their derivatives X', $\dot{x}$ are always positive. Then it is meaningful to speak of the velocity of time $X'(t)$ and acceleration of time $X''(t)$ in the space II at the instant t $(\in k)$, and in the same way of the velocity of time $\dot{x}(T)$ and acceleration of time $\ddot{x}(T)$ in the space I at the instant T $(\in K)$. Two homologous instants $t \in k$ and $T = X(t) \in K$, or $T \in K$ and $t = x(T) \in k$, we shall call *simultaneous*.

Now let oriented straight lines G_{I}, G_{II} be given in the spaces I, II respectively, upon which two points P_{I}, P_{II} are moving. On these straight lines we take fixed points O_{I} and O_{II} respectively, the origin of each line; the instantaneous distances of the moving points P_{I} and P_{II} are measured from these fixed points and are positive and negative in the positive and negative directions of the corresponding straight lines (§ 1.5).

We assume that the motions of the points P_{I}, P_{II} are governed by arbitrary differential equations

$$y'' = q(t)y, \tag{q}$$

$$\ddot{Y} = Q(T)\,Y \tag{Q}$$

where $t \in j$, $T \in J$, as follows: At arbitrary instants $t_0 \in j$, $T_0 \in J$ let us choose the positions of the points P_{I}, P_{II} on the straight lines G_{I}, G_{II} (that is, their distances y_0, Y_0 from the origins O_{I}, O_{II}) and let us choose also their velocities y_0', $\dot{Y}_0$. Then the subsequent motions of the points P_{I}, P_{II} follow the integrals $y(t)$, $Y(T)$ of the differential equations (q), (Q) as determined by the initial values $y(t_0) = y_0$, $y'(t_0) = y_0'$ and $Y(T_0) = Y_0$, $\dot{Y}(T_0) = \dot{Y}_0$. The position of the point P_{I} at any instant $t \in j$ is therefore given by its distance $y(t)$ from the origin O_{I}; moreover, $y(t) > 0$ or $y(t) < 0$ or $y(t) = 0$ according as the point P_{I} lies in the positive or negative direction from the origin O_{I} or is passing through this point, and similarly for the point P_{II}. If the differential equations (q), (Q) are oscillatory, then the points P_{I}, P_{II} are at all times vibrating about the origins O_{I}, O_{II}.

We assume, for definiteness, that $y(t_0) > 0$, $Y(T_0) > 0$. From the theorem of § 24.2, there is precisely one increasing broadest solution X of the differential equation (Qq), which takes the value T_0 at the point t_0 and in its interval of definition k ($\subset j$) transforms the integral Y into the portion of y defined in the interval k. Simultaneously, the function x inverse to X represents in its interval of definition $X(k) = K(\subset J)$ the increasing broadest solution of (qQ), which takes the value t_0 at the point T_0 and which transforms the integral y over the interval K into the portion of Y defined in the interval K. These transformations may be expressed by means of the formula

$$\sqrt[4]{X'(t)}\,y(t) = \sqrt[4]{\dot{x}(T)}\,Y(T). \tag{25.1}$$

We now choose the functions X, x during the time intervals k and K as time functions for the spaces II and I respectively. Moreover, we choose the unit of length in space I at any instant t ($\in k$) as the fourth root of the corresponding velocity in space II, that is to say $\sqrt[4]{X'(t)}$, and analogously we choose that in space II as $\sqrt[4]{\dot{x}(T)}$. Then, by the formula (1), the instantaneous distances of the points $P_{\rm I}$, $P_{\rm II}$ from the origins $O_{\rm I}$, $O_{\rm II}$ are always the same at any instant, that is to say the motion of the points $P_{\rm I}$, $P_{\rm II}$ are the same during the time intervals k, K.

To summarize:

In physical spaces, for appropriate measures of time and length all straight line motions governed by differential equations of the second order are the same.

25.2 Harmonic motion

We now apply the above theory to the case of harmonic motion assuming that the motion of the points $P_{\rm I}$, $P_{\rm II}$ are governed by the differential equations formed with arbitrary constants $\omega > 0$, $\Omega > 0$

$$y'' = -\omega^2 y, \tag{q}$$

$$\ddot{Y} = -\Omega^2 Y \tag{Q}$$

in the time interval $(-\infty, \infty)$.

The initial positions and velocities of the points $P_{\rm I}$, $P_{\rm II}$ we shall choose as follows:

$$y_0 = \frac{c}{\sqrt{\omega}}, \qquad y_0' = 0; \qquad Y_0 = \frac{c}{\sqrt{\Omega}}, \qquad Y_0' = 0 \quad (c = \text{const} > 0).$$

Then the motion of the points $P_{\rm I}$, $P_{\rm II}$ is given by the following integrals of the differential equations (q), (Q):

$$y(t) = \frac{c}{\sqrt{\omega}} \sin\left[\omega(t - t_0) + \frac{\pi}{2}\right], \qquad Y(T) = \frac{c}{\sqrt{\Omega}} \sin\left[\Omega(T - T_0) + \frac{\pi}{2}\right].$$

$$t, T \in (-\infty, \infty).$$

The increasing broadest solution $X(t)$ of the differential equation

$$-\{X, t\} - \Omega^2 X'^2(t) = -\omega^2, \tag{Qq}$$

which takes the value T_0 at the point t_0, and its inverse function $x(T)$ are both linear and have the following forms:

$$X(t) = \frac{\omega}{\Omega}(t - t_0) + T_0; \quad x(T) = \frac{\Omega}{\omega}(T - T_0) + t_0 \quad (t, T \in (-\infty, \infty)).$$

These functions transform the integral Y into y and y into Y, over the interval $(-\infty, \infty)$, and we have

$$\sqrt[4]{\frac{\omega}{\Omega}} \cdot \frac{c}{\sqrt{\omega}} \sin\left[\omega(t - t_0) + \frac{\pi}{2}\right] = \sqrt[4]{\frac{\Omega}{\omega}} \cdot \frac{c}{\sqrt{\Omega}} \sin\left[\Omega(T - T_0) + \frac{\pi}{2}\right]. \tag{25.2}$$

Following the ideas described above, we now take the functions X, x to be our time functions for the spaces II and I respectively in the time interval $(-\infty, \infty)$. The linearity of these functions expresses the linear passage of time in the spaces considered. Moreover, we choose the units of length in the spaces I, II to be constants, having the values $\sqrt[4]{(\omega/\Omega)}$ and $\sqrt[4]{(\Omega/\omega)}$. Then, by (2), the motions of the points P_{I}, P_{II} are the same in the time interval $(-\infty, \infty)$.

To summarize:

Straight line harmonic motions of two points in physical spaces are the same in each space if the time functions are appropriately chosen linear functions and the units of length are appropriately chosen constants.

Complete transformations

This chapter is devoted to questions relating to the existence and generality of complete transformations of two differential equations (q), (Q) and to a study of the structure of the set of such transformations. The relevant theory takes its origin from the results obtained in § 9.5 relating to similar phases of two differential equations, so we shall use the notation which we employed there. In particular, we shall denote the left and right 1-fundamental sequences of the differential equations (q), (Q), when they exist, by

$$(a <)\, a_1 < a_2 < \cdots; \qquad (b >)\, b_{-1} > b_{-2} > \cdots$$

$$(A <)\, A_1 < A_2 < \cdots; \qquad (B >)\, B_{-1} > B_{-2} > \cdots$$

26 Existence and generality of complete transformations

26.1 Formulation of the problem

The starting point of the following study is provided by the existence and uniqueness theorem for solutions of the differential equation (Qq) (§ 24.1) and the remark made on pages 211–2. From this theorem we know that there is precisely one broadest solution $Z(t)$ of the differential equation (Qq) in a certain interval k ($\subset j$) with the initial conditions specified. The range K of this broadest solution $Z(t)$ forms a subinterval of J; $K \subset J$. It is important for our further study to note that, in general, neither does the interval k coincide with j nor does K coincide with J. This means that integrals Y of the differential equation (Q) are in general not transformed by the function $Z(t)$ over their whole domains into integrals y of the differential equation (q) but only portions of one into portions of the other.

We call a *solution* $X(t)$ of the differential equation (Qq) *complete* if its domain of definition k coincides with j and its range K coincides with J. Similarly we speak of *complete transformations* of integrals Y of the differential equation (Q) into integrals y of the differential equation (q) when these are formed from complete solutions $X(t)$ of the equation (Qq) according to formula (23.7). Complete solutions of the differential equation (Qq) are obviously characterized by the property that the corresponding curves pass from one corner of the rectangular region $j \times J$ to the diagonally opposite corner.

By means of complete solutions of the differential equation (Qq), i.e. by complete transformations, integrals of the equation (Q) are transformed into integrals of the equation (q) in their whole domain. Obviously, if X is a complete solution of (Qq) then the inverse solution x of (qQ) is also complete.

We can now formulate the question with which we shall be concerned in this section as follows:

To determine necessary and sufficient conditions for the existence of complete solutions of the differential equation (Qq), *and to determine the number of these solutions.*

26.2 Preliminary

Let us consider two differential equations (q), (Q); their intervals of definition will again be denoted by $j = (a, b)$, $J = (A, B)$.

We know (§ 24.1) that every solution X of the differential equation (Qq) is uniquely determined by two suitable first phases α, $\mathbf{A}$ of the differential equations (q), (Q) by means of the formula

$$\alpha(t) = \mathbf{A}X(t) \qquad (t \in k \subset j). \tag{26.1}$$

We called such phases the generators of X. One of the generators, say α, can be chosen arbitrarily from among the phases of (q); the other, $\mathbf{A}$, is then determined uniquely by this choice of α and the function X.

This holds, in particular, for every complete solution of the differential equation (Qq), in so far as such exist.

26.3 The existence problem for complete solutions of the differential equation (Qq)

The following theorem is fundamental for the study of existence questions relative to complete solutions of the differential equation (Qq):

Theorem 1. Two phases α, $\mathbf{A}$ *of the differential equations* (q), (Q) *generate a complete solution* X *of the differential equation* (Qq) *if and only if they are similar.*

Proof. (a) Let X be a complete solution of the differential equation (Qq) and α, $\mathbf{A}$ be the generating phases. Then we have $X(j) = J$, and for $t \in j$ the relation (1) holds. Clearly, therefore the ranges of the phases α, $\mathbf{A}$ coincide in their intervals of definition j, J so the phases α, $\mathbf{A}$ are similar (§ 9.5).

(b) Let α, $\mathbf{A}$ be similar phases of the equations (q), (Q). Then the ranges of α, $\mathbf{A}$, in their intervals of definition j, J, form the same interval L. It follows that in the interval j the range of the function X constructed from the formula $X(t) = \mathbf{A}^{-1}\alpha(t)$ coincides with J and further that for $t \in j$ the relation (1) holds. The phases α, $\mathbf{A}$ are therefore generators of the complete solution X of the differential equation (Qq). This completes the proof.

From (1) it follows that for $t \in j$, $X'(t) = \alpha'(t)/\dot{\mathbf{A}}X(t)$, and so:

According as the generating phases α, $\mathbf{A}$ of a complete solution X of the differential equation (Qq) are directly or indirectly similar, X represents an increasing or decreasing function in the interval j.

From theorem 1 we deduce that the differential equation (Qq) has complete solutions X if and only if the differential equations (q), (Q) admit of similar phases. This leads, using § 9.6, to the following result:

Theorem 2. The differential equation (Qq) *has complete solutions* X *if and only if the differential equations* (q), (Q) *are of the same character.*

26.4 The multiplicity of the complete solutions of the differential equation (Qq)

We now assume that the equations (q), (Q) are of the same character. By theorem 2, the equation (Qq) therefore admits of complete solutions.

If we apply the theorem of § 9.6, we obtain the following result: Let $t_0 \in j$ be arbitrary, and let $X_0 \in J$ be a number associated directly or indirectly with t_0 with respect to the differential equations (q), (Q). There exist, respectively, increasing or decreasing complete solutions X of the differential equation (Qq), which take the value X_0 at the point t_0. According to the character of the differential equations (q), (Q) and according to whether the numbers t_0, X_0 are singular or not, there is either precisely one complete solution X or a 1- or 2-parameter system of complete solutions of the differential equation (Qq).

More precisely (making use of § 9.6) we have:

There exists precisely one complete solution X, if the differential equations (q), (Q) are general differential equations either of type (1) or of type (m), $m \geqslant 2$, with the numbers t_0, X_0 not singular.

There exist ∞^1 complete solutions X, if the differential equations (q), (Q) are general of type (m) $m \geqslant 2$, and the numbers t_0, X_0 are singular; also if the differential equations (q), (Q) are special of type (1) or of type (m), $m \geqslant 2$, and the numbers t_0, X_0 are not singular; finally if the differential equations (q), (Q) are oscillatory on one side and the numbers t_0, X_0 are not singular.

There exist precisely ∞^2 complete solutions X, if the differential equations (q), (Q) are special of type (m), $m \geqslant 2$, and the numbers t_0, X_0 are singular; also if the differential equations (q), (Q) are oscillatory on one side and the numbers t_0, X_0 are singular; finally if the differential equations (q), (Q) are oscillatory.

This result naturally makes possible also statements relating to the existence and generality of complete solutions of the differential equation (qq), and hence on the possibility and number of complete transformations of integrals of the equation (q) into themselves ($Q = $ q). We observe that the equation (qq) always admits of complete solutions. If (q) is of finite type (m), $m \geqslant 1$, or is oscillatory, then there exist always both increasing and decreasing complete solutions of the differential equation (qq); if however it is oscillatory on one side, then there exist only increasing complete solutions of (qq). We leave it to the reader to prove these results in detail.

27 Structure of the set of complete solutions of the differential equation (Qq)

This section is devoted to a study of the structure of the set of complete solutions of the differential equation (Qq); naturally, this structure depends on the character of the differential equations (q), (Q). In order to keep our study short, we shall first develop a theory applicable to the investigation of all cases, but then only study in full the case of general differential equations (q), (Q) of finite type (m), $m \geqslant 2$.

We consider two differential equations (q), (Q) in the intervals $j = (a, b)$, $J = (A, B)$ and assume that they are of the same character. From § 9.2 this means that (q), (Q) are both general, or are both special of the same finite type (m), $m \geqslant 1$, or are both oscillatory on one side, or finally are both oscillatory.

This is a necessary and sufficient condition for the existence of complete solutions of the differential equation (Qq) (§ 26.3).

27.1 Preliminary

We already have the following information:

1. If $t_0 \in j$, $X_0 \in J$ are arbitrary directly or indirectly associated numbers, then there always exist complete solutions of the differential equation (Qq) which take the value X_0 at the point t_0, i.e. $X(t_0) = X_0$.

Every such complete solution X is obtained from two directly or indirectly similar (first) normal phases α, $\mathbf{A}$ of the equations (q), (Q) with the zeros t_0, X_0, as a solution of the functional equation

$$\alpha(t) = \mathbf{A}X(t). \tag{27.1}$$

The function X increases or decreases according as the phases α, $\mathbf{A}$ are directly or indirectly similar.

We show further that:

2. Let X be a complete increasing or decreasing solution of the differential equation (Qq). Then every two numbers t, $X(t)$, with $t \in j$, $X(t) \in J$ are directly or indirectly associated respectively.

For, let $t_0 \in j$ be arbitrary. We select a normal phase α of (q) with zero t_0. Then there is a phase $\mathbf{A}$ of (Q) similar to α such that the relationship (1) holds in j. The phases α, $\mathbf{A}$ are directly or indirectly similar according as X is increasing or decreasing. From the fact that $\alpha(t_0) = 0$ we have $\mathbf{A}X(t_0) = 0$; consequently, $X(t_0)$ is the zero of $\mathbf{A}$, and on taking account of § 9.5, 2, we deduce that the numbers t_0, $X(t_0)$ are directly or indirectly associated respectively.

From the above results 1 and 2 it follows that:

3. For every number $t \in j$ the values $X(t)$ of all increasing or decreasing complete solutions of the differential equation (Qq) form respectively the set of numbers directly or indirectly associated with t.

27.2 Relations between complete solutions of the differential equation (Qq)

Let us choose, for definiteness, an increasing phase **A** of (Q). Then every complete solution X of the equation (Qq) is determined uniquely from a phase α of the differential equation (q) similar to **A** by a relation such as (1). We shall call the phase α the *generator* of X and say that X is generated by the phase α.

Obviously we have:

1. Two complete solutions X, $\bar{X}$ of the differential equation (Qq) coincide in the interval j if and only if their generators α, $\bar{\alpha}$ coincide.

The mean value theorem gives, for $t \in j$, the relationship

$$X(t) - \bar{X}(t) = \frac{1}{\dot{\mathbf{A}}(T)} [\alpha(t) - \bar{\alpha}(t)], \tag{27.2}$$

in which T is some number lying between $X(t)$ and $\bar{X}(t)$ when $X(t) \neq \bar{X}(t)$.

We see also that:

2. Two complete solutions X, $\bar{X}$ of the differential equation (Qq) are such that at every point $t \in j$ their difference has the same sign as the difference between their generators α, $\bar{\alpha}$; that is $\alpha(t) \geqslant \bar{\alpha}(t) \Rightarrow X(t) \geqslant \bar{X}(t)$ and $\alpha(t) < \bar{\alpha}(t) \Rightarrow X(t) < \bar{X}(t)$.

Moreover,

3. If in the interval J the function $\dot{\mathbf{A}}$ is always greater than some positive constant, then the difference $X(t) - \bar{X}(t)$ is bounded if $\alpha(t) - \bar{\alpha}(t)$ is bounded.

If for example the equation (Q) admits of two independent bounded integrals, then all its integrals are bounded; consequently the amplitudes of all bases of (Q) are bounded and then (by (5.14)) the function $\dot{\mathbf{A}}$ has the property described.

27.3 The structure of the set of complete solutions of the differential equation (Qq) in the case of differential equations (q), (Q) of finite type (m), $m \geqslant 2$

We assume that (q), (Q) are general differential equations of finite type (m), $m \geqslant 2$. Let M be the set of complete solutions of the differential equation (Qq). In what follows, we shall use the term *integral curve* of the differential equation (Qq) to mean the curve $[t, X(t)]$ determined by a complete solution $X \in M$ of the differential equation (Qq).

The set M obviously separates into classes M_1, M_{-1} where M_1 is formed from the increasing and M_{-1} from the decreasing functions. We shall for simplicity concern ourselves only with the set M_1, since the situation in the set M_{-1} is analogous.

1. The region covered by the integral curves of the differential equation (Qq). We set

$$j_\mu = (a_\mu, b_{-m+\mu+1}), \qquad j'_\nu = (b_{-m+\nu+1}, a_{\nu+1}),$$
$$J_\mu = (A_\mu, B_{-m+\mu+1}), \qquad J'_\nu = (B_{-m+\nu+1}, A_{\nu+1});$$
$$\lambda = 1, \ldots, m-1; \qquad \mu = 0, \ldots, m-1; \qquad \nu = 0, \ldots, m-2;$$
$$a_0 = a, \quad b_0 = b; \quad A_0 = A, \quad B_0 = B.$$

Theorem. All integral curves $[t, X(t)]$, $X \in M_1$ *pass through the* $2(m-1)$ *points* $P(a_\lambda, A_\lambda)$, $P(b_{-\lambda}, B_{-\lambda})$ *and their union covers simply and completely the region* D_1 *formed by the union of the open rectangular regions* $j_\mu \times J_\mu, j'_\nu \times J'_\nu$.

All integral curves $[t, X(t)]$, $X \in M_{-1}$ *pass through the* $2(m-1)$ *points* $P(a_\lambda, B_{-\lambda})$, $P(b_{-\lambda}, A_\lambda)$ *and cover simply and completely the region* D_{-1} *formed by the union of the rectangular open regions* $j_\mu \times J_{m-\mu-1}, j'_\nu \times J'_{m-\nu-2}$.

Proof. We restrict ourselves to the proof of the first part of this theorem.

Let $X \in M_1$. From § 27.1, 2, $X(a_\lambda)$ is a number directly associated with a_λ, that is $X(a_\lambda) = A_\lambda$. Let $P(t_0, X_0) \in D_1$, so that $t_0 \in j_\mu$, $X_0 \in J_\mu$, say. Then the numbers t_0, X_0 are directly associated and not singular. Consequently (from § 26.4) there exists precisely one complete solution of the differential equation (Qq) whose value at the point t_0 is precisely X_0. This completes the proof.

2. Normalization of the generators.

Let (u, v) $(uv' - u'v < 0)$ be a principal basis of the differential equation (q) and (U, V) $(UV' - U'V < 0)$ a principal basis of (Q). We assume that u and v respectively are left and right 1-fundamental integrals of (q) and that U and V are respectively such integrals of (Q).

We choose a number r $(= 1, \ldots, m-1)$ and further choose a normal phase $\mathbf{A}$ of the basis (U, V) with the zero A_r and another normal phase $\bar{\mathbf{A}}$ with the zero B_{-r}. The boundary characteristic of $\mathbf{A}$ is $(A_r; -r\pi, (m - r - \frac{1}{2})\pi)$ and that of $\bar{\mathbf{A}}$ is $(B_{-r}, -(m - r - \frac{1}{2})\pi, r\pi)$.

Let $P(a_r)$ be the phase bunch (§ 7.10) formed by those normal phases which vanish at the point a_r of the 1-parameter basis system $(\rho u, v)$ of (q) with $\rho \neq 0$.

For every number ρ $(\neq 0)$ we shall denote by α_ρ the normal phase of the basis $(\rho u, v)$ which is included in the phase bunch $P(a_r)$. We know that $P(a_r)$ breaks up into two sub-bunches, one of which, $P_1(a_r)$, consists of increasing phases and the other, $P_{-1}(a_r)$, of decreasing phases.

Every normal phase $\alpha_\rho \in P_\varepsilon(a_r)$ has the boundary characteristic $(a_r; -r\pi\varepsilon, (m - r - \frac{1}{2})\pi\varepsilon)$ $(\varepsilon = \pm 1)$; consequently the phase α_ρ is directly similar to $\mathbf{A}$ in the case $\varepsilon = 1$ and indirectly similar to $\bar{\mathbf{A}}$ in the case $\varepsilon = -1$. Conversely, every phase of the differential equation (q) which is directly similar to $\mathbf{A}$ or indirectly similar to $\bar{\mathbf{A}}$ has the above boundary characteristic; we deduce that it is included in the sub-bunch $P_1(a_r)$ or $P_{-1}(a_r)$ respectively.

Hence the phases of the differential equation (q) which are directly similar to $\mathbf{A}$ are precisely the elements of the sub-bunch $P_1(a_r)$; the phases of (q) which are indirectly similar to $\bar{\mathbf{A}}$ are precisely the elements of the sub-bunch $P_{-1}(a_r)$.

It follows that the increasing complete solutions of the differential equation (Qq) are (for the above choice of the phase $\mathbf{A}$) generated by the elements in the sub-bunch

$P_1(a_r)$ while the decreasing complete solutions of (Qq) are (for the above choice of the phase $\bar{\mathbf{A}}$) generated by the elements of the sub-bunch $P_{-1}(a_r)$.

3. Properties of the structure of the set M_ε.

Let I_1 and I_{-1} be the intervals comprising all positive and all negative numbers respectively, and let $I = I_1 \cup I_{-1}$.

Corresponding to every number $\rho \in I$ we denote by X_ρ the complete solution of the differential equation (Qq) which is generated by the normal phase $\alpha_\rho \in P_1(a_r)$ using the phase $\mathbf{A}$, or by the normal phase $\alpha_\rho \in P_{-1}(a_r)$ using the phase $\bar{\mathbf{A}}$.

Let $\boldsymbol{K}$ be the mapping $\rho \to X_\rho$ of I on M. Obviously, the mapping $\boldsymbol{K}$ maps the interval I_ε on the set M_ε ($\varepsilon = \pm 1$).

The mapping $\boldsymbol{K}$ *is simple*; this follows from § 27.2, 1 and § 7.12, 2.

For $\rho, \bar{\rho} \in I_\varepsilon$ and $\rho < \bar{\rho}$, in the interval j_μ or j'_ν respectively we have the relations

$$X_\rho < X_{\bar{\rho}} \quad \text{or} \quad X_\rho > X_{\bar{\rho}}.$$

This follows from § 27.2, 2 and § 7.12, 3.

The set M_ε admits of the following ordering relation $\prec$: for $X, \bar{X} \in M_\varepsilon$, we have $X \prec \bar{X}$ if and only if, in every interval j_μ or j'_ν the relation $X < \bar{X}$ or $X > \bar{X}$, respectively, holds.

The mapping $\boldsymbol{K}$ *is order-preserving with respect to this ordering.*

We assume that the values of the function $\mathbf{A}$ lie between positive bounds λ, Λ;

$$\lambda \leqslant \dot{\mathbf{A}} \leqslant \Lambda \quad (\lambda, \Lambda > 0).$$

Then we have (from (2)) the following relationship holding in the interval j for every two elements $X_\rho, X_{\bar{\rho}} \in M_\varepsilon$

$$\lambda |X_\rho - X_{\bar{\rho}}| \leqslant \alpha_\rho - \alpha_{\bar{\rho}}| \geqslant \Lambda |X_\rho - X_{\bar{\rho}}|.$$

From the first inequality and (7.32) it follows that the difference $X_\rho - X_{\bar{\rho}}$ is bounded.

In the set M_ε we define a metric, d, by means of the formula

$$d(X_\rho, X_{\bar{\rho}}) = \sup_{t \in j} |X_\rho(t) - X_{\bar{\rho}}(t)|.$$

In the interval I_ε we take the Euclidean metric. We now show that:

The mapping $\boldsymbol{K}$ *is homeomorphic.*

Proof. From (2) and the relations (7.33), (7.34) we have

$$d(X_\rho, X_{\bar{\rho}}) \leqslant \frac{1}{2\lambda} \frac{|\rho - \bar{\rho}|}{\sqrt{\rho\bar{\rho}}},$$

$$\frac{|\rho - \bar{\rho}|}{1 + \rho\bar{\rho}} \leqslant \tan\left[\Lambda \cdot d(X_\rho, X_{\bar{\rho}})\right].$$

The first relation shows that $\boldsymbol{K}$ is continuous at every point $\bar{\rho} \in I_\varepsilon$; the second relationship shows the continuity of the mapping $\boldsymbol{K}^{-1}$ at every point $X_{\bar{\rho}} \in M_\varepsilon$, which completes the proof.

27.4 Canonical forms of the differential equation (q)

We now make use of the theory of complete transformations to express the differential equation (q) in certain canonical forms. For simplicity we shall call a function $X(t)$, $t \in j = (a, b)$ a *canonical phase function* if, throughout the interval j, $X \in C_3$ and $X' > 0$ and when, moreover, the numbers $C = \lim_{t \to a+} X$, $D = \lim_{t \to b-} X$ are as set out in one of the following five cases:

$$\begin{aligned} &\text{I. (a)}\ C = 0, \quad D = (m - \tfrac{1}{2})\pi, \quad m \geqslant 1, \text{ integral;} \\ &\quad\ \text{(b)}\ C = 0, \quad D = m\pi, \qquad\quad m \geqslant 1, \text{ integral;} \\ &\text{II. (a)}\ C = 0, \quad D = \infty; \\ &\quad\ \text{(b)}\ C = -\infty, \quad D = 0; \\ &\quad\ \text{(c)}\ C = -\infty, \quad D = \infty. \end{aligned}$$

Theorem. The carrier q of every differential equation (q), $j = (a, b)$ *can be represented by means of one of the canonical phase functions X defined in the interval j, in the form*

$$q(t) = -\{X, t\} - X'^2(t). \tag{27.3}$$

According as the differential equation (q) *is of the following types*

I. *of finite type* (m), $m \geqslant 1$ *and*
  (a) *general or*
  (b) *special,*

or

II. (a) *right oscillatory or*
  (b) *left oscillatory or*
  (c) *oscillatory,*

then the function X has the corresponding property 1. (a)–II. (c).

Proof. We shall confine ourselves, as an example, to the proof in the case I. (a).

Let (q) be a general differential equation of finite type (m), $m \geqslant 1$. The differential equation (Q) with $Q = -1$ in the interval $J = (0, (m - \frac{1}{2})\pi)$ is also general of finite type (m). For the function $\mathbf{A}(T) = T$ $(T \in J)$ is obviously a first phase of the differential equation (Q) and its boundary values are $C = 0$, $D = (m - \frac{1}{2})\pi$. We have therefore $O(\mathbf{A}|J) = (m - \frac{1}{2})\pi$ and our statement follows (§ 7.16).

We see that the differential equations (q), (Q) are of the same character. It follows that there exists a first phase α of the differential equation (q) which is directly similar to $\mathbf{A}$. For this phase, we have $\lim_{t \to a+} \alpha(t) = 0$, $\lim_{t \to b-} \alpha(t) = (m - \frac{1}{2})\pi$. Clearly, α is a canonical phase function with the property I. (a). Now, the solution $X(t)$ of the functional equation $\mathbf{A}(X) = \alpha(t)$, i.e. the function $\alpha(t)$, is a complete solution of the differential equation (Qq), so α satisfies the condition (3), and this completes the proof.

IV Recent Developments of Transformation Theory

Since the appearance, in 1967, of the German edition of this book there has been an extensive development of transformation theory and of parts of the theory of Jacobian differential equations connected with it. In particular, an abstract algebraic model of the transformation theory of oscillatory Jacobian differential equations has been constructed; resulting from this progress, attacks have been made on various problems which, while originating in dispersion theory, are also geometrical in nature.

In this Part, consisting of sections 28 and 29, we shall endeavour to present a survey of these recent developments. Section 28 is devoted to the construction—given in detail as it has not been published elsewhere—of the abstract transformation model mentioned above, and to its realization in the analytical case. Section 29 contains a survey of recent progress in various parts of the theory of Jacobian differential equations, which have an impact on transformation theory; it also reviews the relevant literature.

Most of the references in this Part are to the supplementary bibliography on pp. 248–9 these references are indicated by an asterisk.

28 An abstract algebraic model for the transformation theory of Jacobian oscillatory differential equations

28.1 Structure of the group of second order regular matrices over the real number field

Let $\mathfrak{M}$ be the group of regular second-order matrices (i.e. 2×2 square matrices) over the field R of real numbers. We first observe that the group $\mathfrak{M}$ contains the subgroup $\mathfrak{M}_0$ consisting of all elements of $\mathfrak{M}$ with positive determinant. This is invariant in $\mathfrak{M}$ and has index 2; the factor group $\mathfrak{M}/\mathfrak{M}_0$ consists therefore of two classes, $\mathfrak{M}_0$ and M_1. Moreover, the group $\mathfrak{M}$ contains the subgroup $\mathfrak{U}$ consisting of all unimodular elements.

The centre $\mathfrak{C}$ of $\mathfrak{M}$ consists of all matrices $\mu E = \begin{pmatrix} \mu & 0 \\ 0 & \mu \end{pmatrix}$; here $\mu \in R$ ($\mu \neq 0$) and $E = \begin{pmatrix} 1 & 0 \\ 0 & 1 \end{pmatrix}$ is the unit element of $\mathfrak{M}$; $\mathfrak{C}$ is invariant in $\mathfrak{M}$. Moreover, $\mathfrak{C}$ contains the subgroup $\mathfrak{C}_0$ consisting of all matrices λE with $\lambda > 0$; this subgroup $\mathfrak{C}_0$ is invariant in $\mathfrak{M}$ and has index 2 in $\mathfrak{C}$; the factor group $\mathfrak{C}/\mathfrak{C}_0$ consists of two classes, $\mathfrak{C}_0$, C_1.

Any class $\bar{M} \in \mathfrak{M}/\mathfrak{C}$ comprises all matrices of the form μM where $\mu \in R$ ($\mu \neq 0$) and M is an arbitrary element of $\bar{M}$. The determinants of all matrices of $\bar{M}$ have the same sign, namely sgn det M, and $\bar{M} \subset \mathfrak{M}_0$ or $\bar{M} \subset M_1$ according as $M \in \mathfrak{M}_0$ or $M \in M_1$ respectively.

The class $\bar{M}$ separates into two disjoint sets $\bar{M}_0, \bar{M}_1 \in \mathfrak{M}/\mathfrak{C}_0$; i.e. $\bar{M} = \bar{M}_0 \cup \bar{M}_1$, $\bar{M}_0 \cap \bar{M}_1 = \emptyset$. If $\bar{M}_0$ contains the matrix M then $\bar{M}_1$ contains the matrix $-M$. The set $\bar{M}_\alpha$ ($\alpha = 0, 1$) consists of all matrices of the form $\lambda(-1)^\alpha M$, with $\lambda > 0$. Both sets $\bar{M}_0$, $\bar{M}_1$ are simultaneously contained either in $\mathfrak{M}_0$ or in M_1. In each of these sets M_α there is precisely one unimodular matrix U_α ($\in \mathfrak{U}$), namely

$$U_\alpha = \frac{(-1)^\alpha}{\sqrt{(\text{abs det } M)}} M,$$

and both matrices U_0, U_1 obviously have the same determinant, sgn det M.

28.2 An abstract phase group

Let $\mathfrak{G}$ be an abstract group with the properties 1, 2(a), (b), (c) set out below:

(1) $\mathfrak{G}$ *contains a subgroup* $\mathfrak{G}_0$, *invariant in it, of index* 2.

The factor group $\mathfrak{G}/\mathfrak{G}_0$ thus consists of two classes, $\mathfrak{G}_0$ and G_1. For $a \in \mathfrak{G}$ we write sgn $a' = 1$ or sgn $a' = -1$ according as $a \in \mathfrak{G}_0$ or $a \in G_1$. Thus we have, for $a, b \in \mathfrak{G}$,

$$\text{sgn}\,(ab)' = \text{sgn}\, a' \cdot \text{sgn}\, b' \qquad (28.1)$$

(2) $\mathfrak{G}$ *contains a subgroup* $\mathfrak{E}$ *with the following properties* (a), (b), (c):

(a) *the centre* $\mathfrak{Z}$ *of* $\mathfrak{E} \cap \mathfrak{G}_0$ *is an infinite cyclic group with generator* c, (*i.e.* $\mathfrak{Z} = \{c_\nu\}$, $c_\nu = c^\nu$, $\nu = 0, \pm 1, \pm 2, \ldots$, $c_1 = c$) *and for every two elements* $e \in \mathfrak{E}$, $c_\nu \in \mathfrak{Z}$ *there holds the relation*

$$ec_\nu = c_{\nu \operatorname{sgn} e'} e. \tag{28.2}$$

We note that every element $c_\nu \in \mathfrak{Z}$ commutes with every element $e \in \mathfrak{E} \cap \mathfrak{G}_0$, i.e. $ec_\nu = c_\nu e$, and "anticommutes" with every element $e \in \mathfrak{E} \cap G_1$, that is to say $ec_\nu = c_{-\nu}e$.

It follows from (2) that $\mathfrak{Z}$ is invariant in $\mathfrak{E}$. The group $\mathfrak{Z}$ contains the subgroup $\mathfrak{Z}_0$, consisting of all the elements $c_{2\nu}$, i.e. $\mathfrak{Z}_0 \subset \mathfrak{Z}$; $\mathfrak{Z}_0 = \{c_{2\nu}\}$. $\mathfrak{Z}_0$ is invariant in $\mathfrak{E}$ and has index 2 in $\mathfrak{Z}$; the factor group $\mathfrak{Z}/\mathfrak{Z}_0$ consists of the two classes $\mathfrak{Z}_0$, Z_1.

Every class $\bar{e} \in \mathfrak{E}/\mathfrak{Z}$ consists of all those elements $c_\nu e$, where $\nu = 0, \pm 1, \pm 2, \ldots$ and e is an arbitrary element of $\bar{e}$: $\bar{e} = \{c_\nu e\}$. Since $\operatorname{sgn} c'_\nu = 1$, it follows that $\operatorname{sgn} (c_\nu e)' = \operatorname{sgn} e'$, and consequently $\bar{e} \subset \mathfrak{G}_0$ or $\bar{e} \subset G_1$ according as $e \in \mathfrak{G}_0$ or $e \in G_1$. The class $\bar{e}$ separates into two disjoint sets $\bar{e}_0$, $\bar{e}_1 \in \mathfrak{E}/\mathfrak{Z}_0$, (i.e. $\bar{e} = \bar{e}_0 \cup \bar{e}_1$, $\bar{e}_0 \cap \bar{e}_1 = \varnothing$). If $\bar{e}_0$ contains the element e ($\in \mathfrak{E}$) then $\bar{e}_1$ contains the element $c_1 e$ ($\in \mathfrak{E}$). The set $\bar{e}_\alpha$ ($\alpha = 0, 1$) coincides with the totality of elements $c_{2\nu+\alpha}e$: $\bar{e}_\alpha = \{c_{2\nu+\alpha}e\}$, and both sets $\bar{e}_0$, $\bar{e}_1$ are simultaneously contained either in $\mathfrak{G}_0$ or in G_1.

In what follows we shall make use of the following notation:

For $e \in \mathfrak{E}$ we denote by $\bar{e}$ that class of $\mathfrak{E}/\mathfrak{Z}_0$ which contains the element e; that is, $e \in \bar{e} = \mathfrak{Z}_0 e = \{c_{2\nu}\}e = \{c_{2\nu}e\} = \mathfrak{E}/\mathfrak{Z}_0$. Naturally, we also have $\mathfrak{Z}_0 e = e\mathfrak{Z}_0 = e\{c_{2\nu}\} = \{ec_{2\nu}\} \in \mathfrak{E}/\mathfrak{Z}_0$. Similarly, for $M \in \mathfrak{M}$ we denote by $\bar{M}$ that class of $\mathfrak{M}/\mathfrak{C}_0$ which contains the matrix M: $M \in \bar{M} = \mathfrak{C}_0 M = \{\lambda E\}M = \{\lambda M\} \in \mathfrak{M}/\mathfrak{C}_0$ ($\lambda > 0$), and at the same time $\mathfrak{C}_0 M = M\mathfrak{C}_0 = M\{\lambda E\} = \{M\lambda\} \in \mathfrak{M}/\mathfrak{C}_0$.

(b) *There exists an isomorphic mapping* $\mathscr{T}$ of $\mathfrak{E}/\mathfrak{Z}_0$ on $\mathfrak{M}/\mathfrak{C}_0$, *with the following properties* (i), (ii):

(i) *For* $e \in \bar{e} \in \mathfrak{E}/\mathfrak{Z}_0$, $M \in \bar{M} = \mathscr{T}\bar{e} \in \mathfrak{M}/\mathfrak{C}_0$, *we have* $\operatorname{sgn} e' = \operatorname{sgn} \det M$.

(ii) $\mathscr{T}\mathfrak{Z}_0 = \mathfrak{C}_0$, $\mathscr{T}Z_1 = C_1$.

The isomorphism $\mathscr{T}$ induces a homomorphic mapping $\mathscr{H}$ of $\mathfrak{E}$ on $\mathfrak{U}$, defined as follows: for $e \in \bar{e} \in \mathfrak{E}/\mathfrak{Z}_0$, we have $\mathscr{H}e = \mathfrak{U} \cap \mathscr{T}\bar{e}$; thus $\mathscr{H}e$ is the unimodular matrix contained in the class $\mathscr{T}\bar{e} \in \mathfrak{M}/\mathfrak{C}_0$. Obviously, $\mathscr{H}$ is a mapping *onto* $\mathfrak{U}$; we see that it is homomorphic by the following argument: from $e_1, e_2 \in \mathfrak{E}$ it follows that

$$\begin{aligned}\mathscr{H}e_1\mathscr{H}e_2 = (\mathfrak{U} \cap \mathscr{T}\bar{e}_1)(\mathfrak{U} \cap \mathscr{T}\bar{e}_2) = \mathfrak{U} \cap \mathscr{T}\bar{e}_1 \cdot \mathscr{T}\bar{e}_2 = \\ = \mathfrak{U} \cap \mathscr{T}(\bar{e}_1\bar{e}_2) = \mathscr{H}(e_1 e_2),\end{aligned}$$

and on taking account of (ii) we obtain

$$\mathscr{H}c_{2\nu+\alpha} = (-1)^\alpha E \qquad (\nu = 0, \pm 1, \pm 2, \ldots, \alpha = 0,1). \tag{28.3}$$

We also have, for $e \in \mathfrak{E}$,

$$\mathscr{T}\bar{e} = \{\lambda \mathscr{H} e\}, \qquad (\lambda > 0). \tag{28.4}$$

Finally, we observe that the union of the $\mathscr{T}$-maps of every two classes $\bar{e}_0, \bar{e}_1 \in \mathfrak{E}/\mathfrak{Z}_0$, whose union forms a class $\bar{e} \in \mathfrak{E}/\mathfrak{Z}$, (i.e. $\bar{e}_0 \cup \bar{e}_1 = \bar{e} \in \mathfrak{E}/\mathfrak{Z}$), represents a class $\bar{M} \in \mathfrak{M}/\mathfrak{C}$:

$$\mathscr{T}\bar{e}_0 \cup \mathscr{T}\bar{e}_1 = \bar{M} \in \mathfrak{M}/\mathfrak{C}$$

(c) *The normalizer* $\mathfrak{N}_{\mathfrak{E}}$ *of* $\mathfrak{E}$ *in* $\mathfrak{G}$ *coincides with* $\mathfrak{E}$: $\mathfrak{N}_{\mathfrak{E}} = \mathfrak{E}$.

We give the name *abstract phase group* to a group $\mathfrak{G}$ with the above properties 1, 2(a), (b), (c). A subgroup $\mathfrak{E}$ of this with the properties 2(a), (b), (c) we designate a fundamental subgroup of $\mathfrak{G}$. The elements of $\mathfrak{G}$ are called *abstract phases*, but for brevity we generally omit the attribute "abstract" in this connection.

28.3 Linear Vector Spaces

We now introduce the notation $\bar{A} = \mathfrak{G}/_r\mathfrak{E}$.

Let $\bar{a} \in \bar{A}$. If $a \in \bar{a}$ is given arbitrarily, then every element of $\bar{a}$ has the form ea, for a unique $e \in \mathfrak{E}$; conversely ea, with e an arbitrary element of $\mathfrak{E}$, represents an element of $\bar{a}$.

Our object now is to associate simply with every element $\bar{a} \in \bar{A}$ a linear vector space $\mathrm{L}_{\bar{a}}$ of dimension 2 over the field R; this gives us a system $\bar{\mathrm{L}}$ of two-dimensional linear vector spaces over R with, naturally, card $\bar{\mathrm{L}}$ = card $\bar{A}$. Every basis of $\mathrm{L}_{\bar{a}}$ is an ordered pair of elements, $U, V \in \mathrm{L}_{\bar{a}}$; this will frequently be written in the matrix form $\begin{pmatrix} U \\ V \end{pmatrix}$. If B is a given basis of $\mathrm{L}_{\bar{a}}$ then every basis of $\mathrm{L}_{\bar{a}}$ has the form $M\mathrm{B}$, for a uniquely determined matrix $M \in \mathfrak{M}$; conversely $M\mathrm{B}$, for any matrix $M \in \mathfrak{M}$, represents a basis of $\mathrm{L}_{\bar{a}}$.

Now we assume that between phases, on the one hand, and the bases of the linear vector spaces $\mathrm{L}_{\bar{a}} \in \bar{\mathrm{L}}$ on the other hand, there are the following relations:

For every class $\bar{a} \in \bar{A}$ *and* $\mathrm{L}_{\bar{a}} \in \bar{\mathrm{L}}$ *we have the following properties:*
With every phase $a \in \bar{a}$ *there is associated a system* $\bar{\mathrm{B}}_a$ *of bases of* $\mathrm{L}_{\bar{a}}$ *such that*
(a) *the individual bases of* $\bar{\mathrm{B}}_a$ *are constant positive multiples of any one of them,*
(b) *for* $e \in \mathfrak{E}$ *we have* $\bar{\mathrm{B}}_{ea} = \mathscr{H}e \cdot \bar{\mathrm{B}}_a$.

From (a) it follows that, given any $\mathrm{B} \in \bar{\mathrm{B}}_a$, we have $\bar{\mathrm{B}}_a = \{\lambda \mathrm{B}\}$, $\lambda > 0$, while (b) gives, on taking account of (3),

$$\bar{\mathrm{B}}_{c_\nu ea} = (-1)^\nu \bar{\mathrm{B}}_{ea} \qquad (\nu = 0, \pm 1, \pm 2, \ldots).$$

Moreover, for $\mathrm{B} \in \bar{\mathrm{B}}_a$ and $M \in \mathfrak{M}$,

$$M\mathrm{B} \in \bar{\mathrm{B}}_{ea}, \tag{28.5}$$

in which $e = \mathscr{T}^{-1}\bar{M}$.

It follows from the above assumptions that to every $a \in \bar{a}$ there corresponds a basis system $\bar{\mathrm{B}}_a$ of $\mathrm{L}_{\bar{a}}$ with the properties (a), (b). Conversely, every basis system $\bar{\mathrm{B}}$ of $\mathrm{L}_{\bar{a}}$, whose elements differ from one another by a constant positive multiple, coincides with a basis system $\bar{\mathrm{B}}_a$. For, given arbitrary elements $b \in \bar{a}$, $\mathrm{B} \in \bar{\mathrm{B}}$, $\mathrm{B}_b \in \bar{\mathrm{B}}_b$ there correspond elements $M \in \mathfrak{M}$, $e = \mathscr{T}^{-1}\bar{M}$ such that $\mathrm{B} = M\mathrm{B}_b \in \bar{\mathrm{B}}_{eb}$ (from (5)) and consequently $\bar{\mathrm{B}} = \bar{\mathrm{B}}_a$ $(a = eb)$.

If $\mathrm{B} \in \bar{\mathrm{B}}_a$, then we call a a *phase* of B; we also say that B admits of or possesses the phase a, and express this by the notation $\mathrm{B} = \mathrm{B}_a$. If b is also a phase of B, then there exists an $e \in \mathfrak{E}$ with $b = ea$ and we have $\bar{\mathrm{B}}_a = \bar{\mathrm{B}}_{ea} = \mathscr{H}e\bar{\mathrm{B}}_a$. Consequently $\mathscr{H}e = E$ and also $e = c_{2\nu}$ for some appropriate ν $(= 0, \pm 1, \pm 2, \ldots)$; as a consequence, $b = c_{2\nu}a \in \mathfrak{Z}_0 a$; conversely, every element of $\mathfrak{Z}_0 a$ represents a phase of B. Clearly, every basis of a system $\bar{\mathrm{B}}_a$ admits of the same phases and these are precisely the elements of the class $\mathfrak{Z}_0 a$. This class, which is obviously an element of $\mathfrak{G}/_r\mathfrak{Z}_0$, is called the *phase system* of the basis B.

28.4 Quasinorms

We now assume that *with every basis* B *of* $\mathrm{L}_{\bar{a}}$ *there is associated a non-zero real number* $\|\mathrm{B}\|$, *known as the quasinorm of* B, *satisfying the conditions:*

(a) $\operatorname{sgn} \|\mathrm{B}\| = -\operatorname{sgn} a'$,
(b) $\|M\mathrm{B}\| = \det M \cdot \|\mathrm{B}\|$,

where a is a phase of B *and M an arbitrary matrix of* $\mathfrak{M}$.

We now show that

1. *A basis* B *is uniquely determined if we are given one of its phases and its quasinorm.* For, let one phase a and the quasinorm $\|\mathrm{B}\|$ of a basis B be specified. Let us choose a basis $\mathrm{B}_a \in \bar{\mathrm{B}}_a$. Then we have $\mathrm{B} = \lambda\mathrm{B}_a$ for some $\lambda > 0$, and condition (b) above then shows that $\|\mathrm{B}\| = \lambda^2\|\mathrm{B}_a\|$, whence $\lambda = \sqrt{(\operatorname{abs}\|\mathrm{B}\|/\operatorname{abs}\|\mathrm{B}_a\|)}$. Hence B is uniquely determined as
$$\mathrm{B} = \sqrt{\left(\frac{\operatorname{abs}\|\mathrm{B}\|}{\operatorname{abs}\|\mathrm{B}_a\|}\right)}\,\mathrm{B}_a. \tag{28.6}$$

2. *Corresponding to every $k \in R$ $(k \neq 0)$ with* $\operatorname{sgn} k = -\operatorname{sgn} a'$, *every basis system* $\bar{\mathrm{B}}_a$ *contains precisely one basis* B *with quasinorm k, namely*
$$\mathrm{B} = \sqrt{\left(\frac{\operatorname{abs} k}{\operatorname{abs}\|\mathrm{B}_a\|}\right)}\,\mathrm{B}_a,$$
in which B_a *is an arbitrarily chosen basis in* $\bar{\mathrm{B}}_a$.

For, let $\mathrm{B} \in \bar{\mathrm{B}}_a$ and $\|\mathrm{B}\| = k$; then $\mathrm{B} = \lambda\mathrm{B}_a$ $(\lambda > 0)$, so $k = \lambda^2\|\mathrm{B}_a\|$, whence necessarily (since $\operatorname{sgn} k = \operatorname{sgn}\|\mathrm{B}_a\|$), $\lambda = \sqrt{(\operatorname{abs} k/\operatorname{abs}\|\mathrm{B}_a\|)}$. This establishes the assertion.

In particular, every basis system $\bar{\mathrm{B}}_a$ contains precisely one basis with the quasinorm $-\operatorname{sgn} a'$. This we call the *unit basis* in $\bar{\mathrm{B}}_a$ and use for it the notation $\mathscr{B}_a$; that is,
$$\mathscr{B}_a = (1/\sqrt{\operatorname{abs}\|\mathrm{B}_a\|})\,\mathrm{B}_a.$$

28.5 Kummer transformations of bases

Let us take a basis B of $L_{\bar{a}}$ and an element $x \in \mathfrak{G}$; with these we are going to associate another basis, of some suitable vector space $L_{\bar{b}}$ which we shall denote by $B \circ x$.

We choose a phase a of B *and specify, by definition,*

$$B \circ x \in \bar{B}_{ax}, \quad \|B \circ x\| = \|B\| \operatorname{sgn} x'.$$

The basis $B \circ x$ thus admits of the phase ax and has quasinorm $\|B\| \operatorname{sgn} x'$; these specifications serve to determine it uniquely, by § 28.4, 1. Since $\bar{B}_{(c_{2\nu}a)x} = \bar{B}_{c_{2\nu}(ax)} = \mathscr{H} c_{2\nu}\bar{B}_{ax} = \bar{B}_{ax}$, the basis $B \circ x$ is independent of the choice of a.

Let $\bar{b}$ be the class of $\mathfrak{G}/_r\mathfrak{E}$ containing the element ax, i.e. $ax \in \bar{b} \in \mathfrak{G}/_r\mathfrak{E}$, so $B \circ x$ is a basis of $L_{\bar{b}}$. Let us choose $b \in \bar{b}$ and $B_b \in \bar{B}_b$. Clearly, there is a unique element $e \in \mathfrak{E}$ with the property that

$$ax = eb$$

and we have

$$\bar{B}_{ax} = \bar{B}_{eb} = \mathscr{H} e\bar{B}_b = \{\lambda \mathscr{H} eB_b\}, \qquad (\lambda > 0).$$

Since $\operatorname{sgn} \|B\| \operatorname{sgn} x' = -\operatorname{sgn} a' \operatorname{sgn} x' = -\operatorname{sgn} (ax)'$, the system $\bar{B}_{ax}$ contains precisely one basis with quasinorm $\|B\| \operatorname{sgn} x'$, which coincides with $B \circ x$ (see § 28.4, 1). Using (6), we obtain

$$B \circ x = \sqrt{\left(\frac{\operatorname{abs} \|B\|}{\operatorname{abs} \|B_b\|}\right)} \mathscr{H} eB_b. \tag{28.7}$$

The basis $B \circ x$ *may thus be represented explicitly by the formula* (7).

The operation $\circ$ thus starts from any basis B of $L_{\bar{a}}$ and any element $x \in \mathfrak{G}$ and associates with them the basis $B \circ x$ of $L_{\bar{b}}$; we call this operation the *Kummer transformation of* B *with x*, and $B \circ x$ is itself called the *Kummer transform of* B *with x*. Obviously,

$$B \circ x = \sqrt{(\operatorname{abs} \|B\|)} \mathscr{H} e\mathscr{B}_b, \quad \mathscr{B} \circ x = \mathscr{H} e\mathscr{B}_b$$

are special cases of (7).

The only properties of the Kummer transformation which we here need to emphasize are the following:

Let B be a basis of $L_{\bar{a}}$, x, y arbitrary elements of $\mathfrak{G}$ and $M \in \mathfrak{M}$. *Then*

1. $B \circ (xy) = (B \circ x) \circ y$.
   For, let a be a phase of B; then $B \circ (xy)$ admits of the phase $a(xy)$ and has the quasinorm $\|B\| \operatorname{sgn} (xy)'$. But $(B \circ x) \circ y$ admits of the phase $(ax)y$ and has the quasinorm $(\|B\| \operatorname{sgn} x') \operatorname{sgn} y'$, so our statement follows by § 28.4, 1.

2. $(MB) \circ x = M(B \circ x)$.
   For, MB has the phase ea, with $e = \mathscr{T}^{-1}M$ (by (5)), hence $(MB) \circ x$ has the phase $(ea)x$ and the quasinorm of $(MB) \circ x$ is $\|MB\| \operatorname{sgn} x' = \det M\|B\| \operatorname{sgn} x'$. But $M(B \circ x)$ has the phase $e(ax)$ and quasinorm $\|M(B \circ x)\| = \det M\|B \circ x\| = \det M\|B\| \operatorname{sgn} x'$, and the statement follows by § 28.4, 1.

28.6 Kummer transformations of elements

Now we extend the concept of Kummer transformation to the transformation of elements of $L_{\bar{a}}$.

Let $Y \in L_{\bar{a}}$ and $x \in \mathfrak{G}$ be arbitrary elements of the sets indicated. We select a basis B of $L_{\bar{a}}$; we can then represent Y uniquely with respect to this basis B by means of appropriate coordinates β_1, β_2—that is to say, in the form

$$Y = (\beta_1, \beta_2)\mathrm{B}.$$

We now write, by definition,

$$Y \circ x = (\beta_1, \beta_2)(\mathrm{B} \circ x).$$

Thus the operation $\circ$ serves to transform Y into an element $Y \circ x$ of the vector space $L_{\bar{b}}$ containing the basis $\mathrm{B} \circ x$. We call this operation $\circ$ the *Kummer transformation of* Y *with* x, and $Y \circ x$ itself the *Kummer transform of* Y *with* x.

Next we show that $Y \circ x$ *is independent of the choice of* B. For, let C be another basis of $L_{\bar{a}}$; then we have

$$Y = (\gamma_1, \gamma_2)\mathrm{C}$$

for uniquely determined $\gamma_1, \gamma_2 \in R$. Moreover, there is an $M \in \mathfrak{M}$ such that $\mathrm{B} = M\mathrm{C}$ and we have

$$(\gamma_1, \gamma_2)\mathrm{C} = Y = (\beta_1, \beta_2)\mathrm{B} = (\beta_1, \beta_2)M\mathrm{C},$$

hence

$$(\gamma_1, \gamma_2) = (\beta_1, \beta_2)M.$$

Then, on taking account of § 28.5, 2,

$$\begin{aligned} Y \circ x = (\gamma_1, \gamma_2)(C \circ x) &= [(\beta_1, \beta_2)M](\mathrm{C} \circ x) = \\ &= (\beta_1, \beta_2)[M(\mathrm{C} \circ x)] = (\beta_1, \beta_2)[(M\mathrm{C}) \circ x] = \\ &= (\beta_1, \beta_2)(\mathrm{B} \circ x), \end{aligned}$$

which shows that $Y \circ x$ is independent of the choice of B.

28.7 Abstract dispersions

This paragraph introduces the concept of *abstract dispersions*; sub-paragraphs (A) and (B) are preliminary to the definition in sub-paragraph (C) of such dispersions and a study of some of their properties. For convenience, the main results are numbered as Lemmas 1 to 6.

Throughout, by the term "classes" we mean elements of $\mathfrak{G}/_r\mathfrak{E}$. The subgroup of $\mathfrak{G}$ conjugate with $\mathfrak{E}$ with respect to $a \in \mathfrak{G}$ will be denoted by $\mathfrak{G}_a$, thus $\mathfrak{G}_a = a^{-1}\mathfrak{E}a$.

(A) *Kummer complexes:*

Let a, b, A, B be elements of $\mathfrak{G}$.

Lemma 1. The relationship

$$A^{-1}\mathfrak{E}a = B^{-1}\mathfrak{E}b \qquad (28.8.)$$

holds if and only if $B = EA$, $b = ea$, *where* $E, e \in \mathfrak{E}$.

Proof. (i) from $B = EA$, $b = ea$ and $E, e \in \mathfrak{E}$ it follows that

$$B^{-1}\mathfrak{E}b = A^{-1}(E^{-1}\mathfrak{E}e)a = A^{-1}\mathfrak{E}a.$$

(ii) From (8) we have

$$(BA^{-1})\mathfrak{E} = \mathfrak{E}(ba^{-1})$$

and also

$$BA^{-1} = e_0(ba^{-1}), \quad \text{where} \quad e_0 \in \mathfrak{E}. \tag{28.9}$$

This gives

$$(ba^{-1})^{-1}\mathfrak{E}(ba^{-1}) = \mathfrak{E}.$$

Clearly, ba^{-1} is contained in the normalizer $\mathfrak{N}_{\mathfrak{E}}$. On taking account of § 28.2, (2) (c) it follows that $ba^{-1} = e \in \mathfrak{E}$ and, from (9), $BA^{-1} = E \in \mathfrak{E}$ ($E = e_0 e$). We thus have $B = EA$, $b = ea$ and the proof is complete.

It follows from Lemma 1 that to every ordered pair of classes $\bar{A}$, $\bar{a}$ there corresponds a well-defined subset of $\mathfrak{G}$, namely $K(\bar{A}, \bar{a}) = A^{-1}\mathfrak{E}a$, this subset being independent of the choice of the elements A, a in the classes $\bar{A}$, $\bar{a}$; we call this subset $K(\bar{A}, \bar{a})$ the *Kummer complex of* $\bar{A}$, $\bar{a}$.

Lemma 2. The Kummer complex $K(\bar{A}, \bar{a})$ *is characterized by the property of being the unique common element of the two partitions* $\mathfrak{G}/_l\mathfrak{U}_a$, $\mathfrak{G}/_r\mathfrak{U}_A$.

Proof. Obviously we have

$$A^{-1}\mathfrak{E}a = (A^{-1}a)(a^{-1}\mathfrak{E}a) = (A^{-1}a)\mathfrak{U}_a,$$
$$A^{-1}\mathfrak{E}a = (A^{-1}\mathfrak{E}A)(A^{-1}a) = \mathfrak{U}_A(A^{-1}a),$$

hence

$$K(\bar{A}, \bar{a}) \in \mathfrak{G}/_l\mathfrak{U}_a \cap \mathfrak{G}/_r\mathfrak{U}_A$$

Thus $K(\bar{A}, \bar{a})$ occurs as an element in both partitions $\mathfrak{G}/_l\mathfrak{U}_a$, $\mathfrak{G}/_r\mathfrak{U}_A$. But by a known result in group theory ([2*]), the hypothesis of § 28.2, (2)(c) implies that these partitions contain precisely one common element, which thus coincides with $K(\bar{A}, \bar{a})$. We thus have the relation

$$K(\bar{A}, \bar{a}) = \mathfrak{G}/_l\mathfrak{U}_a \cap \mathfrak{G}/_r\mathfrak{U}_A$$

and the proof is complete.

In particular, taking $\bar{A} = \bar{a}$, we see that the Kummer complex of $\bar{a}$, $\bar{a}$ coincides with the subgroup $\mathfrak{U}_a$, i.e. $K(\bar{a}, \bar{a}) = \mathfrak{U}_a$ ($a \in \bar{a}$). This permits us to write, conveniently, $\mathfrak{U}_{\bar{a}}$ in place of $\mathfrak{U}_a$.

(B) *The centre* $\mathfrak{Z}_{\bar{a}}$ *of* $\mathfrak{U}_{\bar{a}} \cap \mathfrak{G}_0$

Now let us consider the centre $\mathfrak{Z}_{\bar{a}}$ of the subgroup $\mathfrak{U}_{\bar{a}} \cap \mathfrak{G}_0$ ($\subset \mathfrak{G}$).

Lemma 3. For $a \in \bar{a}$ *we have*

$$\mathfrak{Z}_{\bar{a}} = a^{-1}\mathfrak{Z}a.$$

Proof. (i) Let $f \in \mathfrak{Z}_{\bar{a}}$; then $f = a^{-1}e_0a$ for some appropriate element $e_0 \in \mathfrak{E}$, sgn $e_0' = 1$. Moreover, f commutes with every element $x \in \mathfrak{U}_{\bar{a}} \cap \mathfrak{G}_0$, i.e. every x of the form $x = a^{-1}ea$, where $e \in \mathfrak{E}$, sgn $e' = 1$. Hence

$$xf = (a^{-1}ea)(a^{-1}e_0a) = a^{-1}(ee_0)a,$$
$$fx = (a^{-1}e_0a)(a^{-1}ea) = a^{-1}(e_0e)a;$$

consequently $ee_0 = e_0e$. Hence $e_0 \in \mathfrak{Z}$ and, finally, $f = a^{-1}e_0a \in a^{-1}\mathfrak{Z}a$.

(ii) Let $f \in a^{-1}\mathfrak{Z}a$, i.e. $f = a^{-1}e_0a$ where $e_0 \in \mathfrak{Z}$ is some appropriate element. Then for every $e \in \mathfrak{E}$, sgn $e' = 1$ and we have $ee_0 = e_0e$, whence $(a^{-1}ea)(a^{-1}e_0a) = (a^{-1}e_0a)(a^{-1}ea)$. Thus f commutes with every element $x \in \mathfrak{U}_{\bar{a}} \cap \mathfrak{G}_0$, i.e. $xf = fx$, and the proof is complete.

From Lemma 3, the centre $\mathfrak{Z}_{\bar{a}}$ of $\mathfrak{U}_{\bar{a}} \cap \mathfrak{G}_0$ is an infinite cyclic group with generators $a^{-1}ca$, $a^{-1}c^{-1}a$, i.e. $\mathfrak{Z}_{\bar{a}} = \{a^{-1}c_\nu a\}$, $c_\nu = c^\nu$, $\nu = 0, \pm 1, \pm 2, \ldots$, $c_1 = c$. The individual elements $f_\nu = a^{-1}c_\nu\, a \in \mathfrak{Z}_{\bar{a}}$ depend on the particular element $a \in \bar{a}$ as follows: for $b = ea$, $e \in \mathfrak{E}$ we have $b^{-1}c_\nu b = a^{-1}e^{-1}(c_\nu e)a = a^{-1}e^{-1}(ec_{\nu\,\mathrm{sgn}\,e'})a = a^{-1}c_{\nu\,\mathrm{sgn}\,e'}\,a = f_{\nu\,\mathrm{sgn}\,e'}$.

(C) *Abstract general and special dispersions: abstract central dispersions.*

(1) We shall apply the term *abstract general dispersions of $\bar{A}$, $\bar{a}$* to the elements of $K(\bar{A}, \bar{a})$, but omit the attribute "abstract" when convenient. A general dispersion x of $\bar{A}$, $\bar{a}$ thus has the form $x = A^{-1}ea$ where A, a are arbitrary elements of $\bar{A}$, $\bar{a}$ and e is an appropriate element of $\mathfrak{E}$.

Let $\bar{A}$, $\bar{a}$ be arbitrary classes and $\mathrm{L}_{\bar{A}}$, $\mathrm{L}_{\bar{a}}$ the corresponding linear vector spaces.

Lemma 4. The Kummer transform $\mathrm{B} \circ x$ *of a basis* B *of* $\mathrm{L}_{\bar{A}}$ *with an element* $x \in \mathfrak{G}$ *is a basis of* $\mathrm{L}_{\bar{a}}$ *if and only if* x *is a general dispersion of* $\bar{A}$, $\bar{a}$.

Proof. Let $\mathrm{B} = \mathrm{B}_A$, $A \in \bar{A}$ and $a \in \bar{a}$.

(i) Assume that $x \in \mathfrak{G}$ has the above property; then $Ax = ea$, $e \in \mathfrak{E}$ and consequently $x = A^{-1}ea \in K(\bar{A}, \bar{a})$.

(ii) Let $x \in K(\bar{A}, \bar{a})$, that is $x = A^{-1}ea$, $e \in \mathfrak{E}$. Then $Ax = A(A^{-1}ea) = ea \in \bar{a}$, whence $\mathrm{B} \circ x = \mathrm{B}_{ea}$ is a basis of $\mathrm{L}_{\bar{a}}$. This proves the Lemma.

A corollary of Lemma 4 is that *the Kummer transform of an element* $Y \in \mathrm{L}_{\bar{A}}$ *with a general dispersion* x *of* $\bar{A}$, $\bar{a}$ *is an element of* $\mathrm{L}_{\bar{a}}$, *i.e.* $Y \circ x \in \mathrm{L}_{\bar{a}}$.

(2) We shall apply the term (*abstract*) *special dispersions of* $\bar{a}$ (or merely *dispersions of* $\bar{a}$) to the elements of $K(\bar{a}, \bar{a})$. A special dispersion x of $\bar{a}$ has thus the form $x = a^{-1}ea$ where a is an arbitrary element of $\bar{a}$ and e an appropriate element of $\mathfrak{E}$.

Lemma 5. The Kummer transform $\mathrm{B} \circ x$ *of a basis* B *of* $\mathrm{L}_{\bar{a}}$ *with an element* $x \in \mathfrak{G}$ *is itself a basis of* $\mathrm{L}_{\bar{a}}$ *if and only if* x *is a special dispersion of* $\bar{a}$.

Clearly, this is a special case of Lemma 4. In particular, *the Kummer transform of an element* $Y \in \mathrm{L}_{\bar{a}}$ *with a special dispersion* x *of* $\bar{a}$ *is itself an element of* $\mathrm{L}_{\bar{a}}$, *i.e.* $Y \circ x \in \mathrm{L}_{\bar{a}}$.

(3) We shall apply the term (*abstract*) *central dispersions of* $\bar{a}$, more briefly, *central dispersions*, to the elements of $\mathfrak{Z}_{\bar{a}}$.

Given a fixed element a in the class $\bar{a}$, and the generating element c of $\mathfrak{Z}$, we can associate with every central dispersion $f \in \mathfrak{Z}_{\bar{a}}$ a unique integer ν, known as the *index* of f, by means of the formula $f = a^{-1}c^\nu a$; we then write $f = f_\nu$. For a different choice of a or c, the index ν of f either remains unchanged or else changes sign to become $-\nu$. Hence, clearly, the parity of the index of every central dispersion is independent of the choice of the elements a, c.

Lemma 6. The Kummer transform $B \circ x$ *of a basis* B *of* $L_{\bar{a}}$ *with an element* $x \in \mathfrak{G}$ *coincides with* B *or* $-B$ *if and only if* x *is a central dispersion of* a *with even or odd index, respectively, i.e.*

$$B \circ f_\nu = (-1)^\nu B.$$

Proof. (i) Let $f = f_\nu \in \mathfrak{Z}_{\bar{a}}$. Then $af_\nu = a(a^{-1}c_\nu a) = c_\nu a$, and hence, using (28.3), $B \circ f_\nu = \mathscr{H}c_\nu B = (-1)^\nu B$.

(ii) Let $x \in \mathfrak{G}$ and $B \circ x = (-1)^\alpha B$, $\alpha = 0, 1$. Then $ax = ea$ and $\mathscr{H}e = (-1)^\alpha E$ $(B = B_a)$. Hence $e = c_{2\mu+\alpha}$, μ integral, and also $x = a^{-1}c_{2\mu+\alpha}a = f_{2\mu+\alpha}$.

It follows from Lemma 6 that *the Kummer transform of an element* $Y \circ L_{\bar{a}}$ *with a central dispersion* f_ν *of* $\bar{a}$ *coincides with either* Y *or* $-Y$ *in the sense that*

$$Y \circ f_\nu = (-1)^\nu Y, \qquad (\nu = 0, \pm 1, \pm 2, \ldots) \tag{28.10}$$

28.8 Realization of the abstract model in terms of analytical transformation theory

Our object now is to obtain a realization of the above abstract model in terms of the transformation theory of oscillatory differential equations (q). To do this, we must interpret the model elements, considered in 28.1–28.6 above, in this realization.

First, however, we introduce the symbol $\mathscr{B}(t)$ to stand for the particular basis of the differential equation (-1)—that is, the equation $y'' = -y$, $t \in j = (-\infty, \infty)$—formed by the elements $\sin t$, $\cos t$. Then for any phase $e(t)$ of (-1) we have, by formula (21.10), the relationship

$$\frac{1}{\sqrt{|e'(t)|}}\mathscr{B}e(t) = \mathscr{H}e\mathscr{B}(t),$$

where $\mathscr{H}e$ is a uniquely determined unimodular matrix; $\det \mathscr{H}e = \operatorname{sgn} e'(t)$.

In our realization of the abstract model, the various elements must be interpreted as follows:

- $\mathfrak{G}$: the phase group described in § 10.1.
- $\mathfrak{G}_0$: the subgroup comprising the increasing elements of the phase group.
- $\mathfrak{E}$: the fundamental subgroup defined in § 10.3.
- $\mathfrak{Z}$: the infinite cyclic group with generator $t + \pi$, that is, $\{t + \nu\pi\}$, $(\nu = 0, \pm 1, \pm 2, \ldots)$.
- $\mathfrak{Z}_0$: the infinite cyclic group $\{t + 2\nu\pi\}$.
- $\mathscr{T}$: the operator such that for every phase $e(t)$ of (-1) and $\bar{e} = \{e(t) + 2\nu\pi\}$, we have $\mathscr{T}\bar{e} = \{\lambda\mathscr{H}e\}$, $\lambda > 0$.
- $\bar{a}$: the set of all first phases of the differential equation (q_a) with $q_a(t) = -\{a, t\} - a'^2(t)$, $a(t) \in \bar{a}$.
- $L_{\bar{a}}$: the integral space of (q_a).
- $\bar{B}_a$: $\left\{\dfrac{\lambda}{\sqrt{|a'(t)|}}\mathscr{B}a(t)\right\}$, $\quad \lambda > 0$.
- $\|B\|$: the Wronskian of B.

We leave it to the reader to verify that the properties assumed for the various elements in the abstract model do in fact hold in the realization. With regard to the properties of § 28.2, 2(c), see [2*].

Now we show that *the operation* $\circ$ *applied to a basis* $\mathrm{B}(t) = (U(t), V(t))$ *of an oscillatory differential equation* (q) *in the interval* $j = (-\infty, \infty)$, *and an arbitrary phase function* $x(t)$, *may be realized by means of the Kummer transformation*

$$(U, V) \circ x = \left(\frac{Ux}{\sqrt{|x'|}}, \frac{Vx}{\sqrt{|x'|}}\right). \tag{28.11}$$

For brevity, in the formula (11), and in what follows, we omit the variable t.

Let a be a proper first phase of B, b be a phase of (q_{ax}) and e be that phase of (-1) determined uniquely by the relation $ax = eb$; finally, let B_b be a basis of (q_{ax}) with proper first phase b.

Then we have, in the first place, the formula

$$\mathrm{B} = \frac{\sqrt{\mathrm{abs}\,\|\mathrm{B}\|}}{\sqrt{|a'|}}\mathscr{B}a, \tag{28.12}$$

and on taking account of (12) and (7), the further relationships

$$\begin{aligned}
\frac{1}{\sqrt{\mathrm{abs}\,\|\mathrm{B}\|}}\,\mathrm{B} \circ x &= \mathscr{H}e\,\frac{1}{\sqrt{\mathrm{abs}\,\|\mathrm{B}_b\|}}\,\mathrm{B}_b = \\
&= \frac{1}{\sqrt{|b'|}}\,\mathscr{H}e\mathscr{B}b = \frac{1}{\sqrt{|b'|}}\,\frac{1}{\sqrt{|e'b|}}\,\mathscr{B}\,eb = \\
&= \frac{1}{\sqrt{|(eb)'|}}\,\mathscr{B}eb = \frac{1}{\sqrt{|(ax)'|}}\,\mathscr{B}ax = \\
&= \frac{1}{\sqrt{|x'|}}\,\frac{1}{\sqrt{|a'x|}}\,\mathscr{B}ax \\
&= \frac{1}{\sqrt{\mathrm{abs}\,\|\mathrm{B}\|}}\,\frac{1}{\sqrt{|x'|}}\,\mathrm{B}x.
\end{aligned}$$

This establishes the formula (11). In particular, *the operation* $\circ$ *applied to an integral* Y *of* (q) *and a phase function* x *is realized by the Kummer transformation*

$$Y \circ x = \frac{Yx}{\sqrt{|x'|}}. \tag{28.13}$$

Finally, we make the remark that when the concept of abstract dispersions is realized on the above lines, known results from the analytic theory of dispersions reappear; for instance, formula (13.10).

29 A survey of recent results in transformation theory

In this section we shall survey recent developments in the field of Jacobian differential equations which have reference to the theory of transformations. We shall be particularly concerned with the theory of dispersions and its applications, especially in the field of central dispersions, and with the generalization of parts of the theory of transformations.

29.1 General dispersions

The theory of general dispersions of two differential equations (q), (Q) has been extended in connection with the study of the phase group $\mathfrak{G}$ (§ 10 and § 20.7). In particular a new characterization of such dispersions has been discovered, and all the general dispersions common to two pairs of differential equations (q_1), (Q_1) and (q_2), (Q_2) determined.

We denote by $I(Qq)$ the subset of $\mathfrak{G}$ comprising all general dispersions of (q), (Q). Then, as we know, $I(qq)$ and $I(QQ)$ are the subgroups conjugate with $\mathfrak{E}$ with respect to arbitrary phases α, $\mathbf{A}$ of (q), (Q) respectively. The fresh information relating to the characterization of general dispersions of (q), (Q) lies in the theorem that the elements $\xi \in I(Qq)$ are characterized by the relationship $\xi^{-1}I(QQ)\xi = I(qq)$. [2*]

With regard to common general dispersions of the above-mentioned pairs o equations, such common dispersions have been shown to exist if and only if all phases of the differential equations (Q_{12}), (q_{12}) are contained in one and the same element of the smallest common covering of the two partitions $\mathfrak{G}/_l\mathfrak{E}$, $\mathfrak{G}/_r\mathfrak{E}$; the carriers Q_{12}, q_{12} are determined by the fact that their phases include the functions $\mathbf{A}_1\mathbf{A}_2^{-1}$, $\alpha_1\alpha_2^{-1}$ formed from arbitrary phases α_1, $\mathbf{A}_1$, α_2, $\mathbf{A}_2$ of q_1, Q_1, q_2, Q_2. [3*]

Further problems involving general dispersions arise in connection with properties of the structure of the phase group $\mathfrak{G}$, and are discussed in [4*] and [5*]. In particular, it has been shown that different concepts from the theory of general dispersions, apparently remote from each other, are related to the centre $\mathfrak{Z}$ of the subgroup of $\mathfrak{E}$ comprising the increasing elements of $\mathfrak{G}$. For instance, the subgroup $\mathfrak{H}$ of the elementary phases (§ 10.4) is the normalizer of $\mathfrak{Z}$ in $\mathfrak{G}$. [5*]

A comparison theorem for general dispersions was given in [39*].

29.2 Dispersions of the 1st and 2nd kinds

New advances have been made in [8*] in the theory of dispersions of the 1st and 2nd kinds. The subgroup $\mathfrak{P}$ ($\subset I(qq)$) comprising the increasing dispersions of $I(qq)$ is the union of three disjoint subsets D_1, D_2 and D_3. The first of these, D_1, consists of the single dispersion $\xi(t) \equiv t$ ($t \in j = (-\infty, \infty)$) which, of course, is a dispersion for every oscillatory differential equation (q). The set D_2 comprises those dispersions for which $\xi(t) \neq t$ ($t \in j$), and the final set, D_3) comprises all other dispersions ξ. The properties of these functions in the various cases are examined; a particular problem studied is to find the conditions under which a dispersion ξ determines the carrier q uniquely. It transpires that this is the case if and only if the dispersion belongs to D_3. Another question considered is the extent to which the subgroup generated by certain dispersions of $\mathfrak{P}$ serves to determine the carrier q. This involves complicated study and can only be answered in special cases; for instance, a continuous one-parameter subgroup in $\mathfrak{P}$ essentially determines q uniquely. These researches can be extended to cover dispersions of the 2nd kind. New problems arise if we consider simultaneous dispersions of the 1st and 2nd kinds; complicated relations occur among these. The only such result to be mentioned here is that the differential equation (q) is determined uniquely "in the general case" by its fundamental dispersions of the 1st and 2nd kinds.

29.3 Central dispersions

In the topic of central dispersions, attention has been directed mainly to central dispersions of the first kind in connection with problems of a different sort. The applicability of central dispersions of all four kinds rests on the fact that, in the first place, they admit of clear geometrical and analytical interpretations, and in the second place that they connect the values taken by integrals and their derivatives (of the relevant Jacobian differential equations) at conjugate and hence distant points; this makes it possible to treat problems of a global character.

This paragraph describes (A) some boundary value problems (B) asymptotic behaviour (particularly boundedness) of integrals and (C) some geometrical studies.

(A) *Boundary value problems*

(i) Consider an oscillatory differential equation $y'' = q(t, \lambda)y$, with $q(t, \lambda) \in C_0$, $t \in j = (-\infty, \infty)$ for every $\lambda > 0$. For such an equation every value λ_0 of λ represents the n-th eigenvalue of the problem

$$y(t_0, \lambda_0) = y(\phi_n(t_0, \lambda_0), \lambda_0) = 0; \qquad (t_0 \in j)$$

here, of course, $\phi_n(t, y)$ denotes the n-th central dispersion of the first kind ($n = 1, 2, \ldots$). [15*]

In the following cases

$$q(t, \lambda) = \lambda q(t), \qquad q(t) < 0,$$
$$q(t, \lambda) = q(t) + \lambda, \qquad q(t) < 0,$$

and in particular for periodic functions q, it has been possible to find bounds for the elements occurring in the boundary value problem, these bounds being partly of a novel structure. [10*], [14*]

(ii) Using phase theory as the basic tool, in [17*] a necessary and sufficient condition is obtained for the existence of a complete generalized Liouville transformation, which takes solutions of

$$\ddot{Y} = (\lambda R(T) + Q(T))\,Y \qquad \text{on } (A, B)$$

into solutions of

$$y'' = (\lambda r(t) + q(t))y \qquad \text{on } (a, b)$$

over their whole domains of definition.

As a special case of this theory, a study is made in [18*] of differential equations $\ddot{Y} = \lambda R(T)\,Y$ and $y'' = (q(t) + \lambda)y$, with periodic carriers, in $t \in (-\infty, \infty)$.

(iii) The two-parameter boundary value problem of F. M. Arscott:

$$y'' + [q(t;\lambda, \mu) + r(t)]y = 0, \quad y(a) = y(b) = y(c) = 0$$

and the special case

$$y'' + [\lambda q(t) + \mu s(t) + r(t)]y = 0, \quad y(a) = y(b) = y(c) = 0,$$

with $q(t;\, \lambda, \mu) \in C_0$ and $q(t) \in C_0$ respectively, for $t \in [a, c)$ and $b \in (a, c]$, was examined in [9*]. For given positive integers n_1, n_2 with $n_1 < n_2$, there exist (under fairly general conditions) eigenvalues of λ, μ such that b is the n_1-th and c the n_2-th zero of $y(t)$ subsequent to a.

(B) *Behaviour of integrals*

We consider an oscillatory differential equation (q), $y'' = q(t)y$, $t \in (a, b)$. Using the theory of central dispersions a study has been made of the connection between, on the one hand, the distribution of zeros of solutions as $t \to b-$ and, on the other hand, the boundedness or the asymptotic behaviour of these solutions.

(i) Let ϕ, $\bar{\phi}$ be respectively the fundamental dispersions of the first kind of (q) and ($\bar{\text{q}}$) and let ϕ_n be the n-th iterate of ϕ ($n = 1, 2, \ldots$). The following results have been obtained:

Every solution of (q) is bounded on $[t_0, b)$ if and only if $\phi_n'(t)$ is bounded on $[t_0, \phi(t_0)]$ for all positive integers n. [58], [19*]

Every solution of (q) belongs to $L^2[t_0, b)$ (i.e. is such that $\int_{t_0}^{b} y^2(\sigma)\, d\sigma < \infty$) if and only if [21*]

$$\sum_{n=0}^{\infty} \int_{t_0}^{\phi(t_0)} \phi_n'^2(\sigma)\, d\sigma < \infty.$$

If $\phi'(t)$ is bounded above on $[t_0, b)$ away from 1, i.e. $\phi'(t) \leqslant k < 1$, where k is a constant, then $b < \infty$ and every solution y of (q) tends to zero as $t \to b-$; every such solution also belongs to $L^p[t_0, b)$ for every $p > 0$ (i.e. $\left[\int_{t_0}^{b} |y(\sigma)|^p\, d\sigma\right]^{1/p} < \infty$. [19*]

If $\phi'(t) \leqslant 1$ on $[t_0, b)$ then every solution of (q) is bounded on $[t_0, b)$. [19*]

If $\phi'(t) = 1$ on $[t_0, b)$ then $b = \infty$ and every solution of (q) is periodic. [20*]

If $\phi'(t) \geqslant$ constant > 1 on $[t_0, b)$, then $b = \infty$ and every non-trivial solution of (q) is unbounded on $[t_0, b)$. [19*]

(ii) Now let (q) and ($\bar{q}$) have the same dispersion ϕ. Then if any of the following statements is true for (q), the same is true for ($\bar{q}$):

(a) every solution is bounded on $[t_0, b)$.
(b) every solution is periodic on $[t_0, b)$.
(c) every solution belongs to $L^2[t_0, b)$.
(d) one non-trivial solution belongs to $L^2[t_0, b)$ but every non-trivial solution which is not a constant multiple of this solution does not belong to $L^2[t_0, b)$.
(e) no non-trivial solution belongs to $L^2[t_0, b)$.

Moreover, if (d) holds, then every solution of both (q) and ($\bar{q}$) belonging to $L^2[t_0, b)$ has the same zeros [19*] and [21*].

In the papers quoted above there are generalizations of these results to L^p classes and also some comparison theorems.

(iii) In [22*], [56] the theory of phases and central dispersions is used to construct all stable periodic differential equations (q), $q \in C_0$, $t \in (-\infty, \infty)$.

Differential equations (q), $q \in C_0$, $t \in (-\infty, \infty)$ with only periodic solutions are shown in [20*] to be characterized by the relation $\phi_n(t) = t +$ constant. A criterion for the periodicity of all solutions of (q) is given in [57] and an explicit form for all such differential equations can be found in [23*]. In [5*] there is an explicit formula for all differential equations (q) with $\phi_n(t) = t + m\pi$, m, n positive integers.

(iv) The first formula (5.5) can be used to construct all differential equations (q) with $\phi_n(t) = t +$ constant, even for carriers which are defined as generalized derivatives of one-sided continuous functions [14*]. The methods used can be extended to the case of linear differential equations of the n-th order [13*]. The geometrical significance of $r(t)$ can be applied to characterize periodic solutions of $y'' + p(t)y - cy^{-3} = 0$ by means of the Liapunov resolvent of $y'' + p(t)y = 0$ [12*].

(v) Oscillatory differential equations (q) in which the derivative $q'(t)$ is a monotonic function of order n, (i.e. $(-1)^i q^{(i+1)}(t) \geqslant 0$, $i = 0, 1, \ldots, n$; $n \geqslant 1$; $t > 0$) are studied in [37*], [38*]. For a differential equation of this kind the zeros t_k of an integral form a monotonic sequence of order n (i.e. $(-1)^i \Delta^{i+1} t_k \geqslant 0$, $k = 0, 1, 2, \ldots$). Theorems have been discovered governing connections between zeros of the integrals of such differential equations and their derivatives, some of which have applications to Bessel functions.

(C) *Geometric studies*

In [20*] a study is made of the close connection between differential equations (q), $q \in C_0$, $t \in (-\infty, \infty)$ with $\phi_n(t) = t +$ constant, and closed (but not necessarily simple closed) curves. This paper also introduces a generalization of the (centro) affine length of the arc of a curve u between the points with parameters t_0 and t_1; the formula is

$$^{c}\Lambda_{t_0}^{t_1}(u) = \operatorname{sgn}\,[u, u'] \int_{t_0}^{t_1} \frac{|[u', u'']|^c}{|[u, u']|^{3c-1}}\, d\sigma$$

or

$$^{c}_{*}\Lambda_{t_0}^{t_1}(u) = \int_{t_0}^{t_1} \frac{\operatorname{sgn}\,[u', u''] \cdot |[u', u'']|^c}{|[u, u']|^{3c-1}}\, d\sigma$$

where $[u, u'] = u_1(\sigma)u_2'(\sigma) - u_1'(\sigma)u_2(\sigma)$, etc. These formulae generalize the formulae of W. Blaschke, O. Borůvka and L. A. Santaló, for which the parameter c takes the particular values $c = \frac{1}{3}$, $c = \frac{1}{2}$, $c = 1$ respectively.

Using integral inequalities for coefficients of differential equations (q) with periodic solutions, derived in [24*] and [25*], the following generalization of the isoperimetric theorem of W. Blaschke and L. A. Santaló was obtained in [25*]:

Let u be a closed plane curve of class C_2 with index n. Let there exist a straight line p through the origin which intersects u in two points symmetrically located with respect to the origin. Let the areas bounded by all simple arcs of u and the line p be the same. Let ${}^c\Lambda(u)$ or ${}^c_*\Lambda(u)$ denote the length of u and let $V(u)$ be the area enclosed by u (the index of u being taken into account). Then

$$ {}^c\Lambda(u) = {}^c_*\Lambda(u) \leqslant 2[V(u)]^{1-2c}(n\pi)^{2c} $$

for $c \in (0, 1]$. If c, $V(u)$ and the index of u are fixed, the maximum length is attained precisely for ellipses with centre at the origin, these being considered as curves of index n.

Further results in centroaffine geometry of plane curves can be found in [6*] and [20*]; results on periodic curvature and closed curves are in [26*], and the impossibility of extending L. A. Santaló's theorem to general simple closed curves is shown in [27*]. A survey of these and further results on differential equations and plane curves is to be found in the lecture notes [28*].

A theorem of W. Blaschke states that a C_2 closed curve has at least three distinct pairs of points with parallel tangents and equal radii of curvature. In connection with this, the following theorem was proved in [15*]:—

Let the central dispersion ϕ_2 of (q) be linear and of the form $\phi_2(t) = t + T$ (hence $q(t + T) = q(t)$), $t \in (-\infty, \infty)$. Further, let $f(t)$ be continuous in an interval $[\tau, \phi_2(\tau)]$. If $f(\tau) = f(\phi_2(\tau))$, and for every integral y of (q) the property

$$ \int_\tau^{\phi_2(\tau)} f(\sigma)y(\sigma)\, d\sigma = 0 $$

holds, then in the interval $[\tau, \phi(\tau))$ there exist at least three distinct values t_i ($i = 1, 2, 3$) such that

$$ f(t_i)\phi'(t_i)^{-\frac{3}{4}} = f(\phi(t_i))\phi'(\phi(t_i))^{-\frac{3}{4}} \quad (\phi = \phi_1). $$

If $\phi(t) = t + T$ and hence $\phi_2(t) = t + 2T$, then this relation implies the existence of at least three distinct values $t_i \in (\tau, \tau + T)$ for which $f(t_i) = f(t_i + T)$.

We have also the further result: let the functions $f(t)$, $g(t)$ be continuous for $t \in (-\infty, \infty)$, $g(t) > 0$, and periodic with period T, where $T \leqslant \phi_2(\tau)$, ϕ_2 being the (general) central dispersion of (q). Also, let every integral y of (q) satisfy the condition

$$ \int_\tau^{\tau+T} f(\sigma)y(\sigma)\, d\sigma = \int_\tau^{\tau+T} g(\sigma)y(\sigma)\, d\sigma = 0. $$

Then the function $f(t)/g(t)$ has at least four relative extrema in the interval $[\tau, \tau + T)$.

A geometrical interpretation of the result of § 15.6 is given in [10*].

29.4 Generalizations

In [14*], [15*] some results from transformation theory of Jacobian differential equations (q) are generalized to differential equations of the form $y'' + Q'(t)y = 0$, $t \in [a, b]$, where Q' is the generalized derivative of a function Q in the sense of distributions.

The integrals of a differential equation (q) form a two-dimensional linear space R of continuous functions. Consequently the Kummer transformation can be regarded as a transformation of one space R_1 into another space R_2. The theory developed in [29*]–[36*] includes the question of how far the transformation theory of differential equations (q) can be taken over to two-dimensional spaces of continuous functions.

A study of the concept and the properties of central dispersions in an abstract group is to be found in [1*].

Bibliography

I Earlier works on transformations of ordinary differential equations of the second order

General:

Encyklop. d. Math. Wiss., Leipzig 1899–1916, IIa 4b.

Books:

L. SCHLESINGER: Handbuch der Theorie der linearen Differentialgleichungen. Leipzig 1897.

E. I. WILCZYNSKI: Projective differential geometry of curves and ruled surfaces. Leipzig 1906.

II More recent works relating to the transformation of ordinary linear differential equations of the second order

1 E. BARVÍNEK: О свойстве заменительности дисперсий и решений дифференциального уравнения $\sqrt{(|X'|)}(1/\sqrt{|X'|})'' + q(X)X'^2 = Q(t)$. Publ. Fac. Sci. Univ. Masaryk, No 393 (1958), 141–155.

2 E. BARVÍNEK: O rozložení nulových bodů řešení lineární diferenciální rovnice $y'' = Q(t) \cdot y$ a jejich derivací. Acta Fac. Nat. Univ. Comenian. V, 8–10 Math. (1961), 465–474.

3 E. BARVÍNEK: Dispersiones de la ecuación diferencial $y'' = Q(t)y$ en el caso general. Memorias de la Facultad de Ciencias de la Universidad de la Habana **1** (1964), Ser. Mat., 31–46.

4 E. BARVÍNEK: Algebraic definition of central dispersions of the 1st kind of the differential equation $y'' = Q(t)y$. Czech. Math. J. **16** (1966), 46–62.

5 O. BORŮVKA: О колеблющихся интегралах дифференциальных линейных уравнений 2-ого порядка. Czech. Math. J. **3 (78)** (1953), 199–251.

6 O. BORŮVKA: Sur la transformation des intégrales des équations différentielles linéaires ordinaires du second ordre. Ann. Mat. Pura Appl. **41** (1956), 325–342.

7 O. BORŮVKA: Théorie analytique et constructive des transformations différentielles linéaires du second ordere. Bull. Math. Soc. Math. Phys. R. P. Roumaine **1 (49)** (1957), 125–130.

8 O. BORŮVKA: Sur les transformations différentielles linéaires complètes du second ordre. Ann. Mat. Pura Appl. **49** (1960), 229–251.

9 O. BORŮVKA: Neuere Ergebnisse auf dem Gebiet der linearen Differentialgleichungen 2. Ordnung. II. Magyar Mat. Kongresszus, Budapest 1960; Ergänzungen zu den Vortragsauszügen, 11–12.

10 O. BORŮVKA: Sur la structure de l'ensemble des transformations différentielles linéaires complètes du second ordre. Ann. Mat. Pura Appl. **58** (1962), 317–334.

11 O. BORŮVKA: Transformations des équations différentielles linéaires du deuxième ordre. Séminaire Dubreil-Pisot, **22** (1961), 1–18.

12 O. BORŮVKA: Über einige Ergebnisse aus der Theorie der linearen Differentialtransformationen 2. Ordnung. Heft 13 der Schriftenreihe des Inst. f. Math. Bericht von der Dirichlet-Tagung. Berlin 1963, 51–57.

13 O. BORŮVKA: Transformation of ordinary second-order linear differential equations. Differential Equations and Their Applications. Proceedings of the Conference held in Prague in September 1962. Prague 1964, 27–38.

14 O. BORŮVKA: Sur l'ensemble des équations différentielles linéaires ordinaires du deuxième order qui ont la même dispersion fondamentale. Bul. Inst. Polit. din Iaşi, Ser. nouă, IX (XIII) (1963), 11–20.

15 O. BORŮVKA: Über die algebraische Struktur der Phasenmenge der linearen oszillatorischen Differentialgleichungen 2. Ordnung. Publ. Fac. Sci. Univ. J. E. Purkyně, No 457 (1964), 461–462.

16 O. Borůvka: Sur quelques applications des dispersions centrales dans la théorie des équations différentielles linéaires du deuxième ordre. Arch. Math. (Brno), **1** (1965), 1–20.
17 O. Borůvka: Sur une application géométrique des dispersions centrales des équations différentielles linéaires du deuxième ordre. Ann. Mat. Pura Appl. **71** (1966), 165–187.
18 O. Borůvka: Über die allgemeinen Dispersionen der linearen Differentialgleichungen 2. Ordnung. An. Şti. Univ. Al. I. Cuza, Iaşi, XI_B (1965), 217–238.
19 L. Frank: O diferenciální rovnici $y'' = Q(x)y$, jejíž integrály mají ekvidistantní nulové body. Sborník Vys. Uč. Techn. Brno, 1958, 91–96.
20 M. Greguš: Aplikácia disperzií na okrajový problém druhého rádu. Mat.-Fyz. Čas. Sloven. Akad. Vied, **1** (1954), 27–37.
21 M. Greguš: O niektorých vlastnostiach riešení lineárnej diferenciálnej rovnice homogénnej tretieho rádu. Mat.-Fyz. Čas. Sloven. Akad. Vied, **2** (1955), 73–85.
22 M. Greguš: O niektorých vlastnostiach riešení diferenciálnej rovnice $y''' + Qy' + Q'y = 0$. Publ. Fac. Sci. Univ. Masaryk, No 365 (1955), 1–18.
23 M. Greguš: О некоторых новых краевых проблемах дифференциального уравнения третьего порядка. Czech. Math. J. **7 (82)** (1957), 41–47.
24 M. Greguš: Poznámka o disperziách a transformáciach diferenciálnej rovnice tretieho rádu. Acta F. R. N. Univ. Comenian. IV, 3–5 Math. (1959), 205–211.
25 M. Greguš: O oscilatorických vlastnostiach riešení lineárnej diferenciálnej rovnice tretieho řádu tvaru $y''' + 2A(x)y' + [A'(x) + b(x)]y = 0$. Acta F. R. N. Univ. Comenian. VI, 6. Math. (1961), 275–300.
26 Z. Hustý: Asymptotické vlastnosti integrálů homogenní lineární diferenciální rovnice čtvrtého řádu. Čas. Pěst. Mat. **83** (1958), 60–69.
27 Z. Hustý: O některých vlastnostech homogenní lineární diferenciální rovnice čtvrtého řádu. Čas. Pěst. Mat. **83** (1958), 202–213.
28 Z. Hustý: Über einige Eigenschaften linearer Differentialgleichungen fünfter Ordnung. Publ. Fac. Sci. Univ. J. E. Purkyně, No 432 (1962), 151–176.
29 Z. Hustý: Некоторые колебательные свойства однородного линейного дифференциального уравнения n-ого порядка ($n \geqslant 3$). Czech. Math. J. **14 (89)** (1964), 27–38.
30 Z. Hustý: Die Iteration homogener linearer Differentialgleichungen. Publ. Fac. Sci. Univ. J. E. Purkyně, No 449 (1964), 23–56.
31 Z. Hustý: Asymptotische Formeln für die Lösungen homogener linearer Differentialgleichungen n-ter Ordnung im oszillatorischen Fall. Čas. Pěst. Mat. **90** (1965), 79–86.
32 Z. Hustý: Adjugierte und selbstadjungierte lineare Differentialgleichungen. Arch. Math. (Brno) **1** (1965), 21–34.
33 Z. Hustý: Asymptotické vlastnosti integrálů homogenních lineárních diferenciálních rovnic 2. řádu. Čas. Pěst. Mat. **90** (1965), 487–490.
34 Z. Hustý: Über die Transformation und Äquivalenz linearer Differentialgleichungen von höherer als der zweiten Ordnung. Czech. Math. J., I. Teil **15 (90)** (1965), 479–502; II. Teil **16 (91)** (1966), 1–13; III. Teil **16 (91)** (1966), 161–185.
35 Z. Hustý: Perturbierte homogene lineare Differentialgleichungen. Čas. Pěst. Mat. **91** (1966), 154–169.
36 Z. Hustý: Asymptotische Eigenschaften von Lösungen homogener Differentialgleichungen n-ter Ordnung. Math. Nachr. (Im Druck).
37 J. Chrastina: O splynutí základních centrálních dispersí 3. a 4. druhu diferenciální rovnice $\ddot{y}(t) + Q(t)y(t) = 0$. Čas. Pěst. Mat. **87** (1962), 188–197.
38 J. Chrastina: K teorii dispersí rovnice $\ddot{y}(t) + Q(t)y(t) = 0$. Publ. Fac. Sci. Univ. J. E. Purkyně, No 454 (1964), 265–273.
39 M. Laitoch: Расширение метода флоке для определения вуда фундаметальной системы решений дифференциального уравнения второго порядка $y'' = Q(x)y$. Czech. Math. J. **5 (80)** (1955), 164–174.
40 M. Laitoch: Sur une théorie des critères comparatifs sur l'oscillation des intégrales de l'équation différentielle $u'' = P(x)u$. Publ. Fac. Sci. Univ. Masaryk, No 365 (1955), 255–266.

41 M. LAITOCH: Совпадение центральных дисперсий 1-го и 2-го рода, соответствующих дифференциальному уравнению второго порядка $y'' = Q(x)y$. Czech. Math. J. **6** (**81**) (1956), 365–380.

42 M. LAITOCH: O jistých řešeních funkční rovnice $F[\varphi(x)] - F(x) = 1$. Čas. Pěst. Mat. **81** (1956), 420–425.

43 M. LAITOCH: O ortogonalitě řešení lineární diferenciální rovnice druhého řádu $y'' = q(x)y$. Sborník Vys. školy Pedagog. Olomouc, VI, **3** (1959), 7–22.

44 M. LAITOCH: Трансформация решений линейных дифференциальных уравнений. Czech. Math. J. **10** (**85**) (1960), 258–70.

45 M. LAITOCH: Über die Nullstellenanzahl der Lösungen der Differentialgleichung $y'' = Q(t)y$. Acta Univ. Palackianae Olomucensis, **3** (1960), 5–9.

46 M. LAITOCH: К проблеме ортогональных систем с весом. Acta Univ. Palackianae Olomucensis, **3** (1960), 11–18.

47 M. LAITOCH: L'équation associée dans la théorie des transformation des équations différentielles du second ordre. Acta Univ. Palackianae Olomucensis, F. R. N. **12** (1963), 45–62.

48 J. MAŘÍK & M. RÁB: Asymptotische Eigenschaften von Lösungen der Differentialgleichung $y'' = A(x) \cdot y$ im nichtoszillatorischen Fall. Czech. Math. J. **10** (**85**) (1960), 501–521.

49 J. MORAVČÍK: Poznámka k transformácii riešení lineárnych diferenciálnych rovníc. Acta F. R. N. Univ. Comenian. VI, 6 Math. (1960), 327–339.

50 J. MORAVČÍK: O fundamentálnom systéme normálnych riešení iterovanej diferenciálnej rovnice štvrtého rádu, Acta F. R. N. Univ. Comenian. VII, 12 Math. (1963), 675–680.

51 J. MORAVČÍK: O zobecnení Floquetovej teórie pre lineárne diferenciálne rovnice obyčajné n-tého rádu. Čas. Pěst. Mat. **91** (1966), 8–17.

52 F. NEUMAN: Sur les équations différentielles linéaires du second ordre dont les solutions ont des racines formant une suite convexe. Acta Math. Acad. Hung., XIII (1962), 281–287.

53 F. NEUMAN: Sur les équations différentielles linéaires oscillatoires du deuxième ordre avec la dispersion fondamentale $\phi(t) = t + \pi$. Bul. Inst. Polit. din Jaşi, X (XIV) (1964), 37–42.

54 F. NEUMAN: Construction of second order linear differential equations with solutions of prescribed properties. Arch. Math. (Brno), **1** (1965), 229–246.

55 F. NEUMAN: Note on the second phase of the differential equation $y'' = q(t)y$. Arch. Math. (Brno), **2** (1966), 57–62.

56 F. NEUMAN: Note on bounded non-periodic solutions of second-order differential equations with periodic coefficients. Math. Nachr. **39** (1969) 217–22.

57 F. NEUMAN: Criterion of periodicity of solutions of a certain differential equation with a periodic coefficient. Ann. Mat. Pura Appl. **75** (1967) 385–396.

58 F. NEUMAN: Relation between the distribution of the zeros of the solutions of a 2nd order linear differential equation and the boundedness of these solutions. Acta Math. Acad. Hung. **19** (1968) 1–6.

59 J. PALÁT: Пример к теории преобразования и дисперсий О. Борувка. Acta Univ. Palackianae Olomucensis, F. R. N. **12** (1963), 63–68.

60 M. RÁB: Poznámka k otázce o oscilačních vlastnostech řešení diferenciální rovnice $y'' + A(x)y = 0$. Čas. Pěst. Mat. **82** (1957), 342–348.

61 M. RÁB: Kriterien für die Oszillation der Lösungen der Differentialgleichung $[p(x)y']' + q(x)y = 0$. Čas. Pěst. Mat. **84** (1959), 335–370.

62 M. RÁB: Asymptotische Formeln für die Lösungen der Differentialgleichung $y'' + q(x)y = 0$. Czech. Math. J. **14** (**89**) (1964), 203–221.

63 M. RÁB: Les développements asymptotiques des solutions de l'équation $(py')' + qy = 0$. Arch. Math. (Brno), **2** (1966), 1–17.

64 S. ŠANTAVÁ: Об основных свойствах интегралов систем двух дуфференциальных уравнений первого порядка. Publ. Fac. Sci. Univ. Masaryk, No 369 (1955), 1–21.

65 S. ŠANTAVÁ: Transformace integrálů systému dvou diferenciálních lineárních rovnic 1. řádu. Sbor. Voj. Akad. A. Z. Brno, **8** (**50**) (1959), 3–14.

66 V. ŠEDA: Niekolko viet o lineárnej diferenciálnej rovnici druhého rádu Jacobiho typu v komplexnom obore. Čas. Pěst. Mat. **88** (1963), 29–58.
67 V. ŠEDA: O niektorých vlastnostiach riešení diferenciálnej rovnice $y'' = Q(x) \cdot y$, $Q(x) \not\equiv 0$ je celá funkcia. Acta F. R. N. Univ. Comenian. IV, 3–5 Math. (1959), 223–253.
68 V. ŠEDA: Transformácia integrálov obyčajných lineárnych diferenciálnych rovnic druhého rádu v komplexnom obore. Acta F. R. N. Univ. Comenian. II, 5–6 Math. (1958), 229–254.
69 V. ŠEDA: On the properties of linear differential equations of the second order in the complex domain. Differential Equations and Their Applications. Proceedings of the Conference held in Prague in September 1962. Prague 1964, 179–186.
70 V. ŠEDA: Применение главной теоремы конформного отображения в теории линейных дифференциальных уравпений 2-ого порядка. Čas. Pěst. Mat. **89** (1964) 10–27.
71 V. ŠEDA: Несколько теорем о линейном дифференциальном уравнении второго порядка типа Якоби в комплексной области. Čas. Pěst. Mat. **88** (1963), 29–58.
72 V. ŠEDA: Исправление работы Несколько теорем о линейном дифференциальном уравнении второго порядка типа Якоби в комплексной области. Čas. Pěst. Mat. **89** (1964), 359–361.
73 V. ŠEDA: Über die Existenz der linearen Differentialgleichungen zweiter Ordnung im komplexen Gebiet, welche den Differentialgleichungen mit konstanten Koeffizienten ähnlich sind. Acta F. R. N. Univ. Comenian. X, 3 Math. **12** (1965), 31–40.
74 V. ŠEDA: Über die Transformation der linearen Differentialgleichungen *n*-ter Ordnung. I. Čas. Pěst. Mat. **90** (1965), 385–412.
75 M. ŠVEC: Sur les dispersions des intégrales de l'équation $y'' + Q(x)y = 0$. Czech. Math. J. **5 (80)** (1955), 29–60.
76 M. ŠVEC: Eine Eigenwertaufgabe der Differentialgleichung $y^{(n)} + Q(x, \lambda)y = 0$. Czech Math. J. **6 (81)** (1956), 46–68.
77 S. TRÁVNÍČEK: О преобразованиях решений систем двух линейных дифференциальных уравнений первого порядка. Acta Univ. Palackianae Olomucensis, F. R. N. **9** (1962), 151–162.
78 S. TRÁVNÍČEK: Об одном использовании функции флоке. Acta Univ. Palackianae Olomucensis, F. R. N. **12** (1963), 64–74.
79 S. TRÁVNÍČEK: О преобразованиях решений систем лниейных дифференциальных уравнений первого порядка. Acta Univ. Palackianae Olomucensis, F. R. N. **18** (1965), 5–23.

III Books to which reference is made

80 W. BLASCHKE: Vorlesungen über Differentialgeometrie, II. Berlin 1923.
81 O. BORŮVKA: Grundlagen der Gruppoid- und Gruppentheorie. Berlin 1960.

IV. Further books on ordinary differential equations

F. M. ARSCOTT: Periodic differential equations, Pergamon, 1964.
R. BELLMAN: Stability theory of differential equations, McGraw-Hill, 1953.
F. BRAUER and J. A. NOHEL: Qualitative theory of ordinary differential equations, W. A. Benjamin, 1969.
L. CESARI: Asymptotic behaviour and stability problems in ordinary differential equations, Springer, 1959.
E. A. CODDINGTON and N. LEVINSON: Theory of ordinary differential equations, McGraw-Hill, 1955.
L. COLLATZ: Eigenwertaufgaben mit technischen Anwendungen, Akad. Verlag, Leipzig, 1963.
L. COLLATZ: The numerical treatment of differential equations, Springer, 1960.
P. HARTMAN: Ordinary differential equations, Wiley, 1964.
E. L. INCE: Ordinary differential equations, Dover, 1926.
J. HORN and H. WITTICH: Gewöhnliche Differentialgleichungen, 6th edn. Berlin, 1960.

E. Kamke: Differentialgleichungen I, 5th edn., Leipzig, 1964.
E. Kamke: Differentialgleichungen; Lösungsmethoden und Lösungen I, 8th edn., Leipzig, 1967.
S. Lefschetz: Lectures on differential equations, Princeton, 1948.
S. Lefschetz: Differential equations; geometric theory, Princeton, 1962.
W. Magnus and S. Winkler: Hill's Equation, Interscience, 1966.
F. Murray and K. Miller: Existence theorems for differential equations.
I. G. Petrovskii: Ordinary differential equations, Prentice-Hall, 1966 (Trans. from Russian).
L. S. Pontrjagin: Ordinary differential equations, Pergamon, 1962.
G. Sansone and R. Conti: Non-linear differential equations, Pergamon, 1965 (Trans. from Italian).
C. A. Swanson: Comparison and oscillation theorems of ordinary differential equations Academic Press, 1968.
V. V. Stepanov: Lehrbuch der Differentialgleichungen, 3rd edn., Berlin, 1967.
F. G. Tricomi: Differential Equations, Blackie, London 1961.

J. H. Barrett: Oscillation Theory of Ordinary Linear Differential Equations. Advances in Mathematics, 3 1969, pages 415–509 (survey article).

Supplementary Bibliography

1* E. Barvínek: Fundamental central dispersion in a simple sustem <𝔊>. Arch. Math. (Brno) **7** (1971) (in press).

2* O. Borůvka: Über eine Charakterisierung der allgemeinen Dispersionen linearer Differentialgleichungen 2. Ordnung. Math. Nachr., **38** (1968), 261–266.

3* O. Borůvka: Sur les solutions simultanées de deux équations differentialles de Kummer. Bull de la Soc. des Sci. math. de Roumanie. IV-ème Congrès des Math. d'expression latine et Commémoration d'Elie Cartan. 1969. (In press.)

4* O. Borůvka: Algebraic elements in the transformation theory of 2nd order linear oscillatory differential equations. Acta Fac. Rerum Nat. Univ. Comen. (Bratislava) Mathematica XVII (1967), 27–36.

5* O. Borůvka: Sur quelques propriétés de structure du groupe des phases des équations différentielles linéaires du deuxième ordre. Revue Roum. de Math. p. et appl. XV (1970, Nr. 9. 1345–1356.

6* O. Borůvka: Eléments géométriques dans la théorie des transformations des équations différentielles linéaires et ordinaires du deuxième ordre. Atti del Convegno Internazionale di Geometria Differenziale. Bologna, 1967, 1–12.

7* O. Borůvka: Théorie des transformations des équations différentielles linéaires du deuxième ordre. Rend. di Mat. **26** (1967), 187–246.

8* J. Chrastina: On dispersions of the 1st and 2nd kind of differential equation $y'' = q(x)y$. Publ. Fac. Sci. Univ. J. E. Purkyně, Brno, No. 508 (1969), 353–377.

9* M. Greguš, F. Neuman and F. M. Arscott: Three-point boundary value problems in differential equations. Jour. Lond. Math. Soc. 2nd series **3**, (1971) (in press).

10* H. Guggenheimer: Some geometric remarks about dispersions. Arch. Math. (Brno), **4** (1968), 193–199.

11* H. Guggenheimer: Hill equations with coexisting periodic solutions, II. Comm. Math. Helv., **44** (1969), 381–384.

12* H. Guggenheimer: An application of Floquet theory. Boll. Un. Mat. It. (4) **2** (1969), 205–207.

13* H. Guggenheimer: Homogeneous linear differential equations with only periodic solutions. Israel J. Math. (In press).

14* H. Guggenheimer: Geometric theory of differential equations, I. Second order linear equations. (In press.)

15* H. Guggenheimer: Geometric theory of differential equations, II. Analytic interpretation of a geometric theorem of Blaschke. (In press.)

16* H. Guggenheimer: Notes on Geometry. Arch. Math. (Brno), **5** (1969), 125–130.

17* F. Neuman: On the Liouville Transformation. Rend. di Mat. **3** (1970), 133–139.

18* F. Neuman: On the Coexistence of Periodic Solutions. J. Diff. Equat. **8** (1970), 277–282.

19* F. Neuman: A Role of Abel's Equation in the Stability Theory of Differential Equations. (To appear in Aequat. Math.).

20* F. Neuman: Centroaffine Invariants of Plane Curves in Connection with the Theory of the Second-Order Linear Differential Equations. Arch. Math. (Brno), **4** (1968), 201–216.

21* F. Neuman: L^2-Solutions of $y'' = q(t)y$ and a Functional Equation. (To appear in Aequat. Math.)

22* F. Neuman: On Bounded Solutions of a Certain Differential Equation. Acta Fac. Rerum Nat. Univ. Comen. (Bratislava) Mathematica XVII (1967), 213–215.

23* F. NEUMAN: An Explicit Form of the Differential Equations $y'' = q(t)y$ with Periodic Solutions. Ann. di Mat. p. ed appl. **85** (1970), 295–300.

24* F. NEUMAN: Extremal Property of the Equation $y'' = -k^2y$. Arch. Math. (Brno) **3** (1967), 161–164.

25* F. NEUMAN: Closed Plane Curves and Differential Equations. Rend. di Mat. **3** (1970), 423–433.

26* F. NEUMAN: Periodic Curvatures and Closed Curves. Rend. Acad. Naz. dei Lincei. **48** (1970) 494–498.

27* F. NEUMAN: A Note on Santaló's Isoperimetric Theorem. (To appear in Revista de Mat. y Fis. Univ. Tucuman).

28* F. NEUMAN: Linear Differential Equations of the Second Order and Their Applications. Lectures notes of the Univ. of Waterloo, Ontario, Canada, also to appear in Rend. di Mat.

29* K. STACH: Die allgemeinen Eigenschaften der Kummerschen Transformationen zweidimensionaler Räume von stetigen Funktionen. Publ. Fac. Sci. Univ. J. E. Purkyně, Brno, 1966, No 478.

30* K. STACH: Die vollständigen Kummerschen Transformationen zweidimensionaler Räume von stetigen Funktionen. Arch. Math. (Brno) **3** (1967), 117–138.

31* K. STACH: Die Kummerschen Transformationen in Räumen mit abgeschlossenen Phasen. Teil A: Allgemeine Eigenschaften. Arch. Math. (Brno) **4** (1968), 141–156.

32* K. STACH: Die Kummerschen Transformationen in Räumen mit abgeschlossenen Phasen. Teil B: Transformationen in Räumen der gegebenen Klasse. Arch. Math. **5** (1969) 61–73.

33* K. STACH: Die Konstruktion eines zweideimensionalen Raums von stetigen Funktionen zur gegebenen Phasenfunktion. Sborník prací VŠB (Ostrava) 2 (1968), 27–31.

34* K. STACH: Die Kategorie der Phasenfunktionen. Publ. Fac. Sci. Univ. J. E. Purkyně, Brno, No. 508 (1969), 379–386.

35* K. STACH: Die Äquivalenz in der Kategorie der Phasenfunktionen. (In press.)

36* K. STACH: Eineallgebraische Struktur der allgemeinen Kummerschen Transformationen. (In press.)

37* J. VOSMANSKÝ: The Monotonicity of Extremants of Integrals of the Differential Equation $y'' + q(t)y = 0$. Arch. Math. (Brno) **2** (1966), 105–111.

38* J. VOSMANSKÝ: Monotonic Properties of Zeros and Extremants of the Differential Equation $y'' + q(t)y = 0$. Arch. Math. (Brno) **6** (1970), 37–74.

39* V. ŠEDA: A Comparison Theorem in the Theory of the Second-Order Linear Differential Transformations. Arch. Math. (Brno) **5** (1969), 7–17.

40* Y. KRBILÀ: Об определении неколеьлюшего днфференчнального уравнения $y'' = q(t)y$ второй глперболической фаэой. Arch. Math. (Brno) **5** (1969), 1–6.

Author and Subject Index